ASHEVILLE-BUNCOMBE TECHNICAL INSTITUTE
STATE
DEPT. OF COMMUNITY COLLEGES
LIBRARIES

DISCARDED

NOV 2 1 2024

MODERN ASPECTS OF ELECTROCHEMISTRY

No. 7

CONTRIBUTORS TO THIS VOLUME

H. H. BAUER
Department of Chemistry
University of Kentucky
Lexington, Kentucky

B. E. CONWAY
Commonwealth Visiting Professor (1969-1970)
Universities of Southampton and Newcastle-on-Tyne
Southampton, England

A. R. DESPIĆ
University of Belgrade
Belgrade, Yugoslavia

P. ELVING
Department of Chemistry
University of Michigan
Ann Arbor, Michigan

P. J. HERMAN
Department of Chemistry
University of Michigan
Ann Arbor, Michigan

V. V. LOSEV
Karpov Institute of Physical Chemistry
Moscow, USSR

J. I. PADOVA
Israel Atomic Energy Commission
Soreq, Israel

K. I. POPOV
University of Belgrade
Belgrade, Yugoslavia

MODERN ASPECTS OF ELECTROCHEMISTRY

No. 7

Edited by

B. E. CONWAY

Department of Chemistry
University of Ottawa
Ottawa, Ontario

and

J. O'M. BOCKRIS

Electrochemistry Laboratory
John Harrison Laboratory of Chemistry
University of Pennsylvania, Philadelphia, Pennsylvania

PLENUM PRESS · NEW YORK · 1972

Library of Congress Catalog Card Number 54-12732
ISBN 0-306-37647-4

© 1972 Plenum Press, New York
A Division of Plenum Publishing Corporation
227 West 17th Street, New York, N.Y. 10011

United Kingdom edition published by Plenum Press, London
A Division of Plenum Publishing Company, Ltd.
Davis House (4th Floor), 8 Scrubs Lane, Harlesden, NW10 6SE,
London, England

All rights reserved

No part of this publication may be reproduced in any
form without written permission from the publisher

Printed in the United States of America

Preface

Despite reductions in the level of research activity in most fields which, for reasons of economic decline, have taken place in the U.S. during the last year or two, world progress in the fundamental aspects has continued actively. An important aspect of such recent work has been the use of nonaqueous solvents in studies on the constitution of the double-layer and electrochemical reactions. Interpretation of the behavior of electrode interfaces in such solvents demands more knowledge of the solvation properties of ions in nonaqueous media. Chapter 1 by Padova on "Ionic Solvation in Nonaqueous and Mixed Solvents" gives an up to date review of the present state of knowledge in this field, together with tabulations of data that are likely to be of quantitative value in further investigations of both homogeneous and heterogeneous electrochemistry in such media.

Electrochemical studies of cathodic processes in nonaqueous solvents have, in recent years, revealed the role of solvated electrons. These are of interest in new approaches to reductive electro-organic synthesis. Similarly, the generation of hydrated electrons in photo-cathodic processes is of great interest. In Chapter 2, by Conway, the conditions under which solvated electrons can arise in electrode processes are critically examined and the electro-organic reactions that have been investigated are reviewed. The supposed electro-generation of hydrated electrons in the water solvent and as intermediates in cathodic hydrogen evolution is shown to be unlikely.

Returning to questions concerned with the double-layer itself, a useful critical appraisal of the significance of measurements and measured quantities concerned with adsorption and the structure of electrode solution interphases is given by Bauer, Herman, and Elving in Chapter 3.

The subject of electrocrystallization and metal dissolution has for many years been a topic of central interest both in fundamental and applied electrochemistry. The possibility of controlling phase growth or dissolution by electric potential control is little appreciated outside the subject of electrochemistry yet it enters into many processes of technological interest not least of which, in a "negative" way, is corrosion. Chapter 4 presents a thorough analysis of various aspects of this topic under the title "Transport Controlled Deposition and Dissolution of Metals" by Despic and Popov. Special attention is given to problems of leveling and dendritic growth. The large number of diagrams and photographs will enhance the value of this chapter both to metallurgists and electrochemists.

Finally, a contribution from the important Russian school of electrochemistry (at the Karpov Institute) is given in Chapter 5 by Losev, who examines the "Mechanisms of Stepwise Electrode Processes on Amalgams." Such studies, which to a large extent are specially his own, allow idealized examination of corrosion-type processes under conditions where various steps in complex reaction schemes can be characterized and their role in the kinetics elucidated. Simultaneous anodic and cathodic processes are involved, the latter naturally involving H.

B. E. Conway
J. O'M. Bockris

Breakers Club
Bermuda
January 1971

Contents

Chapter 1

IONIC SOLVATION IN NONAQUEOUS AND MIXED SOLVENTS
J. I. Padova

Chapter 2

SOLVATED ELECTRONS IN FIELD- AND PHOTO-ASSISTED PROCESSES AT ELECTRODES

B. E. Conway

Chapter 3

CRITICAL OBSERVATIONS ON THE MEASUREMENT OF ADSORPTION AT ELECTRODES

H. H. Bauer, P. J. Herman, and P. J. Elving

Chapter 4

TRANSPORT-CONTROLLED DEPOSITION AND DISSOLUTION OF METALS

A. R. Despić and K. I. Popov

Chapter 5

MECHANISMS OF STEPWISE ELECTRODE PROCESSES ON AMALGAMS

V. V. Losev

1

Ionic Solvation in Nonaqueous and Mixed Solvents

J. I. Padova

Israel Atomic Energy Commission
Soreq, Israel

I. INTRODUCTION

"Since the beginnings of quantitative physical chemistry, the study of electrolyte solutions has occupied a central position and constituted the early basis of electrochemistry."[1] Most investigations have been conducted with water as the solvent and the primary influence of solvation in determining the properties of aqueous solutions of electrolytes has already been stressed.[2]

Early work in *non*aqueous electrochemistry was confined to the extension of theories applied to aqueous solutions. However, in recent years, there has been considerable interest in the behavior of electrolytes in nonaqueous and mixed solvents with a view to investigating changes in the solvation of ions. The testing of electrostatic theories of ion association in media of varying dielectric constant[3] has occupied many researchers, and a steady increase of thermodynamic data as well as theoretical interpretations of the present knowledge on nonaqueous media[4] have appeared in the recent literature. Correspondingly, recent work on electrode processes in nonaqueous media enhances the importance of ionic studies in such solvents.

The structure of solutions of electrolytes may be inferred from the investigation of both reversible and irreversible phenomena. Both approaches will be considered here, although the present discussion will be concerned mainly with thermodynamic behavior in solutions of electrolytes.

It has recently been proposed[5,6] that a differentiation be made between protic and dipolar aprotic solvents. Protic solvents, such as fluoroalcohols, hydrogen fluoride, methanol, formamide, and of course water, are strong hydrogen-bond donors. Dipolar aprotic solvents are highly polar but are no more than very weak hydrogen donors. Common dipolar aprotic solvents are dimethylformamide, dimethylacetamide, dimethylsulfoxide, hexamethylphosphoramide, acetone, nitromethane, nitrobenzene, acetonitrile, benzonitrile, sulfur dioxide, propylene carbonate, sulfolane, and dimethylsulfone. This distinction is made only for solvents of dielectric constant greater than 15 because of extensive aggregation in solvents of lower dielectric constant.

Frank's classification of solutes[7] according to the relative magnitudes of excess mixing functions with water does not quite agree with the above, since ketones are classed together with alcohols. Parker[5] suggested that there are four types of strong solute–solvent interaction that contribute to solvation phenomena: electrostatic (ion–dipole, dipole–dipole), π-complex-forming, hydrogen-bonding, and structure-making or -breaking. Accordingly, in terms of the concept of hard and soft acids and bases,[8] protic solvents are "hard" since they exhibit general hydrogen bonding with small anions, and dipolar aprotic solvents are "soft," as they have a mutual polarizability interaction with large polarizable anions. Structure-making is defined as lengthening the molecular reorientation time, while shortening it is termed structure-breaking[9] or negative solvation. In many cases, the mechanism by which solvation occurs is unknown; however, we shall define solvation, unless otherwise specified, as the total ion–solvent interaction at infinite dilution.

II. THERMODYNAMICS OF SOLVATION

The term solvation will be used in this discussion to describe the total ion–solvent interaction at infinite dilution. The transfer of a pair of gaseous ions into a solvent characterizes the thermodynamic process of ionic solvation, which may be written as follows:

$$C^+(g) + A^-(g) \rightarrow C^+(s) + A^-(s) \qquad \Delta Y_s^\circ \qquad (1)$$

where C^+A^- is the ion pair and ΔY_s° represents the change in the thermodynamic property considered.

The enthalpy, free energy, and entropy of solvation of an electrolyte are usually determined from a specific thermodynamic cycle. For instance, the enthalpy of solvation ΔH_s° may be obtained by a Born–Haber cycle in which the solution process involves the sublimation and the dissociation of the crystal lattice followed by dissolution of the ions at infinite dilution (to avoid any ion–ion interaction), and

$$\Delta H_{s0} = -U_0 + \Delta H_s^\circ \tag{2}$$

where ΔH_{s0} and U_0 are the enthalpy of solution and the crystal lattice energy, respectively.

The molal volume of the solvated ions V_s was shown[74] to be given by

$$V_s = \phi_v + nV_0 \tag{3}$$

where ϕ_v is the apparent molal volume of the electrolyte, V_0 is the molal volume of the solvent, and n is the solvation number characteristic of the electrolyte. By definition, the molal volume of the solvated ions may be expressed equally well as[109]

$$V_s = V_{\text{in}} + nV_s^\circ \tag{4}$$

where V_{in} is the intrinsic ionic volume of the electrolyte[75] and V_s° is the molal volume of the solvent in the solvation shell.

III. DETERMINATION OF THERMODYNAMICS OF SOLVATION

1. Heats of Solvation

(i) Methods

Methods for the determination of heats of solvation have been reviewed by Conway and Bockris.[10] Recently, however, a new method, based on a mass spectrometric study of ion–solvent interaction in the gas phase, has been put forward.[11]

The mass-spectrometric gas-phase studies are based on measurement of the relative concentrations of the clustered ionic species $A^+ \cdot nS$, $A^+ \cdot (n+1)S$, etc. Consider the ion A^+ produced in the gas phase by some form of ionizing radiation or by thermal means. If the atmosphere surrounding the ion contains the vapor of a polar molecule (solvent S), a number of clustering reactions will occur, e.g.,

$$A^+ + S \rightarrow A^+ \cdot S \qquad (0, 1)$$

$$A^+ \cdot S + S \rightarrow A^+ \cdot 2S \qquad (1, 2)$$

$$A^+ \cdot (n - 1)S + S \rightarrow A^+ \cdot nS \qquad (n - 1, n)$$

At equilibrium,

$$\Delta G^\circ_{0,n} = \Delta G^\circ_{0,1} + \Delta G^\circ_{1,2} + \cdots + \Delta G^\circ_{n-1,n} \tag{5}$$

$$\Delta G^\circ_{n-1,n} = -R \ln\left(\frac{P_{A^+ \cdot nS}}{P_{A^+ \cdot (n-1)S + S}}\right) = RT \ln K_{n-1,n} \tag{6}$$

where P_x is the partial pressure of species x. From equation (6), the shell structure will be revealed since the value of $\Delta G_{n-1,n}$ becomes discontinuous whenever a shell is completed. The total free energy of solvation can be obtained from the relation

$$\Delta G_s = \Delta G_{n-1,n} - \Delta G_{\text{evap}}(\text{S}) \tag{7}$$

From measurements of this type taken at different temperatures, ΔH and ΔS can be evaluated. This method looks promising, although to date it has only been tried in water, methanol, and ammonia.

(ii) Experimental Results

Heats of solvation obtained from extrapolated heats of dilution at 25°C in various solvents are given in Tables 1–5. In some cases, solvation enthalpies were obtained from transfer data.

(iii) Ionic Enthalpies of Solvation

A modification of the method proposed by Verwey[26] was applied to the determination of ionic contributions to the enthalpy of solvation in formamide.[17] A plot of the solvation enthalpies of, for example, the lithium halides, as functions of the reciprocal crystal radii of the halide ions may be expressed by the straight line

$$\Delta H_s(\text{LiX}) = A_{\text{Li}} + \left(\frac{B}{r_x}\right) \tag{8}$$

The values of the ionic radii are taken from Ahrens'[27] tabulations:

F = 1.33 Å, Cl = 1.81 Å, Br = 1.96 Å, I = 2.20 Å

Table 1
Enthalpies of Solvation ΔH_s° (in kcal mole^{-1}) of Halides in Methanol

Salt	$-\Delta H_S^\circ$	Ref.	Salt	$-\Delta H_S^\circ$	Ref.
LiCl	214.0	78, 81	$LiClO_4$	188.0	78
LiBr	206.1	81	$NaClO_4$	161.1	78
LiI	199.0	81	$Mg(ClO_4)_2$	583.4	78
NaCl	187.9	81	$Ca(ClO_4)_2$	506.7	78
NaBr	180.8	81	$Su(ClO_4)_2$	475.8	78
NaI	173.7	78, 81	$Ba(ClO_4)_2$	443.3	78
KCl	167.80	81	$Pb(ClO_4)_2$	484.6	78
KBr	167.7	81	$LiNO_3$	205.0	80
KI	152.9	81	$NaNO_3$	179.0	80
RbCl	161.6	13, 81	NH_4Br	157.3	80
RbBr	153.7	81	NH_4NO_3	155.5	80
RbI	146.6	81	$AgNO_3$	189.6	80
CsCl	152.9	13, 78, 81			
CsBr	145.0	78			
CsI	137.8	13, 78, 81			
$ZnCl_2$	686.6	79			
$CaCl_2$	559.6	81			
$SrCl_2$	524.4	81			
$BrCl_2$	504.8	81			

Assuming that the contribution of the anion to the enthalpy of solvation vanishes for $1/r_x = 0$, the constant A_{Li} is identified with the experimental ionic solvation enthalpy of the cation. From this value for Li^+ and the molal solvation enthalpy of the salt, the experimental ionic solvation enthalpies of the halides may be deduced.

Table 2
Enthalpies of Solvation $-\Delta H_s^\circ$ (in kcal mole^{-1}) of Perchlorates in Ethanol, Propanol, and Butanol

Salt	Ethanol	Propanol	Butanol	Ref.
$LiClO_4$	187.0	186.6	186.6	404
$Pb(ClO_4)_2$	478.1	476.4	475.5	404
$Mg(ClO_4)_2$	582.3	576.4	575.2	405
$Ca(ClO_4)_2$	500.3	4 6.7	490.1	405
$Sr(ClO_4)_2$	466.9	464.6	463.6	405
$Ba(ClO_4)_2$	435.4	433.1	432.6	405

Table 3
Enthalpies of Solvation (in kcal mole^{-1}) of Salts in Ethanol, Ethylene Glycol, and Formic Acid

Salt	Ethanol	Ref.	Ethylene glycol	Ref.	Formic acid	Ref.
LiCl	−217.0	12	—	—	—	—
NaCl	−185.6	12	—	—	−186.0	13
NaBr	−178.0	12	—	—	—	—
NaI	—	—	—	—	—	—
KCl	−166.7	25	−173.2	15	−170.0	13
KBr	−159.4	25	—	—	—	—
KI	—	—	—	—	—	—
RbCl	—	—	−154.9	15	—	—
CsCl	—	—	—	—	−164.0	13
CsI	—	—	—	—	−157.0	13
$NaClO_4$	−159.4	16	—	—	−140.0	13

The same procedure has been applied to the alkali salts, and in all cases a linear relation was obtained. The slopes of the plots were the same within 0.2% except in the case of Cs^+. No reason could be found for this deviation.

The ionic contributions to solvation enthalpies in formamide and water are listed in Table 6.

In the treatment of data on propylene carbonate solutions, Wu and Friedman[28] tried to apply the following Latimer–Pitzer–Slansky[29] formula for the free energy of solvation of an ion of charge e_i [see Section (iv)] and Pauling crystal radius r_i in a solvent of dielectric constant ε:

$$\Delta G_i = -e_i^2[1 - (1/\varepsilon)]/2(r_i + \delta_i) \tag{9}$$

where δ_i is a parameter depending on the sign of the ion charge and on the solvent. This is a modification of the Born equation for the solvation energy in which the parameter δ_i depends on variations in the dielectric saturation, structure changes, and other effects.[10]

Differentiation of equation (9) gives the corresponding enthalpy of solvation:

$$\Delta H_i = \Delta G_i\left[1 - \frac{T\partial\varepsilon/\partial T}{\varepsilon(\varepsilon - 1)} + \frac{T\partial\delta/\partial T}{r_i + \delta_i}\right] \tag{10}$$

Table 4
Enthalpies of Solvation (in kcal mole^{-1}) of Halides in Formamide (1), *N*-Methylformamide (2), *N*-Methylacetamide (3), *N*-*N*-Dimethylformamide (4)

Salt	(1)	Ref.	(2)	Ref.	(3)	Ref.	(4)	Ref.
LiF	−240.2	17	—	—	—	18	—	23
LiCl	−210.6	17	−214.3	19	—	18	−215.7	23
LiBr	−240.6	17	—	—	−198.7	18	−212.5	23
LiI	−195.7	17	−198.5	18	—	18	−198.5	24
NaF	−216.5	17	—	—	—	18	—	—
NaCl	−187.9	17	−182.0	19	—	18	—	—
NaBr	−181.0	17	−181.0	19	—	18	−184.0	23
NaI	−171.9	17	−172.7	18, 19	−171.6	18	−178.5	23
KF	−196.7	17	−196.1	18	−195.4	18	—	—
KCl	−168.1	17	−168.5	18, 19	−167.6	18	—	—
KBr	−161.2	17	−162.2	18	−161.1	18	−165.3	23
KI	−152.1	17	−154.3	18	−153.3	18	−159.1	24
Rbf	−190.8	17	—	—	—	—	—	—
RbCl	−162.1	17	—	—	—	—	—	—
RbBr	−155.2	17	—	—	—	—	—	—
RbI	−146.3	17	−148.2	18	−147.0	18	—	—
CaF	−180.5	17	—	—	—	18	—	—
CsCl	−154.3	17	−154.3	19	—	18	—	—
CsBr	−147.2	17	—	—	—	18	—	—
CsI	−138.1	17	−139.6	18	−138.4	18	—	—

Table 5
Enthalpies of Solvation (in kcal mole^{-1}) of Salts in Acetone (1), Acetonitrile (2), Dimethylsulfoxide (3), and Propylene Carbonate (4)

Salt	(1)	Ref.	(2)	Ref.	(3)	Ref.	(4)	Ref.
LiCl	—	—	—	—	−214.3	22	−217.1	21
LiBr	—	—	—	—	−208.4	22	−198.9	21
LI	—	—	—	—	−201.2	22	−192.5	21
NaCl	—	—	—	—	−189.1	20	−181.1	21
NaBr	—	—	—	—	−183.2	20	−176.1	21
NaI	−176.2	16	−172.6	14	−176.0	20	−169.7	21
KCl	—	—	—	—	−170.7	20	−163.1	21
KBr	—	—	—	—	−164.8	22	−158.6	21
KI	—	—	—	—	−157.6	20	−152.2	21
CsCl	—	—	—	—	−155.6	20	−151.2	21
CsBr	—	—	—	—	−149.7	20	−146.0	21
CsI	—	—	—	—	−142.6	20	−136.6	21
$LiClO_4$	—	—	—	—	—	—	−120.0	21
$NaClO_4$	−167.0	16	—	—	—	—	−161.4	21

Assuming that $\partial\delta/\partial T$ is negligible, the linearity of equation (10) is as good as can be expected when applied to the transfer enthalpy of cations from water to propylene carbonate, but not to the solvation enthalpies of the salts. However, assuming[21] that for the largest monatomic ions the modified Born equation (of Verwey)

$$\Delta H_i = B/(r_i + \delta) \tag{11}$$

with $\delta = 0$, yields the electrostatic part of the enthalpy of transfer ΔH_{tr} from water to propylene carbonate, the contributions of Cs^+ and I^- were found to be 0.6 and 0.5 kcal mole^{-1}, respectively. Assuming furthermore that the other structural contributions are also nearly the same for Cs^+ and I^- Wu and Friedman conclude that

$$\Delta H_{tr}(Cs^+) = \Delta H_{tr}(I^-) = \tfrac{1}{2}\Delta H_{tr}(Cs^+ + I^-) = -3.6 \quad \text{kcal mole}^{-1}$$

Similar assumptions for formamide,[17] N-methylformamide,[18,19] and N-methylacetamide[18] give the values -2.9, -3.6, and -3.0 kcal mole^{-1}, respectively, for the ionic enthalpies of transfer of Cs^+. In dimethylsulfoxide, the tetraphenylarsonium cation and the tetraphenylboride anion are assumed to have equal enthalpies of transfer[20] since the main difference between the two ions is the sign of their respective charges, which lie buried within similar organic envelopes. Indeed, the same pattern of anionic behavior as in the previous cases was observed[30] for ionic enthalpies of transfer.

The ionic enthalpies of solvation were calculated and the values are listed in Table 7.

Table 6
Experimental Ionic Enthalpies of Solvation [in kcal (g-ion)$^{-1}$]

Ion	$-\Delta H_s$(water)	$-\Delta H_s$(formamide)
Li^+	115.0	128.5
Na^+	89.3	105.2
K^+	69.1	85.3
Pb^+	63.1	79.6
Co^+	54.7	71.4
F^-	128.7	111.3
Cl^-	95.4	82.5
Br^-	87.6	75.8
I^-	77.2	66.8

Table 7

Ionic Enthalpies of Solvation* (in kcal mole^{-1}) in Formamide (1), *N*-Methylformamide (2), *N*-Methylacetamide (3), Propylene Carbonate (4), and Dimethylsulfoxide (5)

Ion	(1)	(2)	(3)	(4)	(5)
Li^+	−132.4	−134.4	−135.6	−128.5	−138.1
Na^+	−108.7	−108.8	−108.3	−105.6	−113.1
K^+	−88.8	−89.2	−89.9	−87.2	−93.6
Rb^+	−82.9	−83.7	−83.2	−83.8	—
Cs^+	−74.9	−75.6	−75.0	−75.6	−79.8
Cl^-	−81.5	−81.6	−80.0	−77.8	−76.4
Br^-	−79.4	−80.1	−78.3	−77.5	−76.6
I^-	−67.0	−67.1	−67.1	−67.7	−66.6
F^-	−109.3	−108.3	−107.7	—	—

*The enthalpies were calculated from absolute ionic hydration enthalpies[2] and ionic transfer enthalpies.

2. Free Energies of Solvation

The standard free energy of solvation may be defined as the change in free energy experienced by a pair of ions being transferred from the gaseous state into a specified solvent under standard conditions of fugacity and activity. The free energies of formation of pairs of cations and anions in the gaseous state may be calculated, thereby enabling us to obtain the solvation free energy from the measured free energy of solution.

(i) Solubility Method

The standard free energy of solution for a solute is

$$\Delta G^\circ_{sol} = -RT \ln K \qquad (12)$$

where K is the equilibrium constant for the equilibrium between the solid salt and the ions in solution,

$$CA \leftrightharpoons C^+ + A^-$$

In dilute solutions, it is possible to introduce the Debye–Hückel expression for the mean activity coefficients and to obtain the following expression for the free energy of solution:

$$\Delta G^\circ_{sol} = 2.303RT - \nu \log m - \log(\nu_+^{\nu_+} \nu_-^{\nu_-} + A d_0^{1/2} m^{1/2}) \qquad (13)$$

where d_0 is the density of the solvent, A is the Debye–Hückel limiting slope, and ν is the sum of the number of positive ions ν^+ and negative ions ν^-.

(ii) EMF Method

The emf of a chemical cell is directly related to the changes in the free energy involved in the solvation of the ions. In cells that are reversible with respect to both anions and cations, such as

$$\text{M(Hg)}|\text{MX}|\text{AgX} - \text{Ag} \qquad (14)$$

the standard free energy of solution of the salt $\Delta G^\circ_{\text{sol}}$ is obtained from the standard electrode potential of the cell E°, where

$$-\Delta G^\circ_{\text{sol}} = nFE^\circ \qquad (15)$$

$\Delta G^\circ_{\text{sol}}$ may then be combined with energies of dissociation, sublimation, ionization and the lattice energy to obtain the free energy of solvation.

(iii) Medium Effect Method

Using electrochemical cells with transport, the difference of the standard free energies is obtained directly as the free energy of transfer. This may be achieved by combining, e.g., two amalgam cells of the type considered previously, in which the electrolyte is in different media, such as water and another solvent:

$$\text{Ag} - \text{AgX}|\text{MX(water)}|\text{M(Hg)}|\text{MX(solvent)}|\text{AgX–Ag} \qquad (16)$$

The net effect of the cell reaction corresponds to the transfer of MX from one solvent (water) to the other solvent. The standard potential of this double cell is given by the difference in the standard potentials of the aqueous cell ${}_wE^\circ_{\text{MX}}$ and the nonaqueous cell ${}_SE^\circ_{\text{MX}}$. This is, by definition, the medium effect,[38]

$$\ln \gamma_0 = ({}_wE^\circ_{\text{MX}} - {}_SE^\circ_{\text{MX}})/RT \qquad (17)$$

By referring both standard states to the pair of ions in the gaseous state, it may be shown that, in fact, the medium effect represents the difference between the solvation and hydration free energies.

Using these three methods, Izmailov[33] determined and calculated solvation free energies in various solvents. The values thus obtained are listed in Tables 8 and 9 together with more recent values.

Table 8
Solvation Free Energies $-\Delta G_s^\circ$ (in kcal mole^{-1} at 25°C) of Acids and Salts in Nonaqueous Solvents

Electro-lyte	NH_3 (Ref. 33)	N_2H_4 (Ref. 82)	CH_3OH (Ref. 33)	C_2H_5OH (Ref. 33)	C_4H_9OH (Ref. 83)	CH_3COCH_3 (Ref. 82)	$C_5H_{11}OH$ (Ref. 82)	CH_3CN (Ref. 82)	HCOOH (Ref. 33)
HCl	346.2	348.0	325.1	324.3	323.0	322.0	322.5	313.0	323.6
HBr	341.8	345.5	319.8	318.5	320.0	320.0	316.5	311.5	314.0
HI	338.1	324.0	312.9	310.3	309.5	309.0	311.0	306.5	305.5
LiCl	188.9	187.0	187.7	184.8	186.0	185·0	186.0	178.5	194.3
LiBr	186.5	184.5	182.3	179.7	183.0	183.0	180.0	177.0	184.7
LiI	180.9	173.0	175.4	171.5	172.5	172.0	174.5	172.0	176.2
NaCl	164.2	165.5	164.2	160.6	160.0	150.0	155.5	154.5	177.8
NaBr	161.8	163.0	159.9	156.8	157.0	148.0	149.5	153.0	168.2
NaI	156.2	152.5	152.2	148.1	146.5	137.0	144.0	148.0	158.7
KCl	—	148.5	—	—	140.5	138.0	140.0	139.0	159.3
KBr	—	146.0	—	—		136.0	134.0	137.5	149.7
KI	136.8	135.5	135.2	131.5	127.0	125.0	128.5	132.5	141.2
RbCl	—	143.5	—	—	131.5	132.0	135.5	184.5	156.3
RbBr	—	141.0	—	—	128.5	130.0	129.5	133.0	146.7
RbI	130.6	130.5	—	125.5*	118.0	119.0	124.0	128.0	138.2
CsCl	130.9	135.0	131.4	—	128.5	127.0	127.5	125.5	143.6
CsBr	128.5	132.5	127.4	—	125.5	125.0	121.0	124.0	134.7
CsIs	122.8	122.0	120.0	117.0*	115.0	113.0	115.5	119.0	126.2
AgCl	199.7	202.0	178.9	178.6	—	—	—	179.0	197.1
AgBr	196.3	199.5	173.6	173.6	—	—	—	177.5	188.7
AgI	190.6	189.0	166.7	164.6	—	—	—	172.0	180.2
$CaCl_2$	493.6	507.0	—	—	—	—	—	478.8	465.9
$ZnCl_2$	666.0	647.5	624.6	617.5	—	—	—	600.0	644.9
$CdCl_2$	677.0	590.5	560.4	558.4	—	—	—	540.0	567.0

* Ref. 43.

Table 9
Solvation Free Energies $-\Delta G_s^\circ$ (in kcal mole^{-1}) at 25°C of Salt in Nonaqueous Solvents

	Solvent		
Electrolyte	DMF (Ref. 31)	NMF (Ref. 135)	Nitromethane (Ref. 413)
LiF	219.8	—	—
LiCl	193.5	203.2	—
LiBr	190.3	195.0	—
LiI	185.6	—	—
NaF	193.1	—	—
NaCl	166.8	178.9	—
NaBr	163.6	170.7	—
NaI	158.9	—	—
KF	175.5	—	—
KCl	149.2	160.8	—
KBr	146.0	152.6	—
KI	141.3	—	—
CsCl	135.8	139.7	—
CsBr	132.6	131.5	—
CsI	127.9	—	—
$LiClO_4$	—	—	185.85
$NaClO_4$	—	—	156.17
$KClO_4$	—	—	134.11
$RbClO_4$	—	—	127.9
$CsClO_4$	—	—	119.1

(iv) Determination of the Ionic Contributions to the Free Energy of Solvation

Many approaches to this problem were proposed in efforts to establish a single scale for single ion activities in any solvent,[34,35] to set up well-defined absolute standard potentials,[36] and to calculate liquid junction potentials at an aqueous–nonaqueous interface.[37]

(a) The modification of Born's equation (see Section IV.2) as proposed by Latimer *et al.*[29] was fitted by Strehlow[36] to experimental free energies of solvation ΔG_s in various nonaqueous solvents:

$$-\Delta G_s = \frac{Ne^2}{2}\left(1 - \frac{1}{\varepsilon}\right)\left(\frac{1}{r_+ + R_+} + \frac{1}{r_- + R_-}\right) \tag{18}$$

where ε is the dielectric constant, e the charge of the ion, r_+ and r_- the crystal ionic radii of the cation and anion, respectively, and R_+

and R_- are effective increments to account for dielectric saturation, ion–solvent interaction, etc. The ionic contributions are explicitly given in expression (18) and their values can be calculated. Increments in the radii obtained by Strehlow are summarized in Table 10.

This approach was based[39] on the redox equilibrium between ferrocene and its oxidized form, the ferricinium ion:

$$C_6H_5FeC_6H_5 \leftrightarrows C_6H_5Fe^+C_6H_5 + e^- \tag{19}$$

Since the reduced form is neutral, it was assumed that only the ferricinium ion is solvated. The solvation free energy was found to be the same[40] in every solvent, $\Delta G = -52.82$ kcal mole^{-1}, since the oxidation potential did not vary. It was shown, however, that the above couple may be involved in some specific interaction with water[41] and possibly with the nonaqueous solvent. No physical significance may be assigned to equation (18), since no allowance is made for the use of different radii in the gaseous reference state.

(b) An early attempt was made by Pleskov[42] to correlate the standard potentials in various solvents. Rubidium was chosen as the reference in every solvent because of its large radius and low polarizability. However, the assumption of zero transfer free energy cannot be justified under these circumstances.

(c) Izmailov[33] suggested that the free energy of solvation of ions tends to zero with increase in ionic radius. He combined the sums and differences of solvation free energies of ions, obtained by thermodynamic means, for a series of ions of increasing crystallographic radius and extrapolated to infinite radius. The extrapolated value was taken as the ionic free energy of solvation.

He also proposed a second method[43] based on the similarity of solvation energies of isoelectric cations and anions, and on the

Table 10
The Increments R_+ and R_- of the Modified Born Equation for Various Solvents

	H_2O	CH_3OH	CH_3CN	HCOOH	$HCONH_2$	Sulfolane
R_+	0.85	0.81	0.72	0.78	0.85	0.85
R_-	0.25	0.37	0.61	0.38	0.25	—

decrease of these energies with increase in the principal quantum number n of the vacant orbitals. The function to be extrapolated is obtained by adding the sum of the solvation energies of the hydrogen and halide ions to the difference of the solvation energies of the hydrogen and alkali ions, and dividing by two. The result is then plotted versus $1/n^2$ for various values of n and the desired value $\Delta G_s(M^+)$ obtained.

Similar values of the ionic free energy of solvation were obtained by these two methods.[83] However, these long-range extrapolations involve some uncertainties which are reflected by lack of internal consistency in the single-ion medium effects obtained.

(d) In other similar approaches, it has been tacitly assumed that a rule of corresponding states applies between solvation and hydration. Feakins *et al.*[44,45] showed that for the free energy of transfer ΔG°_{tr} from water to water–methanol mixtures, plots of ΔG°_{tr} versus r_c^{-1} for Li^+Cl^-, Na^+Cl^-, and K^+Cl^- and of ΔG°_{tr} versus r_a^{-1} for H^+Cl^-, H^+Br^-, and H^+I^- were approximately linear (r_c and r_a are the cation and anion radii, respectively). The sets of values obtained from the two equations

$$\Delta G^\circ_{tr}(H^+X^-) = \Delta G^\circ_{tr}(H^+) + ar_a^{-1} \tag{20}$$

$$\Delta G^\circ_{tr}(H^+Cl^-) = \Delta G^\circ_{tr}(Cl^-) + br_c^{-1} \tag{21}$$

were said to be in sufficiently good agreement to suggest that a semi-quantitative description of the system had been found.

Bearing in mind that by definition

$$\Delta G^\circ_{tr}(M^+X^-) = G^\circ_s(M^+X^-) - G^\circ_w(M^+X^-) \tag{22}$$

and, generally, in the absence of ion-pairing,

$$\Delta G^\circ(M^+X^-) = G^\circ(M^+) + G^\circ(X^-) \tag{23}$$

where G°_s and G°_w represent standard free energies of solvation and hydration respectively, equations (20) and (21) reduce to

$$G^\circ_s(X^-) = G^\circ_w(X^-) + ar_a^{-1}$$

$$G^\circ_s(M^+) = G^\circ_w(M^+) + br_c^{-1}$$

These may be looked upon as a particular case of the method of comparative measurements.[46]

De Ligny and Alfenaar[47,48] sought to improve upon the preceding method of extrapolation by considering the free energy of transfer ΔG_{tr} to be composed of an "electric part" ΔG_{el} given at low fields by the Born equation, and a "neutral part" $\Delta G_{neutral}$ corresponding to the contribution of an uncharged particle of the same size as the ion. They thought it likely that for large particles, $\Delta G_{neutral}$ is proportional to the particle cross section (i.e., to r^2) and suggested extrapolating the difference $\Delta G_{tr} - \Delta G_{neutral}$ to $r^{-1} = 0$ to minimize any discontinuity. The results thus obtained are claimed to be very accurate,[48] involving very significant contributions of $\Delta G_{neutral}$ which were calculated from the transfer of isoelectronic rare gases. However, this does not agree with Izmailov's estimate that values for $\Delta G_{neutral}$ are only a few per cent of values of ΔG_s.

The application of a nonelectrolytic correction attributed to structural effects[45] did not improve the linearity of the plots but rather implied a change of sign in the ΔG°_{tr} of large cations.

(e) Another proposal was to choose a reference electrolyte[49] whose solvation free energy could be divided equally between the anion and the cation. The reference electrolyte would have to consist of large, symmetric cations and anions, identical in size, surface charge, density, polarizability, and solvation properties, since it is assumed that the two ions experience equal solvation free-energy changes. Tetraphenylphosphonium tetraphenylborate,[51] tetraphenylarsonium tetraphenylboride,[50] and tri-isoamyl-n-butylammonium tetraphenylborate ($TBABPh_4$)[49,52] are composed of very large ions with a central charged atom buried under insulating layers and are of approximately equal crystallographic size.

The electrolyte $TBABPh_4$, recommended[53] for the evaluation of single-ion conductance in nonaqueous solvents, has a low solubility in many solvents, which makes it convenient for calculating the free energies of solvation without considering any activity correction. However, as mentioned above, this assumed equality of the free energies of solvation is based on the transport properties of the ions, specifically the equality of the Stokes radii in the various solvents. No simple correlation necessarily exists between these transport properties and the solvation free energy of an ion,[54] because of other factors, such as dielectric relaxation and the size and solvation number of the ion in the solvent. It is further claimed that even for a more ideal reference electrolyte, such as tetra-isoamylammonium

Table 11
Ionic Free Energies of Transfer from Water to Methanol (in kcal mole^{-1} at 25°C)

Ion	Ref. 48	Ref. 45	Ref. 36	Ref. 33	Ref. 49	Ref. 50	Ref. 42	Ref. 56
H^+	−1.974	−2.93	0.07	+3.1	1.85	—	0.3	−5.84
Li^+	—	−3.8	—	+1.0	—	—	—	−6.7
Na^+	−1.830	−2.9	−0.55	+3.0	—	—	—	−5.9
K^+	−1.223	−2.0	−0.17	+2.0	1.80	1.7	—	−4.9
Rb^+	—	—	−0.06	2.8	0.89	—	0.0	−5.3
Cs^+	—	—	—	—	—	—	—	−5.5
Ag^+	—	—	—	—	0.78	1.2	−0.3	—
Cl^-	7.768	8.57	5.46	−3.0	2.08	1.9	3.7	11.20
Br^-	7.299	9.05	4.81	−1.0	−0.95	—	—	10.60
Z^-	6.222	7.02	4.03	0.2	—	—	—	9.70
SO_4	—	—	—	—	—	—	—	15.18
Picrate$^-$	—	—	—	—	—	−1.2	—	—
TAB^+	—	—	—	—	−4.30	—	—	—
BPh_4^-	—	—	—	—	−4.30	−4.2	—	—

boride,[55] the problem would still persist, especially in dipolar aprotic solvents, which differentiate strongly between the relative solvation of anions and cations.

It is instructive to compare the values (listed in Table 11) of ionic free energies of transfer from water to methanol obtained by the various methods. The results of Case and Parsons[56] for the real free energies of transfer (Section IV.1) are included for comparison.

The disparity in results is only too well illustrated. Evidently, in the various approaches to evaluation of single-ionic free energies there is generally little agreement on the magnitude and sometimes on the sign of the free energy of transfer. It should be kept in mind that the evaluation of individual ionic thermodynamic properties is not possible by thermodynamic means and each value obtained depends on the assumptions made and theoretical principles employed.

3. Entropies of Solvation

A thorough discussion of the various methods of determination of entropies of solvation has been given by Conway and Bockris.[10] Most of the data available in the literature were obtained either from the thermodynamic relationship

$$(\Delta G - \Delta H)/T = -\Delta S \tag{24}$$

from solubility measurements, or from measurements of changes of emf with temperature. In some cases, the values of the free energies of solvation were substituted in the modified Born equation in order to obtain the entropy of solvation.[31] The data available in the literature are given in Table 12.

The standard entropy of solvation is given by

$$\Delta S_s^\circ = \overline{S_2^\circ} - S_g^\circ \tag{25}$$

where $\overline{S_2^\circ}$ is the standard partial molal entropy of the electrolyte and S_g° is the standard molar entropy of the electrolyte in the gas phase calculated from the Sackur–Tetrode equation.[32] Therefore, $\overline{S_2^\circ}$ is the thermodynamic property to be considered in the estimation of the ionic contributions to the molal entropies.

A generalization of Mischenko's assumption[12] for estimating the ionic heats of solvation was applied to standard partial gram-ionic entropies in nonaqueous solvents.[58–63] The total entropies of the solvated electrolytes were divided into their ionic contributions

Table 12
Standard Entropies of Solvation $-\Delta S_s^\circ$ [in cal (deg mole)$^{-1}$; molal scale]

	Solvent				
Electrolyte	DMF (Ref. 61)	MeOH (Ref. 62)	EtOH (Ref. 62)	NMF (Ref. 135)	HCOOH
LiCl	77.3	86.2	89.8	63.7	—
LiBr	73.7	80.2	83.1	—	—
LiI	71.0	76.5	82.4	—	—
NaCl	71.8	75.3	81.2	48.7	55.7
NaBr	68.2	69.9	74.4	47.7	—
NaI	65.5	65.6	73.7	—	—
KCl	68.3	58.5	73.6	45.4	60.1
KBr	64.7	54.5	66.8	—	—
KI	62.0	—	66.1	—	—
CsCl	62.1	—	—	42.7	88.8
CsBr	58.5	—	—	—	—
CsI	55.8	—	—	—	111.5
$CdCl_2$	—	—	1255[(116)]	—	—

by assigning, by trial and error, a value for the partial gram-ionic entropy $\overline{S_2^\circ}$ for the hydrogen ion in such a manner that for both cations and anions, in a given solvent, it falls on the same line when plotted against $\overline{S_2^\circ}$ of the corresponding ions in water. This treatment carried out for all the solvents listed in Table 12 has confirmed[61] that a linear relationship exists in every case:

$$\overline{S_2^\circ}(\mathrm{X}) = a + b\overline{S_2^\circ}(\mathrm{H_2O}) \tag{26}$$

where a and b are constants characteristic of the solvent and $\overline{S_2^\circ}$ (H_2O) is the absolute entropy of the corresponding ion in water. The values for the ionic entropies are listed in Table 13.

An attempt was made[61] to explain the gradual increase of $\overline{S_2^\circ}$ in the following solvents

$$\mathrm{NH_3} < \mathrm{DMF} \simeq \mathrm{EtOH} < \mathrm{MeOH} < \mathrm{NMF} < \mathrm{F} < \mathrm{H_2O} < \mathrm{D_2O}$$

in terms of structure changes. The entropy of solvation was assumed to be made up of three quantities: (1) An entropy decrease caused by the loss of freedom of the gaseous ions, ΔS_F°; (2) an entropy increase caused by the disordering of the solvent structure, ΔS_D°; and (3) an entropy decrease resulting from the ordering of the solvent molecules around the ions, ΔS_O°. Hence,

$$\overline{S_2^\circ} - S_g^\circ = \Delta S_s = \Delta S_D^\circ + \Delta S_O^\circ + S_F^\circ \tag{27}$$

Assuming ΔS_O° to be approximately constant for solvents consisting of molecules with not too widely different dipole moments

Table 13
Standard Absolute Ionic Entropies of Solvation $-\Delta S_s^\circ$ (mole fraction scale) in Nonaqueous Solvents [in cal (deg mole)$^{-1}$ at 25°C][61]

Ion	NH_3	DMF	EtOH	MeOH	NMF	Formamide
Li^+	53.1	48.5	47.5	42.7	41.6	34.3
Na^+	50.0	43.2	44.6	39.5	31.8	33.2
K^+	45.7	39.6	38.5	33.2	28.5	25.9
RB^+	41.0	—	—	30.0	—	24.7
Cs^+	42.0	35.7	—	28.8	25.8	23.6
Cl^-	40.4	38.6	37.8	32.4	28.1	25.6
Br^-	43.4	35.5	35.7	29.8	27.0	23.0
I^-	39.1	32.4	31.0	26.0	—	18.9

Table 14
Constants of Equation (29)

Ion	k'_2, cal (deg^2 mole)$^{-1}$	C, cal (deg mole)$^{-1}$	Δ*
H^+	0.22 ± 0.04	-47.1	± 3.6
Li^+	0.19 ± 0.02	-39.7	± 2.2
Na^+	0.20 ± 0.02	-32.6	± 2.6
K^+	0.23 ± 0.02	-29.5	± 2.2
Cs^+	0.22 ± 0.02	-20.0	± 2.5
Cl^-	0.19 ± 0.02	-24.6	± 2.5
Br^-	0.23 ± 0.02	-24.5	± 1.8
I^-	0.25 ± 0.02	-20.7	± 2.6

* Δ is the standard deviation between the calculated and the experimental values.

and ΔS_F° to be given by the entropy change on dissolution of a gas to form an ideal solution, it is suggested that

$$\overline{S_2^\circ} = kS_{str} + C \tag{28}$$

where S_{str} is that part of the entropy that arises from the order existing among the solvent molecules, and C is a constant characteristic of the ion, representing its partial molal entropy in a solvent with no internal structure.

Equation (28) was tested by relating S_{str} to the deviation from the ideal boiling point of the solvent, ΔT_{bp}. Linear relationships observed when plotting S_2° versus ΔT_{bp} may be represented by the equation

$$S_2^\circ = k' \Delta T_{bp} + C \tag{29}$$

The values of k' and C are listed in Table 14. Equation (29) is especially useful for estimating entropies of ions when no experimental data are available. However, it was pointed out that ΔT_{bp} is not an entirely satisfactory parameter since it suggests that D_2O is more disordered than water, which does not agree with current views.[63] The values of k' are approximately the same for all the ions. This fact would imply a similar disrupting effect in the vicinity of the ion and does not differentiate between structure changes. The values of C are all negative and point only to reorientation of the solvent molecules.

IV. THEORETICAL ASPECTS OF SOLVATION

1. Real Free Energy of Solvation

The real free energy of solvation of an ion in a solvent is the free energy change that would accompany the transfer of the ion from the gas phase into the solvent passing *through* the surface of the solvent.[64] It is equal to

$$\alpha_i(\mathrm{S}) = G_i + Z_i F \chi(\mathrm{S}) \tag{30}$$

where $\chi(\mathrm{S})$ is the surface potential of the solvent, Z_i is the charge on the ion, F the Faraday, and G_i is a measure of the interaction of the ion with the bulk of the solvent and represents the "chemical" free energy of solvation. The last term represents the electric energy for passage of the ion through the surface of the solvent.

The real free energy of solvation may be determined from measurements of Volta potential differences in Galvanic cells.[2] The quantity measured is the compensation potential[45] between an aqueous and a nonaqueous solution, which gives the real free energy of single ions going from one solvent to another:

$$\alpha_i = \alpha_i^S - \alpha_i^w \tag{31}$$

The absolute real ionic free energies of solvation in nonaqueous solvents reported in the literature were obtained in all cases[56,65–69] by using Randle's results for aqueous solutions.[64] Values are listed in Table 15. There may be some uncertainty in the values listed due to the error introduced by assuming (a) that in aqueous solutions, the ionic activity coefficients (which are thermodynamically not defined), of transfer for K^+ and Cl^-, are equal at 0.1 m, and (b) that in methanol, the activity coefficients of NaCl and HCl are equal.

2. Electrostatic Theory of Solvation

Attempts to calculate the free energy of solvation have been only partially successful and were mostly directed specifically toward the calculation of the free energy of hydration.

The earliest model was that due to Born,[70] who assumed the ion to be a rigid sphere of radius r_i and charge $Z_i e$ in a continuous dielectric medium of dielectric constant ε. The free energy of solvation was calculated as the change in electrostatic free energy ΔG when a charge is transferred to the solvent from a vacuum.

Table 15
Real Free Energies of Ions in Some Nonaqueous Solvents [in kcal (g-ion)$^{-1}$ at 25°C]

Ion	Methanol (Ref. 56)	Methanol (Ref. 67)	Ethanol (Ref. 56)	Butanol (Ref. 56)	Formamide (Ref. 56)	Formic acid (Ref. 56)	Acetonitrile (Ref. 56)
H^+	266.3	265.1	265.9	264.5	264.0	253.8	257.8
Li^+	228.8	127.4	127.1	127.5	—	124.3	121.1
Na^+	104.1	103.9	102.0	101.5	—	102.3	98.8
K^+	86.1	87.1	86.7	82.0	85.0	83.8	83.0
Rb^+	80.8	81.1	80.7	73.0	77.9	80.8	78.2
Cs^+	76.5	71.1	72.7	70.0	—	71.8	70.3
Cl^-	59.5	60.5	58.8	58.5	66.8	68.2	55.7
Br^-	54.3	54.2	53.1	55.5	—	60.2	54.2
I^-	47.5	47.8	46.7	45.0	—	50.7	49.5
Tl^+	87.1	—	87.3	—	55.8	—	—
Ag^+	120.8	—	120.7	—	—	121.8	124.5
Cu^+	—	—	—	—	—	—	154.0
Zn^{2+}	494.8	—	489.2	—	491.6	495.6	477.5
Cd^{2+}	442.8	434.3	439.7	—	438.0	434.6	427.6
Cu^{2+}	507.3	—	513.3	—	507.2	—	519.6
Pb^{2+}	372.5	—	369.3	—	368.3	—	351.6
Ca^{2+}	—	—	—	—	—	368.8	368.8

ΔG is given by the equation

$$\Delta G = -\frac{N(Z_i e)^2}{2r_i}\left(1 - \frac{1}{\varepsilon}\right) \tag{32}$$

A difficulty encountered in the numerical evaluation of this equation involves the choice of values for r_i, which should not be the same in vacuum as in the solvent. In addition, neglecting dielectric saturation caused by the intense field near the ions affects the values of ΔG considerably. In order to improve agreement between experimental and calculated values, empirical corrections were made,[29] but these will not be considered here.

An alternative approach,[84] based on Noyes' formulation[85,86] for aqueous solutions, was proposed to apply to any solvent. The free energy of solvation of an ion, ΔG_s°, was defined by

$$\Delta G_s^\circ = \Delta G_{\text{con}}^\circ - Z_i \Delta G_R^\circ \tag{33}$$

where ΔG°_{con} is the conventional standard free energy of solvation of the ion (and by definition $\Delta G^\circ_{con} = 0$ for H^+) and ΔG°_R is a reference free energy of solvation. The value of the conventional free energy of solvation was obtained from thermodynamic data[57] (ionization potentials, dissociation and sublimation energies, emf, etc.). The solvation energy ΔG°_{con} was assumed to be strictly electrostatic (ΔG°_{el}), the solvation being considered as occurring in three steps: discharge of the gaseous ion, solvation of the neutral species, and the recharging of the solvated neutral species. The actual electrostatic charging energy was represented as a power series expansion in the reciprocal of the radius, the first term being of course that corresponding to the Born equation,

$$\frac{-\Delta G^\circ_{con}}{Z} = \Delta G^\circ_R + \frac{Z_i k}{r} + \frac{Z_i A}{r^2} + \frac{Z_i B}{r^3} + \cdots \tag{34}$$

The constants $A, B, \ldots$ obtained by least-squares fitting of equation (34) are not assumed to have any physical significance but represent the deviations from a continuum dielectric model. They may possibly be interpreted as arising from specific ion–solvent interactions, dielectric saturation and ion–quadrupole interactions (cf. Buckingham[87]). Values obtained for the ionic free energies of solvation in liquid ammonia are given in Tables 16 and 17.

This method of extrapolation to infinite ionic size improves on earlier approaches but is open to the same criticism as that of other methods.

Table 16

Ionic Free Energies of Solvation $-\Delta G^\circ_s/Z_i$ in Liquid Ammonia (in kcal mole^{-1} at 25°C) for Ions Having Inert Gas Structures

Ion	Radius, Å	$-\Delta G^\circ_s/Z_i$	Ion	Radius, Å	$-\Delta G^\circ_s/Z_i$
H^+	—	283.0	Be^{2+}	0.31	299.0
Li^+	0.60	129.6	Mg^{2+}	0.65	235.8
Na^+	0.95	101.6	Ca^{2+}	0.99	195.0
K^+	1.33	81.6	Sr^{2+}	1.13	183.0
Rb^+	1.48	68.0	Br^{2+}	1.35	164.2
Cs^+	1.69	70.9	Al^{3+}	0.50	375.0
Cl^-	1.36	63.2			
Br^-	1.81	58.9			
I^-	1.95	55.2			

Table 17
Ionic Free Energies of Solvation $-\Delta G_s^\circ/Z_i$ in Liquid Ammonia (in kcal mole^{-1})

Ion	Radius, Å	$-\Delta G_s^\circ/Z_i$	Ion	Radius, Å	$-\Delta G_s^\circ/Z_i$
Mn^{2+}	0.80	228.0	Hg^{2+}	1.10	245.4
Fe^{2+}	0.76	243.9	Fe^{3+}	0.64	366.0
Co^{2+}	0.74	254.4	Co^{3+}	0.63	394.2
Ni^{2+}	0.72	266.0	Cu^{+}	0.96	162.5
Cu^{2+}	0.74	270.7	Ag^{+}	1.26	137.2
Zn^{2+}	0.74	259.8			
Cd^{2+}	0.97	242.0	Tl^{+}	1.40	90.7
Sn^{2+}	1.12	192.8	Pb^{2+}	1.20	192.4

A more rigorous approach is based[4,71] on the fundamental equation[72] for the free energy of a dielectric continuum in an electrostatic field:

$$\Delta G = \left(\frac{1}{4\pi}\right)\int\int \mathbf{E}\cdot d\mathbf{D}\,dv \tag{35}$$

where **E** and **D** are vectors representing field strength and dielectric displacement, respectively, in a volume element dv.

If the dielectric constant is assumed to be independent of the field strength, i.e., no dielectric saturation takes place, (35) may be shown to reduce to Born's equation (32) when referred to the standard gaseous state in which the ion has the same radius as in the solvent considered.

There is a danger of misusing the classical free-energy density of a continuous dielectric

$$\delta G = \varepsilon E^2/8\pi \tag{36}$$

in the calculation of total free energy of solvation. Indeed, ΔG has been calculated[73] by integrating the free-energy density $\delta G = \varepsilon E^2/8\pi = \mathbf{E}\cdot\mathbf{D}/8\pi$ over the whole volume v in the equation

$$\Delta G = \left(\frac{1}{8\pi}\right)\int\int \mathbf{E}\cdot\mathbf{D}\,dv$$

with **D** being correctly taken as dependent on the field strength.

However, this has not been assumed when deriving (36) from (35) and the final results are therefore not entirely correct.

Considering an element of volume of solvent dv around an ion to be given by $dv = 4\pi r^2\, dr$, using the differential dielectric constant ε_d defined by*

$$\varepsilon_d = dD/dE \tag{37}$$

and substituting in (35), the basic equation

$$\Delta G = \tfrac{1}{2}\int_{r_e}^{\infty}\int_{0}^{E_r} \varepsilon_d\, d(E^2) r^2\, dr \tag{38}$$

is obtained, where E_r is the field strength at a distance r and r_e is the effective radius of the bare ion in solution.[76] The double integral represents the increase in the electrostatic part of the free energy of the dielectric due to the field of one ion. Since the localization of the energy is a formal description, it may be looked upon as the result of the interaction between the ion and the continuous dielectric. Referring it to the reference gas phase, a general expression was obtained for the free energy of solvation ΔG_s in kcal (g-ion)$^{-1}$,

$$\Delta G_s = \Delta G - (NZ_i r_e^2/2r) \tag{39}$$

The procedure for the computation of the integral in (38) has been described[74,75] and calculations for the free energy of solvation of ions in methanol[76] as a function of the intrinsic radius r_e have been carried out. Good agreement between experimental and calculated values for the medium effect in methanol was reported.[136]

The entropy and enthalpy of solvation may similarly be obtained from the equations

$$\Delta S = \tfrac{1}{2}\int\int \left(\frac{\partial \varepsilon_d}{\partial T}\right)_{P,E} d(E^2) r^2\, dr \tag{40}$$

$$\Delta H = T^2\left[\frac{\partial(\Delta G/T)}{\partial T}\right]_{P,E} \tag{41}$$

3. Thermodynamics of a Dielectric Continuum

Many attempts have been made to treat the thermodynamics of a fluid in an electrostatic field rigorously. However, it was only in 1955, as a result of Frank's[77] clear thermodynamic analysis, that a

*Scalar magnitudes of E and D are normally used since $\mathbf{D}$ and $\mathbf{E}$ lie radially at a spherical ion.

method was found which permits the electrostatic field strength E to be written as a variable of state of the fluid in as general a sense as are the variables of state, pressure P and temperature T. It was possible to extend this treatment to the case where the field changes continuously within the dielectric[4] and to calculate the various thermodynamic properties of ions from the general equation

$$dG^* = -S\,dT + v\,dP - (v\varepsilon_d E/4\pi)\,dE + \sum \mu_i dn_i \tag{42}$$

where G^* is a thermodynamic function analogous to the Gibbs free energy function. It should be stressed, however, that in applying the thermodynamics of a dielectric continuum to the calculation of the various thermodynamic properties of ions, changes in the properties are assigned to the solvent only, although they refer to the ions considered. The role played by the ions as the source of the field strength does not entail any change in their properties either on account of a charging process or the dissolution of gaseous ions, since they are supposed to be initially present in the solution in their final state.

The partial molal volume $\overline{V}°$ of an electrolyte is thermodynamically defined[117] as the actual increase in volume that takes place when 1 g mole of electrolyte is added to an infinite amount of solvent. This change in volume is due to two effects[75]: an increase in volume due to the presence of the electrolyte which occupies an intrinsic volume* V_{in}, and a decrease in volume due to the contraction or electrostriction of the solvent ΔV under the influence of the electric field caused by the ions. The isotonic electrostriction is obtained from (42) by cross-differentiation:

$$v^{-1}\left(\frac{\partial v}{\partial E}\right)_{\mu,T} = \frac{E}{4\pi}\left(\frac{\partial \varepsilon_d}{\partial P}\right)_{E,T} \tag{43}$$

and the total electrostriction is[75]

$$\Delta V = V_{\text{in}} - \overline{V}° = \frac{N_0}{2}\int_{r_e}^{\infty}\int_0^{E_r}\left(\frac{\partial \varepsilon_d}{\partial P}\right)_{E,T} d(E^2)r^2\,dr \tag{44}$$

Integration of the electrostatic volume change is carried out up to r_e, the intrinsic radius of the ion, where by definition, $2.52r_e^3 = V_{\text{in}}$

*This volume may be smaller than that of the ion in the gas phase, due to the contractive pressure of the solvent.

and by using Grahame's relation[118]

$$\varepsilon_d = N_\times^2 + (\varepsilon_0 - N_\times^2)/(1 + bE^2) \tag{45}$$

where $N_\times$ is the refractive index and b is a parameter independent of E and equal to 2.2×10^{-8} for methanol.

The partial molal volume $\overline{V}°$ in methanolic solutions may be calculated from (44) for every r_e. The ionic contributions to the partial molal volume were evaluated[76] taking $\overline{V}° = 6.0$ ml mole^{-1} for the Br^+ ion.[119] Approximately the same value is obtained[120] from the method of Conway, Verrall, and Desnoyers whereby the partial molal volumes of the alkylammonium bromides plotted versus their molecular weights are extrapolated to zero to give the partial molal volume of the bromide counter ion.

It has been found that in methanol,[119] cations are generally larger than in water, whereas anions tend to be smaller. This holds even for large cations such as tetraalkylammonium ions. The difference, however, grows smaller with increasing cation size.

The dependence of the differential dielectric constant ε_d on the distance r from the ion may be seen in Figure 1. In methanol, as in water, the dielectric constant above a certain critical value for r is equal to the bulk dielectric constant (in methanol, $r \doteqdot 12$ Å)

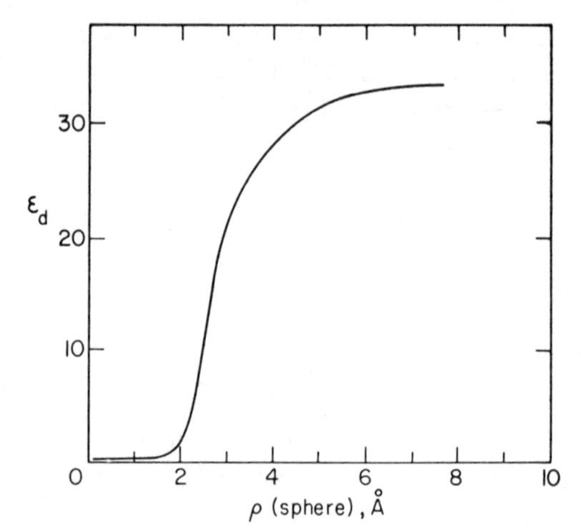

Figure 1. The differential dielectric constant ε_d of methanol as a function of r.

Table 18
Radius of Dielectric Saturation for Nonaqueous Solvents

Solvent	Radius r_0°, Å
Methanol	1.90
Ethanol	2.10
Propanol	2.35
Isopropanol	2.40
Butanol	3.20
Pentanol	3.60
Hexanol	3.80
Glycerol	1.97

and independent of the field strength. If the dielectric constant is less than another critical value r_0°, it is again field-independent. This picture of ions in solution extends the model suggested by Frank and Wen[120] for aqueous solutions to alcoholic solutions as well. The ion is considered to be a small sphere containing the charge and surrounded by three concentric regions. The innermost region is one of "immobilization" (its limit is r_0°, the radius of dielectric saturation[148]), the second is more randomly organized, and the third region corresponds to the bulk solvent.

The dependence of the differential dielectric constant ε_d on r in various solvents was found[121] to be similar to that shown in Figure 1. In Table 18, the radii of dielectric saturation r_0° for various alcohols are given.

The calculation of the electrostatic part of the free energy of the dielectric due to the ions has already been considered. It may also be applied to the evaluation of the "medium effect".[71] The primary medium effect, usually denoted by $\log \gamma^\circ$ and given by Owen[134] as

$$2k \log \gamma^\circ = {}^{w}E_N^\circ - {}^{s}E_N^\circ \tag{46}$$

where ${}^{w}E_N^\circ$ and ${}^{s}E_N^\circ$ are the standard potentials of the pair of ions in water and in another solvent, respectively, represents the effect of transferring a pair of ions from one solvent at infinite dilution to another at infinite dilution where the only effects are the ion–solvent interactions. This is equivalent to the difference between the Gibbs free energies of solvation, or the free energy of transfer. Indeed, the simplest expression of the primary medium effect was given by the Born equation, thereby implying that the standard cell potential is a linear function of the reciprocal of the dielectric constant.[135] This prediction, however, is not borne out by experiment and there is usually considerable departure from a straight line.

The application of the above treatment, which takes into consideration the change of the dielectric constant of the solvent within the ion field, leads to the following expression:

$$\log \gamma^\circ = \frac{1}{2kT}\left[N_\times^2 \int_{r_e}^{\infty} E_r^2 r^2\, dr + \frac{\varepsilon_0 - N_\times^2}{b}\int_{r_0}^{r_0^\circ} \ln(1 + bE_r^2) r^2\, dr\right]_{\text{water}} - \frac{1}{2kT}\left[N_\times^2 \int_{r_e}^{\infty} E_r^2 r^2\, dr + \frac{\varepsilon_0 - N_\times^2}{b}\int_{r_0}^{r_0^\circ} \ln(1 + bE_r^2) r^2\, dr\right]_{\text{solvent}} \tag{47}$$

where N is the refractive index of the solvent, ε_0 is the static dielectric constant, b, r_0, and r_0° are constants characteristic of the solvent, and E_r is the field strength at a distance r from the ion. Calculated values of $\log \gamma^0$ for the transfer of HCl from water to methanol were found to be in reasonable agreement with experimental results.[136]

The calculation of the entropy of solvation should be carried out at a constant chemical potential of the solvent.[77] The change in entropy of the solvent when subjected to an electrostatic field of strength E under these conditions is obtained from equation (42). Cross-differentiation yields

$$(\partial S/\delta E)_{\mu,T} = (vE/4\pi)(\partial \varepsilon_d/\delta T)_{P,E} \tag{48}$$

By integrating around the ion using spherical symmetry, the electrostatic contribution to the entropy of solvation of N ions is found to be

$$\Delta S = \tfrac{1}{2}\int\int (\partial \varepsilon_d/\partial T)_{P,E}\, d(E^2) r^2\, dr \tag{49}$$

This evaluation of the entropy, as well as that of the enthalpy, involves the temperature dependence of certain parameters, which is not yet known. In addition, structural effects certainly contribute significantly to the entropy of solvation, so there is no way to appraise the above calculation for the time being.

It is believed that the most significant consequence of the above treatment is that the picture of an ion in solution as obtained from a continuum theory corresponds quite closely to that obtained by Frank and Wen on molecular grounds and originated by Bernal and Fowler, namely that an ion is a charge contained in a cavity which is surrounded by three concentric regions. The importance of this correspondence between continuum and molecular models whereby the dielectrically saturated layer matches the primary solvated layer,[148] and the other two regions correspond to the secondary solvated layer and the bulk solvent, respectively, should not be overlooked in structural considerations.

4. Structural Approaches

Eley and Pepper[88] attempted to apply an earlier treatment[89] to the solution of ions in methanol. The heat of solvation was evaluated by splitting the process into two stages:

1. A cavity the size of the ion is regarded as being made in the solvent. The internal energy change is generally taken to be 8 kcal, while the reorientation energy of methanol molecules, due to the formation of the hole, is neglected.

2. The ion is removed from the gas phase and placed in the cavity, a process which involves a change in the energy of methanol molecules in the first coordination shell (tetrahedral configuration). The energy of interaction between the primarily solvated ion and the rest of the solvent is given by the Born–Bjerrum equation

$$E = \frac{NZ_i^2 e^2}{2(r_i + 2r_s)}\left(1 - \frac{1}{\varepsilon} - \frac{T}{\varepsilon^2}\frac{\partial\varepsilon}{\partial T}\right) \tag{50}$$

where $2r_s$ is the diameter of a solvent molecule.

Fairly good agreement was obtained with experimental data when the ionic contributions were obtained on the basis of the calculated ratio $\Delta E_{KCl}/\Delta E_{LiCl} = 80/72$. The entropies, however, could not be reproduced by this simple model, which stresses only the electrostatic ion–dipole energy.

Table 19
Ionic Enthalpies of Solvation in Formamide (in kcal mole^{-1} at 25°C)[90]

Ion	Radius, Å	$-\Delta H_s^{tetra}$	$-\Delta H_s^{absolute}$	$-\Delta H_s^{octa}$
Li^+	0.68	72.8	122.9	102.5
Na^+	0.98	65.1	99.6	89.0
K^+	1.33	58.0	79.7	77.3
Rb^+	1.48	55.3	74.0	73.4
Cs^+	1.67	52.5	65.8	68.7
F^-	1.33	94.8	116.9	114.0
Cl^-	1.81	76.7	88.1	92.6
Br^-	1.96	72.4	81.4	87.5
I^-	2.20	66.0	71.4	79.9

It was suggested that the first shell of oriented dipoles may affect solvent molecules further away. A more detailed model[90] based on the Buckingham theory[87] involves a primary solvation layer composed of the nearest solvent molecules rigidly bound in either a tetrahedral or octahedral configuration within a continuum having macroscopic properties of the pure solvent. Within the solvation layers, the energies of interaction between the ion and polarizable spheres with permanent multipole moments and between the spheres themselves are taken into account. The polarization of the solvent due to the solvated entity completes the description of the interaction energy. These detailed interaction energies computed for formamide solutions and the ionic enthalpies of solvation in formamide were calculated for both a tetrahedral and an octahedral model. Values are listed in Table 19. Only the calculated enthalpies for the octahedral coordination are in fairly good agreement with the experimental values. The deviations are probably due to neglect of the orientation effects outside the first solvation layer and to the possibility of a continuous change in the number of solvating molecules. Hydrogen bonding of the anion and the libration of the molecules in the first shell were not taken into account.

V. SOLVATION NUMBERS

Bockris[91] proposed the term "primary solvation number" for the number of molecules near to an ion which have lost their translational degrees of freedom and move as an entity with the ion during its

Brownian motion. Accordingly, the primary solvation number is a definite quantity, while the "secondary solvation" caused by electrostatic interaction beyond the first solvation shell depends on the type of property observed.

Various methods have been suggested for the determination of primary solvation numbers.[10]

(*a*) *The isotope dilution technique.* This method was applied[92] to the study of the exchange of methanol between solvated cations and the solvent. The conditions required for the sucess of the method are that the solvated cation be a specific entity distinguishable from the solvent and that the exchange be not yet complete at the time of isotopic sampling. It was suggested[92,93] that this method measures a "kinetic solvation number" comparable to a primary solvation number. Results obtained are listed in Table 20.

(*b*) *The nuclear magnetic resonance method.* NMR was reported to afford a direct method for the determination of the solvation numbers of ions in nonaqueous solvents.[94] Assuming that there are individual NMR peaks for a solvent nucleus in the bulk and the solvation shell, the solvation number of the cation is determined directly from the ratio of the areas under the absorption peaks. Measurements were reported in methanol,[94–96,101] dimethylsulfoxide,[97] and dimethylformamide.[98–100] In other cases,[102–105] the technique introduced by Jackson *et al.*[106] to differentiate between different CH_3 groups was used. An addition of cupric ion to the solution selectively broadened the peak of the CH_3 group in the bulk, thereby permitting observation of the peak of the CH_3 group in the solvation sphere. The solvation numbers obtained in various solvents are shown in Table 21.

Table 20
Primary Solvation Numbers by the Isotope Dilution Method

Cations	Co^{2+}	Ni^{2+}	Mg^{2+}	Fe^{2+}
n	6	6	6	6 ± 1

Table 21
Primary Solvation Numbers Obtained by NMR*

Ion	Ammonia	Methanol	Ethanol	DMSO	DMF
Mg^{2+}	5(107)	6(102,94)	6(105)	—	—
Co^{2+}	—	6(95)	—	—	6(98)
Ni^{2+}	—	6(95,104)	—	—	—
Be^{2+}	—	—	—	—	4(100)
Zn^{2+}	—	6(101)	—	—	—
Al^{3+}	6(103)	—	—	6(97)	6(99)
Ga^{3+}	—	—	—	—	6(99)

* References are given in parentheses. (See also Table 31.)

(*c*) *Compressibility method.* Passynski[108] suggested that the decrease in compressibility of electrolyte solutions is due to the fully electrostricted solvent molecules. Thus, the solvent molecules within the primary solvation shell are rendered incompressible. If β_0 and β are the compressibilities of the solvent and the solution, respectively, the incompressible volume fraction of the solution[10] is $1 - (\beta/\beta_0)$. The solvation number n_s of the electrolyte at a molar concentration c is given by[109] $n_s = 1000d_0[(1 - \beta/\beta_0)/\beta_0 c M_0] - (\phi_v d_0/M_0) = -\phi_k d_0/M_0\beta_0$, where ϕ_v and ϕ_k are the apparent molar volume and the apparent molar compressibility, respectively, and d_0 and M_0 are the density and molar weight of the solvent.

At infinite dilution, the relation for n_s reduces to

$$n_s^\circ = -\phi_k d_0^\circ/M_0\beta_0$$

Complications, in interpretation of compressibility results, due to structural effects were stressed by Conway and Verrall.[121] Measurements were carried out in methanol,[111–115] ethanol,[111,114,115] butanol,[114,115] and formamide.[110,116] Results may be found in Table 22.

(*d*) *Volume difference method.* The application of density measurements to the calculation of primary solvation numbers[10] is limited to the evaluation of the apparent molal volume of the electrolyte in solution. The volume of the solvated ions must also be known since the following relation holds:

$$V_s^\circ = \phi_v^\circ + n_s^\circ V_0 \tag{51}$$

where V_s°, ϕ_v°, and n_s° are the volume of the solvated ions, the apparent molal volume, and the primary solvation number, respectively, all at infinite dilution; V_0 is the molal volume of the solvent. The molal volume of the solvated ions is usually obtained by the following kinetic methods, which may or may not involve some secondary solvation contribution.

(i) The viscosity method. Viscosity data are usually analyzed using the Jones–Dole equation[122,288]

$$\eta/\eta_0 = 1 + Ac^{1/2} + Bc \qquad (52)$$

where η and η_0 are the viscosity of the solution and the solvent, respectively, A is a theoretical coefficient[123] representing the contribution from interionic forces which may be neglected at concentrations higher than 0.1 M, B represents the contribution of the cospheres of the ions,[124] and c is the molar concentration. It

Table 22
Solvation Numbers from Compressibility Measurements at 25°C*

	Solvent			
Electrolyte	Methanol	Ref.	Ethanol	Ref.
LiCl	4.2	111	2.7	111
LiBr	5.0	111	3.4	111
LiI	5.6	111	3.7	111
NaCl	4.7	111	—	—
NaBr	5.6	111	2.9	111
NaI	6.2	111	3.2	111
KBr	5.2	111	—	—
KI	6.0	111	—	—
CsCl	3.0	112	—	—
NH_4Cl	4.9	111	—	—
NH_4Br	5.0	111	2.5	111
NH_4I	5.5	111	3.3	111
$LiNO_3$	5.3	111	3.4	111
$NaNO_3$	5.9	111	—	—
NH_4NO_3	5.2	111	2.9	111
$MgCl_2$	12	113	—	—
$CaCl_2$	10	113	—	—
$ZnCl_2$	6	113	—	—

* Incomplete data are given in Refs. 110, 114, 115.

has been shown[125,126,132,133] that the coefficient B may be identified with the product σV_s, where σ is a shape factor equal to 2.5 for spheres[129] (cf. Refs. 38, 127, 128).

The solvation number is then obtained from the relation

$$n_s^\circ = (V_s^\circ - \phi_v^\circ)/V_0 \tag{53}$$

Results obtained from data in the literature may be seen in Table 23.

(ii) The mobility method. It was suggested[137] that the radius r_s of the solvated ion may be obtained through the conventional primitive model[138] of Stokes, which gives the relation

$$\lambda_i^\circ = |Z_i|F^2/6\pi\eta_0 r_{si} \tag{54}$$

where λ_i° is the equivalent conductance of the ion i at infinite dilution, Z_i and r_{si} are the charge and radius of the solvated ion, respectively, F is the Faraday, and η_0 is the viscosity of the solvent. It was generally assumed that the volume V_s^s of the solvation shell is

$$V_s^s = (4\pi/3)(r_{si}^3 - r_c^3) \tag{55}$$

where r_c is the crystallographic radius of the ion. The solvation number was then assumed to be

$$n_s = V_s^s/V_0 \tag{56}$$

where V_0 is the molar volume of the solvent.

Supposing even that Stokes's relation holds, the ionic solvated volume should be obtained, because of packing effects,[139,148] from the expression

$$V_{si} = 4.35 r_{si}^3 \tag{57}$$

where V is in ml mole^{-1} and r_{si} in angstroms. The solvation number should be evaluated as above when the apparent molar volume ϕ_v° of the electrolyte is known [equation(3)]. Various corrections both empirical[141,145] and theoretical[140,142–144,148] have been suggested with a view to applying Stokes' relation to all ions (cf. Section VII.3). Nevertheless, here only solvation numbers obtained from relation (54) and which have been correctly evaluated or recalculated will be quoted. The results may be seen in Table 24.

(e) *The activity method.* Robinson and Stokes[131] derived an equation for the rational mean ionic activity coefficient of an electrolyte assumed to be solvated with n moles of solvent per mole

Table 23
Solvation Numbers Obtained from Viscosity and Apparent Molal Volumes

Ion	Solvent	
	Methanol	Formamide
Li^+	—	4.0
Na^+	—	4.0
K^+	6.0	2.0
Rb^+	—	1.5
Cs^+	—	1.5
NH_4^+	—	1.0
Me_4N^+	0	0
Et_4N^+	0	0
Cl^-	1.5	1.5
Br^-	1.0	1.0
I^-	1.0	0
SCN^-	—	0
NO_3^-	—	0

Table 24
Solvation Numbers from Mobility Measurements in Nonaqueous Solvents

Ion	Solvent						
	Methanol	Ethanol	DMSO	Acetone	Formamide	DMF	N-Methyl acetamide
H^+	—	—	—	—	3.5	3.0	—
Li^+	5.0	5.0	4.3	2.9	5.4	3.2	5.1
Na^+	5.0	4.0	2.3	2.6	4.0	3.0	3.5
K^+	4.0	3.0	2.4	2.0	2.5	2.0	3.3
Rb^+	3.0	2.0	—	—	2.3	—	—
Cs^+	3.0	2.0	—	—	1.9	—	2.6
NH_4^+	2.0	2.0	—	1.0	2.0	—	2.7
Bu_4N^+	0	—	0	—	—	—	—
Ba^{2+}	—	—	—	—	7.0	—	9.0
Sr^{2+}	—	—	—	—	6.0	—	8.6
Ca^{2+}		—	—	—	6.0	—	8.6
Mg^{2+}	—	—	—	—	—	—	10.3
Cl^-	1.5	2.0	0.6	1.0	—	0.5	2.1
Br^-	1.0	2.0	0.5	1.0	1.0	0.5	1.7
I^-	1.0	1.0	0.4	—	1.0	0.5	1.5
ClO_4^-	2.0	2.0	0	—	—	1.0	—
NO_3^-	2.0	2.0	0	—	—	—	1.5
SCN^-	—	—	0	—	—	—	1.3

Table 25
Solvation Numbers from the Activity Coefficient Expression

Electrolyte	Solvent		
	Formamide	Methanol	Ethanol
LiCl	5.1	—	—
NaCl	4.2	3.7	—
KCl	3.5	—	—
NaBr	—	3.2	3.3
NaI	3.8	2.2	—
KBr	—	2.7	2.5
RbCl	2.5	—	—
CsCl	1.0	—	—
KNO_3	0	—	—
KI	4.4	—	—

of salt, in terms of the conventional activity coefficients. Solvation numbers obtained[130] by a best fit to this equation are reported in Tables 25 and 26. Their method involved, however, use of a Debye–Hückel expression for the electrostatic activity coefficient at concentrations where it would be inapplicable.

Table 26
Solvation Numbers in Sulfuric Acid at 25°C

Ion	(*)	(†)
Li^+	2.6	2.3
Na^+	3.8	3.0
K^+	2.4	2.1
Ag^+	2.4	2.1
NH_4^+	1.2	1.2
Me_2COH^+	1.5	1.0
$MePhCOH^+$	3.8	1.4
Ph_2COH^+	7.2	1.3
$PhNH_3^+$	1.5	0.8
$Ph_2NH_2^+$	3.8	0.6
Ph_3NH^+	7.2	0.6
Ba^{2+}	11.5	6.5
H_3O^+	2.1	1.8

* Fitted to Robinson and Stokes's Eq. (38).[131]
† Fitted to Gluckauf's equation.[431]

VI. MIXED SOLVENTS

Most of the work on mixed solvents has been concerned with aqueous–organic solvents, with emphasis on changes in thermodynamic behavior of electrolytes with decreasing water content of the mixed solvent.

Equilibrium properties of mixed solvents will be considered in terms of the thermodynamic properties of transfer. The molal free energy of transfer of an electrolyte ΔG°_{tr} may be defined as the difference between the standard molal free energies of an electrolyte in the mixture $_s\overline{G^\circ}$ and in water $_w\overline{G^\circ}$. Thus,

$$\Delta G^\circ_t = {}_s\overline{G^\circ} - {}_w\overline{G^\circ} \tag{58}$$

actually characterizes the change in solvation during transfer from one solvent to another, as we have previously seen. An enthalpy

Table 27
Enthalpies of Transfer from Water to Water–Glycerol Mixtures (in cal mole^{-1})[172]

Mole fraction of glycerol	$NaNO_3$	KNO_3	$RbNO_3$	$CsNO_3$	$Ca(NO_3)_2$	$Sr(NO_3)_2$	$La(NO_3)_3$
0.005	31	67	73	26	172	110	192
0.010	61	118	146	66	327	195	306
0.015	90	174	203	102	465	290	411
0.020	125	213	260	138	583	389	512
0.030	197	293	365	226	799	717	709
0.040	267	373	471	316	975	873	902
0.050	328	451	575	413	1143	1105	1084
0.060	378	530	676	520	1304	1325	1282
0.070	434	608	775	625	1447	1558	1488
0.080	492	704	888	738	—	—	1674
0.090	546	787	986	863	—	—	1852
0.100	612	871	1096	1005	—	—	2012
0.110	679	958	1188	1158	—	—	2269
0.120	737	1043	1282	—	—	—	2339
0.130	792	1125	—	—	—	—	2492
0.140	835	1214	—	—	—	—	2654
0.150	890	—	—	—	—	—	2812
0.160	948	—	—	—	—	—	2923
0.170	1002	—	—	—	—	—	—
0.200	1172	—	—	—	—	—	—
0.220	1302	—	—	—	—	—	—

ΔH°_{tr}, entropy ΔS°_{tr}, and volume $\Delta \overline{V^\circ_{tr}}$ of transfer may be similarly defined. It is more accurate to deal with the thermodynamic transfer properties obtained directly than to obtain them indirectly as a small difference between large values of separately measured properties in the solvents.

The methods for determination of the various thermodynamic properties have already been considered and collected, and the recalculated thermochemical data may be found in Tables 27–29.

1. Structural Changes in Mixed Solvents

It has been suggested that the use of ions as "internal indicators" of the electrochemical properties[158] of partially aqueous solvents should be quite profitable as well as instructive.

The experimental standard free energies of transfer of electrolytes ΔG°_{tr} from water to methanol–water mixtures[44,45,159,160] and

Table 28
Enthalpies of Transfer from Water to Mixed Solvents (in cal mole^{-1})[161]

Electrolyte	Organic solvent	Percent 10	20	43.12	68.3	90
LiCl	Methanol	900	1300	680	−390	−1750
NaCl	Methanol	480	760	680	170	−1270
KCl	Methanol	290	490	375	−640	−2060
RbCl	Methanol	170	290	330	−50	−1370
CsCl	Methanol	160	210	90	−590	−1660
HCl	Methanol	330	560	880	−100	−1250
HBr	Methanol	320	500	570	−660	−2550
HI	Methanol	395	560	190	−850	−3640
LiCl	Dioxane	—	−585	—	—	—
NaCl	Dioxane	—	−330	—	—	—
KCl	Dioxane	—	−295	—	—	—
RbCl	Dioxane	—	−335	—	—	—
CsCl	Dioxane	—	−585	—	—	—
NaCl	Dioxane	—	−330	—	—	—
NaBr	Dioxane	—	−745	—	—	—
NaI	Dioxane	—	−1165	—	—	—
KCl	Dioxane	—	−295	—	—	—
KBr	Dioxane	—	−730	—	—	—
KI	Dioxane	—	−1135	—	—	—

Table 29
Free Energies of Transfer from Water to Mixed Solvents (in cal mole^{-1})

Electrolyte	Organic solvent	Percent							Ref.
		10	20	30	43.12	50	68.3	90	
LiCl	Methanol	341	692	—	1520	—	2640	3980	161
NaCl	Methanol	454	922	—	2066	—	3560	5340	161
KCl	Methanol	484	980	—	2256	—	3970	6540	161
HCl	Methanol	180	353	—	701	—	1329	2748	161
HBr	Methanol	125	219	—	438	—	914	2273	161
HI	Methanol	58	138	—	12	—	233	1460	161
HCl	Glycol	80	—	163	—	238	420	1425	412
LiCl	Dioxane	—	595	—	—	—	—	—	162
NaCl	Dioxane	—	762	—	—	—	—	—	162
KCl	Dioxane	—	765	—	—	—	—	—	162
RbCl	Dioxane	—	789	—	—	—	—	—	162
CsCl	Dioxane	—	801	—	—	—	—	—	162
AgCl	Dioxane	—	619	—	—	—	—	—	162
NaCl	Dioxane	—	762	—	—	—	—	—	162
NaBr	Dioxane	—	576	—	—	—	—	—	162
NaI	Dioxane	—	825	—	—	—	—	—	162
HCl	Dioxane	—	427	—	—	—	—	—	162
HBr	Dioxane	—	241	—	—	—	—	—	162
HI	Dioxane	—	−40	—	—	—	—	—	162

to dioxane–water mixtures[162,163] have confirmed that the application of the Born equation,

$$\Delta G_{tr}^{\circ} = \frac{Ne^2}{2}\left(\frac{1}{\varepsilon_s} - \frac{1}{\varepsilon_w}\right)\left(\frac{1}{r_+} + \frac{1}{r_-}\right) \tag{59}$$

where ε_s and ε_w are the dielectric constants of the mixed solvent and water, respectively, and where it is inherently assumed that the radii r_+ and r_- are constant in any solvent, could do no more than correctly predict the sign of ΔG_{tr}° (cf. Ref. 29). Nevertheless, changes in the standard free energy of transfer of KCl from water to ethanol–water mixtures were calculated[164,165] using the model developed by Stokes[166] for the free energy of hydration of ions with a noble gas structure. This modification of the Born equation accounts for the increase in the size of the ion due to hydration as well as for dielectric saturation in the vicinity of the ions. The increasing discrepancy between calculated and experimental values is attributed to increased solvation by ethanol since, in the calculations, selective hydration is assumed in the ethanol–water mixtures.

The free energy of transfer ΔG_{tr}° for alkali metal chlorides in methanol–water mixtures, up to a concentration of 90 % w/w,[44] varies monotonically with solvent composition. It is possible that the free energies of transfer of NaCl and KCl pass through a maximum at methanol content between 90–100 %. On the other hand, the variation of ΔG_{tr}° with the reciprocal of the cationic radius for the same salts in methanol–water mixtures[45,44] and in dioxane–water mixtures[162,163] is small and monotonic, and has been used as an extrathermodynamic assumption for a rough separation of the transfer properties into ionic contributions. Objections to this procedure have been mentioned earlier, yet after proclaiming the total breakdown of the Born equation for cations, the same form of equation is still used by several authors[167] in their extrapolation by using a $1/r$ plot. This extrapolation yields values of free energy of transfer of cations and anions that are opposite in sign.[162,163]

It is further claimed[162] that all cations are in lower and all anions in higher free-energy states in the mixtures than in water. Grunwald *et al.*[157] explained analogous behavior in a 50 % dioxane–water solvent by the highly characteristic behavior of cation dioxanation and of anion hydration. As a matter of fact, both explanations may be shown to be identical in predicting selective solvation of cations by the organic component and of selective hydration of the anions [cf. discussion following equation (69)]. However, evidence from NMR measurements indicates that cations are preferentially solvated in 50% dioxane–water, implying thereby that the apportionment of the ΔG°_{tr} into its ionic contributions was not carried out properly.

The suggestion was made by several investigators[44,47,48,157,159,162] that two main effects are contributing to each of the transfer functions. First, the ion affects the structure of the solvent; second, it has a potential energy due to the interaction of its own charge with the complex distribution of the charge on the solvent molecules.[44] The structural contribution to the free energy of transfer ΔG°_{tr} is regarded as small (because of a mutual compensation effect between enthalpy and entropy of transfer) as compared to the large electrostatic contribution (cf. Section III.2.iv(e)), which explains why in many systems, the free-energy function is less discriminating than either the enthalpy or entropy functions.[173] However, deviations from linearity have been attributed to the "nonelectrolytic contribution," which is especially significant in the case of H^+I^-.[45] The changes observed in the enthalpies of transfer ΔH°_{tr} with the composition of the mixed solvent are quite different from those presented by the free energy of transfer ΔG°_{tr} and show no uniform trend such as would be expected from any simple electrostatic theory.

A recent careful analysis[158] of methanol–water mixtures has shown that enhancement of the three-dimensional order characteristic of pure water reaches a maximum at a mole fraction of methanol between 0.1 and 0.3, after which a structural breakdown takes place. Plots of ΔH°_{tr} of alkali metal halides in methanol–water mixtures[25,44,45,169–171,174] invariably show a maximum in the so-called structurally critical region,[44] reflecting thereby the well-known effects of ions on the structure of the solvent. This is further confirmed by the complex variation of ΔH°_{tr} with either cationic or anionic radii. It is suggested[45] that the maximum in ΔH°_{tr} in the region of enhanced solvent structure arises from the resistance set up by the solvent to the creation of ion-ordered solvent, thereby producing a maximum of structure-breaking and a minimum of structure-making, as confirmed by the entropy contribution.

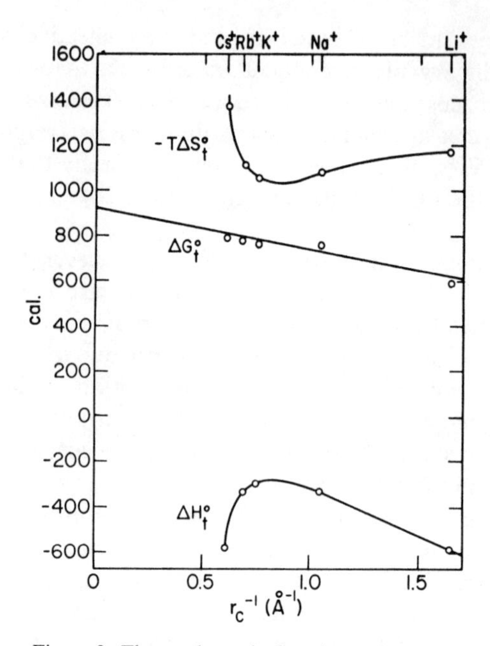

Figure 2. Thermodynamic functions of transfer from water to dioxane–water mixtures.

The difference in behavior of the molar enthalpy and free energy of transfer in dioxane–water[161] mixtures is illustrated in Figure 2, where these properties are plotted as a function of the cation radius.[163] Similar behavior in methanol–water mixtures and the fact that whenever ΔH°_{tr} has been determined for electrolytes in mixed solvents it has been found nearly always[25,169–171,174] to show a maximum at a high mole fraction of water, points toward the existence of such a maximum. The effect is smaller in dioxane–water mixtures than in other binary mixtures.[174] Later data for NaI in dioxane–water mixtures[175,176] indicated only the presence of an inflection at a dioxane mole fraction of about 0.15, which is still within this critical structure region.

Although similar effects were reported for ΔH°_{tr} of alkali metal halides in ethanol–water mixtures[25,179] and HCl in ethylene glycol–water mixtures,[177] no maxima were observed for solutions of $RbNO_3$, $CsNO_3$, KNO_3, $NaNO_3$, $Ca(NO_3)_2$, $Sr(NO_3)_2$, and

$La(NO_3)_3$ in ethylene glycol–water[172] and glycerin–water mixtures.[178] The similarity between the structures of the two solvents is suggested as an explanation for this normal behavior.

It may be concluded that the difficulty in interpreting the experimental evidence for structural changes in mixed solvents resides mainly in the thermodynamic impossibility of unambiguously obtaining the ionic contributions to the properties considered, especially with regard to the sign of the contribution and to which of the ions is in a higher- or lower-energy state. Some progress has been made by use of physical models which will be considered later.

2. Solvation Approach to Mixed Solvents

A specific model, the hard-sphere model of a solvated ion, was suggested[109] for the correlation of apparent properties of electrolytes in solution. The solvated ion is pictured as a spherical cavity in the solvent. A spherical shell just inside the cavity contains solvent, while the charge of the ion is situated inside this shell. The high electrostatic field at the surface of the ion, or the inner surface of the solvation shell, causes dielectric saturation and makes this shell incompressible.[108] The ion–solvent complex is assumed to be made up of one mole of solute and n_s moles of solvent. The molal volume of the solvated electrolyte V_s, irrespective of the solvent, is then[74]

$$V_s^\circ = \phi_v + n_s V_0 \tag{60}$$

and at infinite dilution,

$$V_s^\circ = \phi_v^\circ + n_s^\circ V_0 \tag{61}$$

In mixed solvents, the solvation number n_s° may be assumed to be the same as in aqueous solutions because of selective solvation effects.[146–148] The molal solvated volume thus obtained was shown to be linearly related to the solvent composition and the reciprocal of the dielectric constant of the mixed solvent for a number of salts in ethanol–water and acetone–water mixtures. The treatment of relative viscosity measurements by an extended Mooney equation[149] yielded a very simple relation[125] between the solvated molal volume and the B coefficient of the Jones–Dole equation,[122]

$$V_s = 0.4B \tag{62}$$

thereby permitting the calculation of solvation numbers and solvated molal volumes from viscosity measurements[150] (cf. Ref. 128). Assuming the additivity of molar refractions, the solvated molar refraction was calculated and it was shown[151] that the refractive index of the solvated complex is independent of concentration. The polarizability of solvated ions could be calculated, as well as the electronic contribution to the molar polarization of the solvated salt. Actually, the same model has been inherently used for the ion in the development of electrolyte theories and has properties independent of the concentration, which point to its usefulness as a working entity. However, it is still open to the same criticisms that are directed against the sphere-in-the-continuum model.[3]

3. Thermodynamic Approach to Selective Solvation in Mixed Solvents[4]

Preferential solvation shown by electrolytes in mixed solvents is a direct consequence of the specific interaction between an ion and one of the components of the mixed solvents. It may be defined as the relative change in composition of the mixed solvents in the vicinity of the ions and expressed by

$$(n_1/n_2)/(n_1^\circ/n_2^\circ) = 10^\alpha \tag{63}$$

where n_1/n_2 is the mole ratio of the components in the vicinity of the ion, n_1°/n_2° is the initial mole ratio of the components, and α is called the index of selective solvation.

Many attempts have been made to derive theoretical expressions for (63), either assuming that the mixed solvent is an ideal mixture and the electrostatic contribution is obtained from Born's expression,[152] or assuming an empirical correction for the change in the dielectric constant of the solution,[153] or other experimental corrections.[154–156]

The change in local composition produced by an ion field in a mixture of solvents may be obtained from the fundamental thermodynamic expression (42), which for a two-component system is

$$dG^* = -S\,dT + v\,dP - (v\varepsilon_d\,dE/4\pi)\,dE + \mu_1\,dn_1 + \mu_2\,dn_2 \tag{64}$$

where μ_1 and μ_2 are the chemical potentials of the solvents. A combination of cross-differentiation and use of the Gibbs–Duhem equation yields the following general equation for selective interaction of mixed solvents with ions:

$$\ln\frac{N_1}{N_2} - \ln\frac{N_1^\circ}{N_2^\circ} = \frac{1}{8\pi}\int_v dv \int\left[\frac{\partial\mu_1}{\partial(\ln N_1)}\right]^{-1} \times \left[\left(\frac{\partial\varepsilon_d}{\partial n_1}\right)_{P,T,E,n_2} - \left(\frac{\partial\varepsilon_d}{\partial n_2}\right)_{P,T,E,n_1}\right] d(E^2) \tag{65}$$

In the limiting case where the dielectric constant and the volume v are assumed to be independent of the field strength E, and the binary mixture is an ideal one, the expression (65) may be shown to reduce either to Debye's expression[152] or to that obtained by Frank.[77]

In the case of aqueous–organic solvents, where it has been shown that the activity of water is very nearly given by the mole fraction of water in the solvent, expression (65) may be written in spherical coordinates:

$$\ln\frac{N_1}{N_2} - \ln\frac{N_1^\circ}{N_2^\circ} = -\frac{1}{2RT}\int_{r_e}^{\infty} r^2\,dr \times \int_0^{E_r}\left[\left(\frac{\partial\varepsilon_d}{\partial n_1}\right)_{P,E,n_2} - \left(\frac{\partial\varepsilon_d}{\partial n_2}\right)_{P,E,n_1}\right] d(E^2) \tag{66}$$

The concept of partial free energy of solvation at infinite dilution of an ion $\overline{\Delta G_i}$ in component i of a mixed solvent may be formally introduced in the expression

$$\Delta G = n_1\overline{\Delta G_1} + n_2\overline{\Delta G_2} \tag{67}$$

where ΔG is the free energy of the ion in the mixed solvent, and it may rigorously be shown that

$$\ln\frac{N_1}{N_2} - \ln\frac{N_1^\circ}{N_2^\circ} = -\frac{1}{RT}(\overline{\Delta G_1} - \overline{\Delta G_2}) \tag{68}$$

and

$$\alpha = -(\Delta G_1 - \Delta G_2)/2.3RT$$

By combining the ionic indices of preferential solvation, a mean selective index $\bar{\alpha}$ given by

$$\bar{\alpha} = (v_a\alpha_a + v_c\alpha_c)/v = -(\overline{\overline{\Delta G_1}} - \overline{\overline{\Delta G_2}})/2.3RT \tag{69}$$

Table 30
Values of $\bar{\alpha}$ for Various Solvents as Compared to Water

Salt	Solvent								50% Dioxane–water
	NH_3	N_2H_4	CH_3OH	C_2H_5OH	C_4H_9OH	$C_5H_{11}OH$	$(CH_3)_2CO$	HCOOH	
HCl	−5.58	−2.67	2.02	2.56	2.89	3.12	3.12	2.94	2.98
HBr	−6.73	−3.78	2.02	2.56	2.02	3.19	1.84	4.22	—
HI	−7.58	−2.67	1.47	2.56	2.57	2.20	2.75	4.58	—
$HClO_4$	—	—	—	—	—	—	—	—	0.21
NaCl	1.47	1.09	1.28	1.83	3.07	4.71	6.78	−1.10	5.06
NaBr	0.32	0	1.28	1.83	2.2	4.78	5.50	0.18	—
NaI	−0.55	1.10	0.73	1.83	2.75	3.79	6.42	0.55	—
NaOH	—	—	—	—	—	—	—	—	6.45
$NaNO_3$	—	—	—	—	—	—	—	—	3.59
LiCl	0.92	1.65	1.28	2.19	1.97	2.02	2.38	−0.73	4.21
LiBr	−0.23	0.55	1.28	2.19	1.10	2.09	1.10	0.54	—
LiI	−1.08	1.65	0.73	2.19	1.65	1.10	2.02	0.92	—
KCl	4.76	1.10	1.28	2.38	3.98	4.22	4.95	−0.55	5.06
KBr	3.61	0	1.28	2.38	3.12	4.29	3.67	0.73	4.14
KI	2.76	1.10	0.73	2.38	3.66	3.30	4.58	1.10	3.11
RbCl	2.93	1.28	1.83	2.93	5.63	4.22	5.50	0.55	4.81
RbBr	1.80	0.18	1.83	2.93	4.77	4.29	4.22	1.78	—
RbI	0.92	1.28	1.28	2.93	5.31	3.30	5.13	2.20	—
CsCl	4.76	0.92	1.83	2.67	3.26	3.67	3.85	−1.78	4.58
CaBr	3.57	−0.18	1.83	2.67	2.39	3.74	2.57	−0.52	—
CsI	2.73	0.92	1.28	2.67	2.93	2.75	3.49	−0.13	—
AgCl	−5.50	−6.23	2.20	2.38	—	—	—	−2.20	—
AgBr	−6.65	−7.33	2.20	2.38	—	—	—	0.92	—
AgI	−7.50	−6.23	1.65	2.38	—	—	—	−0.55	—
$CaCl_2$	6.98	2.44	—	—	—	—	—	0.49	—
$ZnCl_2$	−5.99	−3.66	4.40	5.86	—	—	—	1.84	—
$CdCl_2$	−4.03	−4.37	3.98	4.51	—	—	—	0.98	—

is obtained, where $\nu = \nu_a + \nu_c$ and $\overline{\overline{\Delta G_i}}$ is the partial molal free energy of solvation of the electrolyte in component i of the solvent. The property $\overline{\overline{\Delta G_i}} - \overline{\overline{\Delta G_2}}$ is identical with the thermodynamic quantity dF_0/dZ as defined by Grunwald *et al.*,[157] and can be obtained by applying the Backhuis–Roozeboom procedure to the free energy of solvation of the electrolyte per mole of mixed solvent. As a first approximation, the difference $\overline{\overline{\Delta G_1}} - \overline{\overline{\Delta G_2}}$ was evaluated from the molal free energy of solvation of the electrolyte in the pure component i of the mixed solvent as tabulated by Izmailov. The calculated values of the mean index of selective solvation with respect to water are given in Table 30. A positive index indicates that water is preferentially solvated, whereas a negative index shows that a specific interaction arises between the electrolyte and the other solvent. It may be seen from the table that generally salts will be solvated by water preferentially except in the case of mixtures of $NH_3 + H_2O$, $N_2H_2 + H_2O$, and $HCOOH + H_2O$, where, probably, chemical interaction promotes complex formation. The assumption of selective solvation made in interpreting physical and chemical phenomena in mixed solvents[4] is justified by the tabulated data (see Table 30) and relations to salting-out[148] are to be noted.

4. NMR Aspects of Solvation

Derivation of direct evidence for solvation in mixed solvents is complicated by the number of species present and the possibility of preferential solvation. NMR techniques yield direct information concerning the interactions occurring in solutions since the NMR chemical shift of a nucleus is determined primarily by the local electronic environment, which may be perturbed by strong interactions with a neighbor. An ion in solution causes an external perturbation in the electron density around the solvent molecules either by electrostatic interaction or chemical bonding and structural changes. Such changes are observed through a shift in the resonance frequency of the solvent molecules relative to the pure solvent and through the spin–lattice relaxation time.

In water-proton magnetic resonance, a proton involved in hydrogen bonding resonates at a lower applied magnetic field than nonassociated protons.[180] Two processes occur when an electrolyte is added to an aqueous mixed solvent. First, there is a partial destruction of hydrogen bonding, which is more or less compensated by any corresponding ionic solvation. Then, there is a chemical shift to a lower or higher field, which may be interpreted either in terms of structure-forming (solvation) or structure-breaking, respectively. However, NMR shifts reflect the total effect produced by both ions and hence indicate only which effect predominates. A number of factors are involved in determining the chemical shift[2].

High-resolution NMR measurements carried out on electrolyte solutions in dioxane–water mixtures[168] exhibit significant shifts in the water-proton resonance positions. The upfield effect shown by the alkali halides emphasizes structure-breaking by the latter. The low-field shifts characteristic of the alkaline earths are especially large for beryllium, as well as for lanthanum and thorium. In contrast to the behavior of the water-proton resonance, which shows a marked dependence on the cation, the dioxane resonance frequency remains essentially unchanged in all solutions. This may be considered as conclusive evidence for selective solvation of cations by water molecules in mixed dioxane–water solvents, with no indication whatsoever of interactions between dioxane and either of the two ionic species. Further confirmation was provided by line-width variations. However, later, but preliminary NMR chemical shift investigations on aqueous dioxane solutions of $Al(ClO_4)_3$ and $Mg(ClO_4)_2$ seemed to indicate solvation by dioxane.[184] It would appear that the perchlorate ion is solvated by dioxane, since it has been repeatedly shown that aluminium chloride is not solvated by dioxane in dioxane–water mixtures.[181–183]

Similar studies of paramagnetic salts,[185] as well as $AlCl_3$, and $TiCl_4$[182] in aqueous mixtures of tetrahydrofuran, confirm that the water signal is the only one affected to any extent by the addition of the salt. Thus, tetrahydrofuran, although a polar molecule, behaves like dioxane, implying that the cyclic ether linkage does not undergo chemically significant ion–dipole interaction in aqueous solution to any appreciable extent.

The extension of NMR investigations to other aqueous solvent mixtures has produced some unexpected results. In aqueous mixtures of acetone,[181–182] ethanol,[181,182] or acetonitrile,[182] the chemical shifts indicate that water solvates ions extensively in all the mixtures. Interpretation of the water results in aqueous mixtures of N,N-dimethylformamide (DMF), dimethyl sulfoxide (DMSO), and N-methylformauride (NMF) supports the conclusion that the nonaqueous species does solvate added ions. For aluminum chloride solutions in DMF–, DMSO–, and NMF–H_2O mixtures, separate signals corresponding to the bulk and complexed species were observed for the organic components.[181] A specific interaction between $AlCl_3$ and DMSO was suggested in order to explain the absence of enhanced solvating ability of DMSO over $Al(ClO_4)_3$ in mixtures with H_2O.[207] Those NMR measurements that have been extended to a wider range of solvents, e.g., with dimethylacetamide (DMA) and tetramethyl urea (TMU) as additional organic solvent components,[182] have confirmed competitive solvation by the organic component, especially at *higher concentrations and at low water content*. Separate bulk and bound resonance signals were obtained for aluminum chloride,[186,187] aluminum perchlorate, and aluminum nitrate, but not for thorium nitrate, magnesium, and lithium perchlorates in N-methylacetamide (NMA)–water solvents. From the effects of changes in solvent composition and electrolyte concentration, the relative order of solvation of the cations by water and NMA was found to be $Al^{3+} > Th^{4+} > Mg^{2+} > Li^+$, while the order obtained for the anions studied was $Cl^- > NO_3^- > ClO_4^-$. Solvation of both ions and not solvent structure-breaking processes were observed in the various salt solutions studied.[187]

In aqueous pyridine solutions,[188] the effect of alkali chlorides is similar to that observed in dioxane–water mixtures,[168] while the pyridine proton signals are barely affected at all. The alkaline earth chlorides are strongly solvated and, on the basis of chemical shifts, Be^{2+} was stated to behave as a strong Lewis acid and to be competitively solvated by water and pyridine. Anions and tetraalkylammonium salts are said to produce predominantly a structure-breaking effect in the solutions.

A possible explanation for the competitive effect shown by certain organic components with regard to water may be found in terms of the relatively high concentrations of salts and organic species involved in such solutions. Solutions of an electrolyte at infinite dilution in mixed solvents with high water content would presumably be selectively solvated by water as shown by studies of lithium-7 chemical shifts in acetonitrile–water mixtures[190] and sodium-23 chemical shifts in many solvents.[191] Experimental evidence that the $^{35}Cl^-$ chemical shift reflects the solvent environment of the ion in mixtures of CH_3CN–H_2O and DMSO–H_2O appears to support the hydrogen-bonding hypothesis in the former, and in the latter to indicate an even competition between the two solvents for sites in the solvation shell of Cl^-.[221]

In some electrolyte solutions at ambient temperatures, only one proton resonance signal is observed even though the exchange of water molecules is relatively slow.[189] This results from a rapid proton exchange, which averages the signals arising from the bulk and the bonded molecules. However, it was shown[192] that by cooling to a specific temperature, depending on the particular mixed solvent, the proton exchange is slowed to such an extent that separate bulk and associated-water resonances can be observed. Solvation numbers in aqueous mixed solvents may thus be determined in concentrated solutions for both aqueous and organic components by direct integration of the two signals. A series of proton resonances in mixed acetonitrile–water solutions was obtained from differently hydrated aluminum ions.[193]

A different method for the study of preferential solvation based on the broadening of an NMR proton signal resulting from contact with paramagnetic species was recently proposed.[200] This method permits determination of the fraction of the solvation shell of a paramagnetic ion occupied by a given component in a mixed solvent.

The results of further investigations[201] indicate that DMSO and DMF might be good competitors for coordination sites but that dioxane is excluded by water. This was independently checked by a method based on the contact shift experienced by protons on solvent molecules in the coordination spheres of paramagnetic ions.[200] The sensitivity of the ^{59}Co chemical shift to its environment confirmed the above results.[202] More evidence for selective solvation in water–acetone, water–acetonitrile, methanol–acetone, and methanol–acetonitrile mixtures was observed.[203] In salts when there is a large difference between the size of the ions, only the small ion was found to produce a chemical shift in the O—H proton resonance. Since the effects of Et_4NClO_4 on the proton chemical shift is negligible, only perchlorates or tetraalkylammonium compounds were investigated. The ions Na^+, Li^+, Mg^{2+}, Br^-, Cl^-, and I^- all showed a downfield shift in the water line which was assigned to solvation. In proton spin relaxation studies of the Mn^{2+} ion in formic acid,[204] the lengthening of the longer relaxation time with increasing water content was

Table 31
Solvation Numbers in Aqueous and Nonaqueous Solvent Mixtures Determined by NMR

Solvent mixture*	Ref.	Be^{2+}	Mg^{2+}	Al^{3+}	Ga^{3+}	In^{3+}	Ag^{+}	NO_3^-
H_2O	196	6	—	—	6[198]	6[196]	—	—
Acetone		—	—	—	—	—	—	—
H_2O	197	4	6	—	—	—	—	—
Acetone		—	—	—	—	—	—	—
H_2O	199, 183	—	—	6	—	—	—	—
Acetone		—	—	—	—	—	—	—
H_2O	183, 195	—	—	5	—	—	—	—
DMSO		—	—	1	—	—	—	—
H_2O	199	—	—	6	—	—	—	—
$DMSO_2$		—	—	—	—	—	—	—
H_2O	199	—	—	4	—	—	—	—
TMSO		—	—	—	—	—	—	—
H_2O	199	—	—	6	—	—	—	—
$TMSO_2$		—	—	—	—	—	—	—
H_2O	183	—	—	4.5	—	—	—	—
DMF		—	—	1.5	—	—	—	—
H_2O	183	—	—	6	—	—	—	—
Dioxane		—	—	—	—	—	—	—
H_2O	183	—	—	6	—	—	—	—
THF		—	—	—	—	—	—	—
H_2O	193	—	—	6	—	—	—	—
CH_3CN		—	—	—	—	—	—	—
Acetone	194	—	—	—	—	—	—	—
Methanol		—	6	—	—	—		
H_2O	407	—	—	—	—	—	0	2
							↓	↓
Acetonitrile		—	—	—	—	—	4	2

* DMSO, dimethyl sulfoxide. TMSO, tetramethylene sulfoxide. $DMSO_2$, dimethyl sulfone. $TMSO_2$, tetramethylene sulfone. $DMSO_4$, dimethyl sulfate. THF, tetrahydrofuran.

attributed to the possible increase of the distance between the proton and the ion. No complex hydration shell was formed. Other complementary aspects of NMR studies have been reviewed.[205,206].

Direct determinations of solvation numbers were obtained by the above technique in aqueous acetone[183,194,197] and methanol–acetone mixtures[194] where preferential solvation for methanol was shown by magnesium ions, in water mixtures of DMSO,[183,195,197] DMF,[183] tetrahydrofuran,[183,199] tetramethylurea,[183,199] dimethylsulfate, and tetramethylene sulfone.[199] Aluminum perchlorate was singled out for showing some nonaqueous solvent interaction, probably of ClO_4^-, in aqueous acetone, aqueous tetrahydrofuran, and acetone–tetrahydrofuran[129] at low water and rather high electrolyte concentrations. Results obtained are summarized in Table 31.

VII. TRANSPORT PROCESSES

The various kinetic properties of ions such as their viscosity increment, diffusion constants, and limiting ionic conductance are known to reflect, to various extents, the size of the ions in solution. Diffusion and conductance involve essentially the same phenomenon and are therefore related by several equations including the viscosity of the medium. The various methods by which these measured hydrodynamic properties have been interpreted usually involve Kohlrausch's law of independent ion migration and a modified physical model of a sphere in a continuum, with structure-breaking or -forming according to the experimental evidence. The solvation aspects will now be considered.

1. Viscosity

In a very dilute solution, the interstitial solvent in the region between the cospheres of the ions is unmodified and has the same properties as in the pure solvent.[124] The ions are expected to contribute equally toward any change in the viscosity. Jones and Dole[122] suggested that the relative viscosity η/η_0 of strong electrolytes could be represented by the equation

$$\eta/\eta_0 = 1 + Ac^{1/2} + Bc \qquad (70)$$

where A is a constant that may be obtained through the Onsager–

Fuoss theory[209] and B is an empirical constant, either positive or negative, that accounts for the ion–solvent interaction. Negative B coefficients are confined to highly associated solvents such as water,[127] sulfuric acid,[208] glycerol,[211,222–224,226] and ethylene glycol.[211,222,226] They are usually found for monatomic ions with low surface charge density.[217] According to Frank and Evans,[225] the approximate balance between the ordering tendency of the field at the surface of the solvation shell and the resilience of the solvent molecules to stay within their three-dimensional structure, leads to a structural collapse in that region. This effect, which tends to make B more negative, may exist at the periphery of the dielectric saturation region even for structure-making ions.

It was suggested[125] that the term "negative solvation" proposed by Samoilov[226] for disrupted solvent structure should be applied to these systems. This has recently been justified in a thorough investigation on negative hydration by nuclear magnetic relaxation methods[227] and reorientation time in aqueous–organic mixtures.[228] Measurements of relative viscosity for KCl, RbCl, and CsCl in methanol–water mixtures[216] and KI in ethanol–water mixtures[214,215] and in DMSO–water mixtures[214] have definitely proved the existence of negative B coefficients in aqueous mixtures. This was regarded[216] as confirming the structure model of Franks and Ives for highly aqueous alcohol–water mixtures. The existence of a minimum in both positive and negative B coefficients when plotted versus the mole fraction of water in mixed solvents was taken as further evidence strengthening the above theory.[216] However, no such minima were observed in dioxane–water mixtures[218,229,230] except for RbCl and CsCl.[237] In pure nonaqueous solvents, the structure-breaking contribution is probably negligible,[217] as indicated by the large positive B values found in N-methyl formamide,[217] acetone,[231] N-methyl-propionamide,[232] and others.[211]

Evidence for the additivity of ionic contributions appears to be rather conclusive.[128] Additivity is therefore applied to the separate ions and most correlations with other properties in aqueous solutions have been made in terms of ionic B coefficients. In aqueous and mixed solvents, less evidence was produced. For bisulfate salts in sulfuric acid, Gillespie[220] inferred that HSO_4^- anions, being similar in size and character to the solvent (cf. OH^- in water), will have very little effect on the solution structure and therefore assigned the total B to the cation concerned. In DMSO-H_2O mixtures, the suggestion of equal ionic B coefficients for $(i\text{-Aml})_3BuN^+$ and $B(Ph)_4^-$ was based on the assumption that the two ions are of equal size and are unsolvated due to the large symmetric shape with low surface charge density.[210] It was suggested that the close correlation between ionic B coefficients and solvation properties for KCl in aqueous solutions confirmed that equal contribution of the ions in KCl is a good approximation.[128] There is no reason why a similar division should not be tried in other systems. In the first instance, as shown by Gurney,[124] the B coefficients for various salts both in aqueous and nonaqueous solutions change in a predictable manner with the molar entropy of solution. Secondly, ionic B coefficients in aqueous solutions which were computed by equating the contributions made by K^+ and Cl^- with the B coefficients of KCl at 25°C, showed a linear relationship with partial molar ionic entropies. Then, Asmus[234] and Nightingale[235] showed that a single linear relationship may be used to correlate entropies of hydration with ionic B values for both monatomic and polyatomic ions. Since it is usual to link ionic entropies with B coefficients,[217] the same correlation should be made in nonaqueous and mixed solvents.[233] As it was shown that a linear relationship exists between ionic entropies in various solvents [cf. Eq. (61)], it may be assumed that in aqueous mixtures as well as in nonaqueous solvents, the equality of the ionic B coefficients for K^+ and Cl^- should be preserved.[233] Following these suggestions, correlation with entropies of solvation were made and ionic B coefficients could be evaluated.

The shape and size of electrolytes were correlated[125] with the B coefficients using the Einstein equation[129] as extended by a refinement of Vand's equation.[236] Stokes[127,128] assumed a rigid volume for the solvated electrolyte V_s independent of concentration; however, this assumption may be made only at high dilution. The relation

$$B = \sigma V_s \tag{71}$$

was obtained, where σ is the shape factor (cf., p. 34), equal to 2.5 for spheres. Tuan and Fuoss[132] assumed that the large tetraalkylammonium salts have spherical symmetry and tried to find a correlation between partial molal volumes and viscosity measurements through Einstein's equation. Quantitative agreement was

B Viscosity Coefficients in Nonaqueous Solvents at 25°C*

Electrolyte	Methanol	Ethanol	Ethylene-glycol	Glycerol	DMSO	CH_3CN	CH_3NO_2	NMF	Acetone	*N*-methyl propiona-mide
LiCl	—	—	—	—	—	—	—	—	0.382[231]	—
KCl	0.7635[212]	—	—	—	—	—	—	0.615[217]	—	1.37[232,305]
KBr	0.7396[212]	—	—	—	—	—	—	0.584[217]	—	—
KI	0.6747[212,213]	—	0.0327[211]	−0.185[211]	—	—	—	—	—	—
NaI	—	1.15[208]	—	—	—	—	—	—	—	—
NaCl	—	—	—	—	—	—	—	—	—	—
NaCl	—	—	—	—	—	—	—	0.599[217]	—	—
NH_4Cl	0.661[212]	—	—	—	—	—	—	—	—	—
CsI	—	—	−0.080[211]	−0.408[211]	0.68[207]	—	—	—	—	—
NaSCN	—	—	—	—	0.62[210]	—	—	—	—	—
$NaClO_4$	—	—	—	—	0.62[210]	—	—	—	—	—
$NaB(Ph)_4$	—	—	—	—	1.14[210]	—	—	—	—	—
$(i\text{-Am})_3$-$BuNBPh_4$	—	—	—	—	1.57[210]	1.47[403]	—	0.567[217]	—	—
Me_4NBr	0.42[132]	—	—	—	—	—	—	—	—	—
Bu_4NBr	0.84[132]	—	—	—	—	0.93[132]	0.75[132]	—	—	—
Et_4NBr	0.56[255]	—	—	—	—	0.69[132]	—	—	—	—
Pr_4NBr	0.70[255]	—	—	—	—	—	—	—	—	—
Bu_4NBr	0.84[255]	—	—	—	—	0.87[132]	—	—	—	—
Pr_4NI	0.66[255]	—	—	—	—	0.71[132]	—	—	—	—
$Bu_4NB(Ph)_4$	—	—	—	—	—	1.35[132]	—	—	—	—
$Pr_4NB(Ph)_4$	—	—	—	—	—	1.25[132]	—	—	—	—
Bu_4NPi	—	—	—	—	—	1.13[132]	—	—	—	—
Pr_4NPi	—	—	—	—	—	0.90[132]	—	—	—	—
Et_4NPi	—	—	—	—	—	0.85[132]	—	—	—	—
Me_4NPi	—	—	—	—	—	0.78[132]	—	—	—	—

* References are given in parentheses.

found only for the large ions, as might be expected due to their negligible electrostriction. For smaller ions, however, the viscosity data indicated larger radii than did the molar volumes. The difference is due to solvation. Investigations in dioxane–water did not confirm the sphere-in-continuum model but rather a model of water structure in which "icebergs" of bonded molecules exist in a sea of free molecules.[237] Some negative B coefficients were reported, but no minima were found. The model of a preferentially water-solvated sphere has been previously considered [cf. discussion following Eq. (69)] and found to reproduce the various properties satisfactorily.

Since no theoretical calculation of the coefficient B has been proposed[124] (Ref. 238), it may be emphasized that a quantitative theory of the B coefficient should be developed to elucidate the points raised. Values of B for salts in nonaqueous and mixed solvents may be found in Tables 32 and 33A–C.

2. Conductance

During recent years, a substantial body of experimental data on conductance in nonaqueous and mixed solvents has accumulated. The data have usually been analyzed in terms of the Fuoss–Onsager conductance theory,[358] which is basically an application of the interionic attraction theory[359] to the specific model of rigid, charged spheres representing the ions in an electrostatic and hydrodynamic continuum, i.e., the solvent.[3] The conductance equation, which has two parameters, Λ_0, the limiting conductance at infinite dilution, and $å$, the contact distance between the ions, may be able to reproduce the concentration dependence of the equivalent conductance up to a concentration corresponding to $\kappa å \leq 0.2$,[362] where κ is the characteristic Debye–Hückel length. The equation is usually given in the following form:

$$\Lambda = \Lambda_0 + S(\gamma c)^{1/2} + Ec\gamma \log(c\gamma) + (J - B\Lambda_0)c\gamma - K_A c\gamma y_{\pm}^2 \Lambda \quad (72)$$

S and E are functions of ε_0, T, and η_0 of the solvent but independent of $å$, while J is an explicit function of $å$; B is the viscosity coefficient "B" of the Jones–Dole equation; γ is the degree of dissociation; K_A is the association constant; and $y_{\pm}$ is the Debye–Hückel limiting mean molar activity coefficient. For unassociated electrolytes, the last term disappears and $\gamma = 1$.

Table 33A
***B* Viscosity Coefficients in Aqueous Mixtures at 25°C**

Solvent	Percent	Electrolyte						
		LiCl	NaCl	KCl	KI	RbCl	CsCl	CH_3COONa
Methanol	10	0.146*	0.069*	—	—	—	—	0.363†
	20	0.145*	0.064*	−0.028*	—	−0.058*	−0.070*	0.346†
	30	—	—	—	—	—	—	0.353†
	40	0.25*	−0.111*	−0.002*	—	−0.023*	−0.034*	0.362†
Ethanol	0.5	—	—	—	—	—	—	0.323†
	18.5	—	—	—	—	—	—	0.294‡
	29.5	—	0.046†	—	−0.158†	—	—	0.274‡
	40.4	—	—	—	—	—	—	0.247‡
	52.0	—	—	—	—	—	—	0.270‡
Acetone	8.5	—	—	—	—	—	—	0.353‡
	21.2	—	—	—	—	—	—	0.382‡
	30.8	—	—	—	—	—	—	0.380‡
	40.0	—	—	—	—	—	—	0.387‡
	52.0	—	—	—	—	—	—	0.393‡
DMSO	10	—	—	—	—	—	—	0.379‡
	20	—	—	—	—	—	—	0.390†
	30	—	0.139†	—	−0.065†	—	—	0.419†
	40	—	—	—	—	—	—	0.418†
	50	—	—	—	—	—	—	0.415†

* Ref. 217.
† Refs. 214, 219.
‡ Ref. 408.

Table 33B
***B* Viscosity Coefficients in Aqueous Mixtures at 35°C**

		Electrolyte		
Solvent	Per-cent	$BaBr_2$ (Ref. 430)	$BaCl_2$ (Ref. 430)	$SrCl_2$ (Ref. 218)
Dioxane	10	0.284	0.230	0.350
	20	0.440	0.405	0.500
	30	0.675	0.648	0.620

From (72), the contact distance $\mathring{a}$ may be obtained either from J, a_J or from an expression for the association constant K_A,[360] whereby a_K is obtained from the slope of $\log K_A$ versus $1/\varepsilon_0$ or from the dependence of the Walden product $\Lambda_0\eta_0$ on dielectric constant.[141] Discussion of the parameter $\mathring{a}$ will be deferred until the Walden product is considered.

Fuoss's application of a more sophisticated approach to alkali halides in dioxane–water mixtures, which included explicit retention of the Boltzman factor,[361] was found to give more accurate results.[3] However, the derivation of the Fuoss–Onsager equation[361] was questioned in a discussion of Pitt's equation.[362] Accurate experiments should permit a decision to be made between the two theories.[357] Nevertheless, the limiting conductances Λ_0 obtained by the Onsager–Fuoss extrapolation will be considered here since this method has been most widely used.

At infinite dilution, the motion of an ion is limited only by interactions with the surrounding molecules, as there are no other ions within a finite distance. Under such circumstances, the validity of Kohlrausch's law of independent migration of ions is almost axiomatic.[38] Thus, for an electrolyte yielding two kinds of ions,

$$\Lambda_0 = \lambda_1^\circ + \lambda_2^\circ \tag{73}$$

where λ_i° is determined by measuring the transference number t_i and

$$\lambda_i^\circ = t_i^\circ \Lambda_0 \tag{74}$$

The Hittorf method of determining true transference numbers[363] may shed some light on the ion–solvent interaction in mixed solvents. Solvent molecules that are firmly bound to an ion will move with it under the influence of the current. Accordingly, the changes in the

Table 33C
B Viscosity Coefficients in Dioxane–Water Mixtures*

	Percent dioxane												
Salt	18.5	19.1	28.9	30	39.7	41.1	43.8	51.2	54.6	55.3	56	63.3	71.1
LiCl	0.175	—	0.235	—	—	—	0.316	—	—	0.375	—	0.441	—
NaCl	0.143	—	—	0.160	0.192	—	—	—	—	—	0.242	0.277	—
KCl	0.035	—	0.037	—	—	—	—	—	0.130	—	—	0.160	—
RbCl	—	−0.008	0.010	—	—	—	—	—	—	0.092	—	0.138	—
CsCl	—	−0.016	0.032	—	—	—	—	—	0.084	—	—	—	0.161
CsI	—	—	—	—	—	—	0.0	—	—	0.066	—	0.118	—
NaF	—	0.282	—	0.300	—	0.358	—	—	—	—	—	—	—
NaBr	0.093	—	0.108	—	—	0.179	—	—	—	(0.0258)	—	0.285	—
NaI	0.067	—	0.100	—	0.172	—	—	—	—	0.293	—	0.400	—
Me_4NBr	—	—	0.13	—	0.15	—	—	—	0.17	—	—	0.23	—
Et_4NBr	—	—	0.37	—	—	0.31	—	0.42	—	—	—	0.39	—
Pr_4NBr	—	—	0.77	—	—	0.72	—	—	—	—	—	0.59	—
Bu_4NBr	—	—	1.08	—	1.04	—	—	—	—	—	0.84	0.76	—

* Reynolds, Ph.D. Thesis, Yale University (1966); also additional data at other temperatures.

number of moles of solvent at the cathode, for instance, will be given by

$$\Delta N = t_+ n_+ - t_- n_- \quad (75)$$

where t_+ and t_- are the transference numbers of the cation and anion, respectively, and n_+ and n_- are the number of solvent molecules bound to each ion. In ethanol–water[365–371] and dioxane–water mixtures,[372,402] α-methyl glucoside acted as an inert reference for the determination of the composition of the mixed solvent. This method was shown[10] not to yield accurate solvation numbers in aqueous solutions since the results varied with the reference substance chosen,[364] but information on selective solvation of cations by water in mixed solvents is obtainable. Direct determinations of solvent composition in methanol–water mixtures suggested that both Ca^{2+} and Cl^- are selectively solvated by water,[374] while in acetonitrile–water mixtures, Ag^+ and NO_3^- may be preferentially solvated by acetonitrile,[372] and Zn^{2+} is definitely solvated by water.[373] In hydrazine–water mixtures, the ions Zn^{2+} and Cl^- were shown to be selectively solvated by hydrazine.[373] This result is consistent with the free energy of transfer for the Zn^{2+} ion in these solvent pairs (cf. Ref. 4). The use of labeled formic acid was tried for the determination of the solvation number of HCl solutions,[375] assuming that both cation and anion have the same mobility. The results obtained by the above method are in general agreement with those obtained by other methods, at least qualitatively.

(*i*) *Limiting Ionic Conductances*

Reliable values of single ion conductances are very useful for the investigation of ion–solvent interactions. As mentioned above, the division of electrolyte conductances into the limiting ionic conductances ideally requires transference numbers.[239] For several solvents, however, provisional scales of ionic conductances have been based on the assumption that the constituent ions of reference electrolytes, such as tetra-*n*-butyl ammonium tetraphenyl-fluoroborates,[376] tetra-*n*-butylammonium tetraphenyl-borate[377] and especially triisoamyl-*n*-butyl ammonium tetraphenyl-borate[328] and tetraisoamylammonium tetraisoamylboride,[239] have equal mobilities.

Limiting ionic conductances of ions in some pure solvents are listed in Table 34A,B and the basis for the assignment of the ionic contributions is indicated.

Limiting Ionic Conductances in Nonaqueous Solutions at 25°C

Ion	Methanol	Ethanol	Propanol	Butanol	Tetrahydro-furan	Dimeth-oxyethans	Acetonitrile	Nitromethane	DMSO
Li^+	39.08[329]	17.07[329]	—	8.10[262]	36.6[266]	55.7[265]	—	—	11.4[247]
Na^+	45.09[329]	20.37[276]	8.35[261]	—	49.2[266,267]	—	76.9[244]	—	14.54[250]
K^+	47.78[410]	22.2[410]	6.88[329]	—	49.8[266,267]	—	83.6[244]	—	14.7[250]
Cs^+	61.33[329]	26.46[279]	—	—	68.4[266]	53.7[265]	87.3[244]	—	16.1[248]
Rb^+	56.08[329]	—	—	—	—	—	—	—	15.3[249]
Ag^+	50.07[342]	—	—	—	—	—	86.0[409]	—	—
NH_4^+	—	—	—	6.68[262]	—	—	—	—	—
Me_4N^+	68.73[255]	29.65[385]	14.40[385]	9.67[263]	—	—	94.5[242]	54.9[274–4]	18.6[251]
Et_4N^+	60.5[255]	29.27[385]	15.05[385]	10.40[263]	—	—	84.8[242]	47.7	16.5[250]
Pr_4N^+	46.08[255]	22.98[385]	12.19[385]	8.80[263]	—	—	70.3[242]	39.1	13.4[250]
Bu_4N^+	38.94[385]	19.67[385]	0.71[385]	7.84[263]	—	—	61.4[242]	34.1	11.0[250]
$(EtOH)_4N^+$	—	—	—	—	—	—	64.0[242]	—	—
$(i\text{-Am})_4N^+$	35.4[255]	—	—	—	—	—	56.0[239]	—	—
$(i\text{-Am})_3BuN^+$	36.6[255]	18.31[385]	10.17[385]	7.67[263]	40.3[266]	46.3[265]	58.0[245]	32.62[271]	11.05[252]
$(n\text{-Am})_4N^+$	34.8[255]		—		—	—	—	—	—
$(Hepta)_4N^+$	—	14.93[385]	8.29[385]	6.25[263]	—	—	—	—	—
Ph_4As^+	—	—	—	—	—	—	55.8[239]	—	—
Cu^+	—	—	—	—	—	—	64.7[409]	—	—
Me_3S^+	67.60[258]	—	—	—	—	—	97.20[258]	—	—
Et_3S^+	62.0[258]	—	—	—	—	—	89.18[258]	—	—
Pr_3S^+	50.47[258]	—	—	—	—	—	71.6[258]	—	—
$\frac{1}{2}Mg^{2+}$	—	50.9	—	—	—	—	—	—	—
$\frac{1}{2}Zn^{2+}$	48.0[281]	—	—	—	—	—	—	—	—
Cl^-	52.09[255]	21.87[385]	10.45[385]	7.76[263]	—	—	—	62.7[272–4]	24.4[251]
Br^-	56.43[255]	23.88[385]	12.22[385]	8.23[263]	—	—	100.7[245]	62.9[272–4]	24.1[251]
I^-	62.62[255]	27.0[385]	13.81[385]	9.32[263]	—	—	102.1[245]	—	23.8[251]
Picrate	47.07[255]	—	—	—	—	—	77.0[242]	—	—
Ph_4B^-	37.05[255]	—	—	—	40.3[266]	46.3[265]	58.1[265]	33.12[271]	11.05[252,250]
SCN^-	—	—	—	—	—	—	—	—	29.50[252]
ClO_4^-	71.0[342]	—	16.42[385]	11.22[263]	—	—	103.7[244]	—	24.40[250,252]
$(i\text{-Am})_4B^-$	—	—	—	—	—	—	57.6	—	—
BF_4^-	—	—	—	—	—	—	108.5[409,241]	—	—
PF_6^-	—	—	—	—	—	—	104.0[409]	—	—
NO_3^-	61.13[320]	—	—	—	—	—	106.4[409]	—	—

* References are given in parentheses. Additional data in Refs. 243, 246, 253, 254, 256, 259, 260, 264, 269, 275, 277, 278, 280.

Table 34B
Limiting Ionic Conductances in Nonaqueous Solutions*

Ion	Formamide	Formic acid	N-Methyl acetamide (35°C)	NMF (Ref. 304)	N-Methyl propionamide (30°C)	Sulfolane (30°C) (Refs, 300,301)	DMF	Acetone
Li^+	8.5[306]	19.36[307]	5.65[311]	—	—	4.33	25.0[342]	72[325]
Na^+	10.1[306]	20.97[307]	7.19[311]	21.56	5.06[302]	3.61	29.9[342]	75.9[325]
K^+	12.7[306]	23.99[307]	7.28[311]	22.13	5.36[232]	4.05	30.8[342]	75.4[325]
Rb^+	12.8[306]	—	—	—	—	4.16	—	—
Cs^+	13.5[306]	—	8.33[311]	24.39	—	4.27	—	—
NH_4^+	15.6	27.01[307]	—	—	—	4.97	38.7	—
Me_4N^+	12.5	23.62[307]	11.28[308]	—	—	—	38.9	—
Et_4N^+	10.0	—	10.33[311]	26.20	—	3.98	35.6	—
Pr_4N^+	—	—	8.23[308]	—	—	—	29.2	—
Bu_4N^+	6.8	—	7.11[308]	—	—	—	26.2	—
H^+	10.8	79.63[307]	—	—	—	—	—	—
Tl^+	15.8[310]	—	—	—	—	—	38.6[342]	—
Me_3PhN^+	10.7	—	—	—	—	—	—	—
$(i\text{-}Am)_3BuN^+$	—	12.33[307]	—	—	—	—	—	—
Am_4N^+	—	—	6.68[308]	—	—	—	—	—
$(Hex)_4N^+$	—	—	6.50[308]	—	—	—	—	—
$(Hep)_4N^+$	—	—	6.20[308]	—	—	—	—	—
Ag^+	—	—	—	—	—	—	35.2[342]	—
Cl^-	17.1[306]	26.52[307]	10.60[311]	19.70	6.24[302]	9.30	55.1[342,339]	—
Br^-	17.2[306]	28.30[307]	11.72[311]	21.56	7.06[302]	8.92	53.6[342]	—
I^-	16.6[306]	—	13.42[311]	22.76	8.34[302]	7.22	52.3[342]	—
SCN^-	17.2[306]	—	—	—	—	9.64	—	—
NO_3^-	17.4[310]	—	—	—	—	—	57.3[342]	—
$HCOO^-$	15.3[310]	50.05[307]	—	—	—	—	—	—
CH_3COO^-	11.9[310]	—	—	—	—	—	—	—
$PhCO_2^-$	9.8	—	—	—	—	—	—	—
$PhSO_3^-$	10.4	—	—	—	—	—	—	—
ClO_4^-	—	29.35[307]	—	—	—	6.685	52.4[342]	115.3[325]
Ph_4B^-	—	12.33[307]	—	13.08	—	—	—	—
Picrate	—	—	10.93[312]	—	—	—	31.5	—
PF_6^-	—	—	—	—	—	5.95	—	—

3. Stokes' Law and Walden's Rule

If the ions of an electrolytic solution are represented by solvation spheres of radius R_i and charge $Z_i e$, and the solvent is regarded as a continuum of viscosity η_0,[141] then Stokes' law[378] may be shown to yield what is called the primitive model[138] for limiting ionic conductances λ_i°,

$$\lambda_i^\circ = |Z_i|eF/\pi\eta_0 R_i \tag{76}$$

where F is the Faraday equivalent. Substituting numerical values for the constants gives

$$\lambda_i^\circ = 0.820|Z_i|/R_i\eta_0 \tag{77}$$

where λ_i° is in $\text{ohm}^{-1}\ \text{cm}^{-1}\ \text{M}^{-1}$, η_0 is in poises, and R_i is expressed in angstroms. Stokes' law is only valid for a moving sphere which is large compared to the particles of the medium.[379] For bulky organic ions, ionic conductances should therefore be inversely proportional to the ionic radius R_i. If it is assumed that the radius of an ion does not change on going from one solvent to another, the product

$$\lambda_i^\circ\eta_0 = 0.820|Z_i|/R_i \tag{78}$$

is a constant. Expression (78) is known as Walden's rule.[380] On the other hand, effective radii calculated from limiting ionic mobilities have been used to estimate the number of solvent molecules associated with the moving ion, thereby assigning the ion–solvent interaction to a solvation shell present around the ion. Since Stokes' law does not apply to small ions, attempts were made[38] to correct the law by assuming that smaller tetraalkylammonium ions are unsolvated, and their behavior deviates from Stokes' law because of a size deficiency. A correction factor was evaluated and used in drawing an empirical curve from which corrected Stokes radii could be obtained for a given limiting ionic mobility. This procedure was applied to methanol,[343] DMSO,[252] and other solvents[340] (cf. Ref. 144). It was suggested that this approach suffers from one serious flaw. The basic assumption on which this approach rests is that Walden's product is invariant with temperature.[335] Experimental results[381,382] indicate that this assumption is incorrect and that the resulting solvation numbers obtained from this type of correction to Stokes' law are meaningless.

The changes in the Stokes radius R_s were attributed by Born[140] to ion–dipole reorientation that yields an apparent increase in the real radius r of the ion. The following expression was suggested:

$$R_s = r[1 + \tfrac{1}{3}(r_0/r)^4] \tag{79}$$

where r_0 is a characteristic length given by[346]

$$r_0^4 = (33/28)\alpha_0^2(e^2\mu^2/k^2T^2) \tag{80}$$

α is an intermediate value between $1/\varepsilon$ and 1; μ is the dipole moment of the solvent near the ion. The effect of dielectric saturation on R was later discussed by Schmick,[383] but no further improvements in this model were made until Fuoss noticed the dependence of Walden's product $\Lambda_0\eta_0$ on the dielectric constant and considered the effect of electrostatic forces on the hydrodynamics of the system.[141] It was thought that, due to the finite relaxation time of the solvent dipoles, an ion moving through a solvent would orient too few dipoles in the direction of its motion, while behind it, too many dipoles would be still pointing in its general direction. In other words, the electrostatic coupling between the ion and the solvent acts to produce an increase in viscosity and work must be done to orient the dipoles as the ion moves on. These heuristic arguments led to the expression

$$\lambda_i^\circ\eta_0 = Fe|Z_i|/6\pi R_\infty(1 + A/\varepsilon R_\infty^2) \tag{81}$$

from which the classical Stokes radius R_s may be derived as

$$R_s = R_\infty + (A/\varepsilon) \tag{82}$$

R_∞ is the hydrodynamic radius of the ion in a hypothetical medium of infinite dielectric constant where all electrostatic forces vanish, and A is an empirical constant. Experimental results suggest that A increases linearly with R_∞ in hydrogen-bonding solvent mixtures and decreases hyperbolically with R_∞ in aprotic solvent mixtures.

Fuoss's treatment was placed on a firmer basis by Boyd,[142] who showed that the effect of dielectric relaxation on ionic motion can be treated theoretically on a macroscopic basis in a manner consistent with Stokes' law itself and that the dielectric relaxation times for polar solvents lead to an effect of the proposed magnitude in some aprotic solvent mixtures. The equation he obtained was

$$\lambda_i^\circ = \frac{Fe|Z_i|}{6\pi\eta_0 r[1 + (2/27)(1/\pi\eta_0)Z_i^2 e^2 \tau/r_1^4 \varepsilon_0]} \tag{83}$$

where τ, which was assumed to be unaffected by the hydrodynamic disturbance of the ionic motion, is the Debye relaxation time for the solvent dipoles. Finally, Zwanzig,[143] in a more rigorous evaluation of Boyd's derivation, showed that the "ultimate that has been devised"[384] by an application of a continuum model for the solvent to ionic mobilities, yielded for the Walden product the expression

$$\lambda_i^\circ \eta_0 = Fe|Z_i|/[6\pi r + (B/r^3)] \tag{84}$$

where B is a function of the solvent properties only and is given by

$$B = (2e^2/3)(\tau/\eta_0)[(\varepsilon_0 - \varepsilon_\infty)\varepsilon_0^2] \tag{85}$$

ε_∞ is the limiting high-frequency dielectric constant of the solvent.

It may be seen that Born's and Zwanzig's equations are very similar and both may be written in the form[138]

$$\lambda_i^\circ = AR_i^3/(C + R_i^4) \tag{86}$$

from which a maximum value of λ_i°, $\lambda_m^\circ = A(3^{3/4}/4)C^{-1/4}$ conductance units corresponding to the radius $R_m = (3C)^{1/4}$ Å, may be obtained. An excellent discussion of the above effect in aqueous solutions[138] indicates that neither slipping nor streaming through the solvent would improve the model, but the use of a local viscosity might produce an almost perfect fit for halide conductances.

(i) *Nonaqueous Solvents*

For comparing results in different solvents, equation (81) can be used in the linear form suggested by Atkinson and Mori.[341] Rearranging terms and inserting numerical constants gives

$$15.5/\lambda_i^\circ \eta_0 = 18.8r + (15.3 \times 10^{12}/r^3)[(\tau/\eta_0)(\varepsilon_0 - \varepsilon_\infty)/\varepsilon_0^2] \tag{87}$$

To facilitate plotting, equation (87) may be written as

$$L^* = 18.8r + (15.3 \times 10^{12}/r^3)R^* \tag{88}$$

where L^* and R^* are defined by comparison with (87). The quantity r may be obtained both from the intercept and the slope of the line. In order to test Zwanzig's theory,[143] equation (88) was applied to

Table 35
Ionic Radii Calculated from the Intercept and Slope of a Plot of Zwanzig's Equation for Alcohol Solutions at 25°C

Ion	r° (intercept), Å	r° (slope), Å
Li^+	3.0	6.1
Na^+	2.9	7.2
K^+	2.5	7.8
Cs^+	2.0	7.3
Me_4N^+	1.8	7.9
Et_4N^+	2.3	9.4
Pr_4N^+	3.2	15.1
Bu_4N^+	3.8	16.3
Cl^-	2.2	6.3
Br^-	2.1	6.7
I^-	1.9	7.1

methanol, ethanol, and acetonitrile solutions,[341,384] solvents for which accurate conductance[244,255,258,263,382,385] and transference data[239,245] are available. All the plots were found to be straight lines.[341,384] A quantitative test was obtained by comparing the value of r from the slope and the intercept of the straight line obtained from data[384] for various alcohols. These are given in Table 35.

It is quite obvious that the two radii are far from equal, indicating that the theory does not account quantitatively for mobility changes within the homologous series of alcohols. It was further suggested[384] that the relaxation effect is not the predominant factor affecting ionic mobilities and that these mobility differences could be explained qualitatively[386] if the microscopic properties of the solvent, dipole–moment, and free–electron pairs were considered the predominant factors in causing the deviation from Stokes' law.

(ii) Mixed Solvents

The behavior of the limiting ionic conductance–viscosity product in mixed solvents was ascribed to the structural changes that arise from increasing the organic solvent content in the aqueous mixed solvent.[384] The data indicate that some salts show an increase in $\lambda_i^{\circ}\eta_0$ and some a decrease as the dielectric constant decreases. In alcohol–water mixtures, structure enhancement appears to reach a maximum near 20% or 30% alcohol, and at higher alcohol concentrations, the water structure is progressively reduced.[7,158] Similarly, the Walden product increases initially, then decreases steadily as the added alcohol destroys the water structure and the ions lose their excess mobility. This was shown for water mixtures of methanol,[257,387] ethanol,[276,384] propanol,[261] methyl cellosolve,[388] ethylene glycol[314–317,319–230] and tetrahydrofuran.[268,270]

The larger alkali metal and halide ions appear to be better structure-breakers in water-rich solvents than in pure water. For alkali nitrates, a different solvation process for the nitrate ions, which solvate by hydrogen bonding, and alkali ions, which solvate through ion dipole interaction, is suggested[257] as an alternative explanation for the behavior of the $\Lambda_0\eta_0$ curve. The ability of the tetraalkylammonium ion to enforce water structure decreases continuously, leading to an increase in $\Lambda_0\eta_0$ as more organic solvent is added until a maximum is reached, after which a slow decrease takes place. Sometimes, there is no maximum, as for example with Bu_4NBr[318] and Et_4NPi,[389] where only the structure-forming function of the ion is present. A similar situation exists in glycerol–water mixtures,[282–285,388] where the increase of the Walden product has been ascribed to the formation of H-bonded complexes. It was suggested[381] that a negative term be introduced into the equation

$$\eta = \eta_0\{1 + (B/R_\infty^2\varepsilon) - [f(h)/R_\infty^2\varepsilon]\} \tag{89}$$

where $f(h)$ is an implicit function of the number of H-bonds, to adjust the overall effect manifested in the apparent viscosity η.

In dioxane–water mixtures,[268,270,289,292,391–397] for a dioxane content >20–40%, the Walden product was, in general, found to decrease with increase of dioxane in the solvent. It has been suggested[384] that this monotonic decrease might mean that the maximum has already been passed at the dioxane concentrations studied, and that a maximum might occur at about 0.15 mole fraction of dioxane in the solvent, where the viscosity of dioxane–water mixtures shows a sharp minimum.[270] Indeed, a careful perusal of previous and current investigations reveals that at this dioxane concentration, a maximum does occur for potassium chloride,[270,390] sodium bromate,[398] tetrabutylammonium iodide,[398] tetrabutylammonium bromide,[398] thallous nitrate,[399] thallous chloride,[400]

Table 36
Comparison of Radii Obtained from Equation (88) for Dioxane–Water Mixtures

Ion	r (intercept)	r (slope)	r (crystalline)
Li^+	1.69	5.1	0.68
Na^+	1.22	5.3	0.98
K^+	0.60	5.2	1.33
Cs^+	0.64	5.4	1.67

potassium and silver nitrates,[401] and sodium sulfate[287] (cf. Ref. 290). The maximum has been shown to occur even more specifically for the ions Cl^- [286,390] and NO_3^-.[286,296,297] For the perchlorate ion,[290,295] the rapid decrease in the mobility of the ion was tentatively explained by selective solvation of ClO_4^- by dioxane.[286]

The application of Zwanzig's equation to the extensive set of data for dioxane–water mixtures[3] yielded straight lines for the alkali metal ions.[341] However, again (see Table 36), no correlation is found between the radii calculated from the slope and the intercept of plots based on equation (88).

Similar results were obtained for glycerol–water mixtures, where in the case of K^+, the "intercept" radius was 0.58 Å and the "slope" radius 3.0 Å. In ethanol–water mixtures,[384] the Bu_4N^+ relation is linear down to 50% ethanol, yielding r(intercept) = 6.0 Å and r(slope) = 3.1 Å. Then, the equation curves up to that for pure water, with a negative slope. For the Cs^+ ion, on the other hand, a straight line is obtained only at the low ethanol end down to 20 mole% ethanol, and the intercept radius (r = 0.80 Å) and slope radius (r = 3.1 Å) are again in poor agreement.

There is no doubt that the continuum theory is unable to explain the data either in the water-rich region or in the water-poor region, since even in the most favorable case it does not give a quantitative result. It is suggested that the complicated behavior of ionic conductances in aqueous mixtures may be interpreted by the existing continuum theory, taken in conjunction with structural effects.[384] Thus, the solvent composition around the ion must be defined and selective solvation or competition between the solvents ascertained. Then, the interaction of the solvated ions with the solvent must be evaluated (e.g., on the basis of a model similar to Frank and Wen's), in terms of microscopic properties of the solvent whenever possible. The molal entity defined in Section VI.2 might well serve as a first step in this direction.

4. Association

For those solutions that do not conform to the equation

$$\Lambda = \Lambda_0 - Sc^{1/2} + E'c \ln c + Jc \quad (90)$$

the electrolyte is assumed to be associated. With the aid of the mass action law, the conductance function for associated electrolytes may be expressed by (72). This equation involves three arbitrary parameters Λ_0, J, and K_A from each of which a distance of closest approach $\mathring{a}$ may be calculated, a_Λ, a_j, and a_a, respectively. It was felt that for an electrolyte represented by charged spheres in a continuum, the three $\mathring{a}_i$ should be equal.[414] This rests, however, on the approach taken to obtain a theoretical expression for the association constant K_A. Before considering the various expressions that have been proposed, we should bear in mind the definitions of ion pairs. Four classes of ion pairs were defined by Griffith and Symons[415]:

1. Complexes: two or more ions held in contact by covalent bonds.
2. Contact ion pairs: ions in contact but without any covalent bonding between them.
3. Solvent-shared ion pairs: pairs of ions linked electrostatically through a single oriented solvent molecule.
4. Solvent-separated ion pairs: pairs of ions linked electrostatically but separated by more than one solvent molecule.

It will be seen that the general term ion pair has been used alternatively when referring to one of the classes 2, 3, or 4. However, no distinction can be made conductometrically among them.

Attempts to relate the association constant to properties of the solvent and solute were all limited in their application and can only be used under restricted conditions if the results are to have any meaning.[353]

(*a*) *Bjerrum's approach.* Bjerrum[416] formulated a function which described the probability of finding one ion of opposite charge in the immediate vicinity of a given ion as a function of the distance r

between them. He found a minimum in the probability at a characteristic distance q from the central ion. For $r < q$, ions were supposed to be associated into an ion pair. This model actually covers classes 2, 3, and 4 of ion pairs. The Bjerrum equation has been criticized especially for two reasons.[417] It yields different $\mathring{a}$ values for the same electrolyte in different media and it predicts a limiting dielectric constant beyond which no association occurs. It has also been suggested that the parameter $\mathring{a}$ is not correct and depends on the dielectric constant, the temperature, and the nature of the solvent.[423] Also, q is unrealistically large for low-ε solvents.

(*b*) *The Denison and Ramsey* (*DR*) *approach.* Denison and Ramsey calculated free-energy changes ΔG_{IV} in the following cycle:

$$\begin{array}{ccc} \text{ion + ion at infinity in medium} & \xrightarrow{\text{IV}} & \text{ion pair at } r_2 \text{ in medium} \\ \downarrow{\text{I}} & & \uparrow{\text{III}} \\ \text{ion + ion in vacuo} & \xrightarrow{\text{II}} & \text{ion pair in vacuo} \end{array}$$

$$\Delta G_{\text{IV}} = \Delta G_{\text{I}} + \Delta G_{\text{II}} + \Delta G_{\text{III}} \tag{91}$$

The solvation energy changes ΔG_{I} and ΔG_{III} of the free ions and the ion pairs were considered to be approximately the same and were omitted from the calculation. The final expression obtained was

$$K^{-1} = \exp(e^2/akT) \tag{92}$$

where only ions in contact were counted as ion pairs (class 2).

It may be shown[417] that only for $a = 5.1$ Å would the Bjerrum and DR approaches yield the same value for K_A. For smaller ions, the two theoretical values for K_A differ.

(*c*) *The Gilkerson approach.* Gilkerson[313] derived an equation for K^{-1} in terms of free volume. However, it cannot be used to calculate K from experimental data, since it contains E_s, the difference in solvation energy between the ion pair and the free ions.

(*d*) *Coulombic theory.* Fuoss[419] applied a method devised by Boltzmann to obtain the following expression:

$$K_A = (4\pi N a^3/3000)\exp(e^2/a\varepsilon kT) \tag{93}$$

where only ions in contact are counted as pairs. This expression was derived using the familiar continuum model but Fuoss stated that the equation may be corrected for solute–solvent interactions by multiplying by the proper term.

Other expressions were proposed,[353,420–422,424] but will not be considered here since most investigators have used the Fuoss or DR equations. Insofar as the molecular nature of the solvents can be ignored, as in Fuoss's equation, the association constant of a salt AB in a series of solvents may be expected to be a smooth function of the dielectric constant ε of the solvents.[426]

The Coulombic theory predicts that a plot of log K_A versus $1/\varepsilon$ for any given salt should be a straight line whose slope is proportional to $1/\mathring{a}_k$ and intercept equal to log K_A°. $\mathring{a}_k$ may be obtained[336] from either the slope or the intercept; however, the values usually differ considerably.[263] Comparison of the values of $\mathring{a}_j$ and $\mathring{a}_k$ have led to various suggestions as to the class of ion pairing present in the solution.

Free ions in acetone solutions were thought to be extensively solvated, while either solvent-shared or solvent-separated ion pairs were formed[323] depending on external conditions. In LiCl ion pairs, the unsolvated ions are in contact, while in LiI, the ion pair is formed from fully solvated ions. LiBr is an intermediate case,[324] as shown by comparison of the Denison–Ramsey and Stokes radii (cf. Refs. 297, 344, 349). Similarly, in THF, the dissociation constants obtained were interpreted in terms of different ion pairing.[267]

For a variety of salts in dioxane–water mixtures, it was found[426] that the plot of log K_A versus $1/\varepsilon$ was concave downwards in the range $78.5 \geq \varepsilon > 30$, becoming linear in the range $\varepsilon \lesssim 30$. Extrapolation of the linear part of the curve into the region of high ε gave values for K_A which were too high. However, by using various aqueous mixtures of ethylene carbonate and tetramethylsulfone, it was shown[348] that variation of the solvent composition at fixed ε resulted in K_A varying up or down by an order of magnitude. This confirms that the overall ε is not the predominant factor in mixtures, even when the difference E_S between ion–solvent and ion-pair–solvent interactions is taken into account.[349]

For the alcohols, a fairly good straight line is obtained for the tetraalkylammonium halides in EtOH, PrOH, BuOH, and PeOH.[263] Thus, it would seem that salts in alcohols follow the simple exponential law predicted by electrostatics despite the difference in $\mathring{a}_k$ which has already been mentioned. It appears that some factor, such as anionic solvation, could be assumed to be the predominant factor controlling the extent of pairing. However, the extremely high degree of association of tetrabutylammonium perchlorate compared to that in the isodielectric solvent acetone, makes this explanation unlikely.[385]

It is felt that the best alternative would be based on a multistep association process involving classes 2, 3, and 4 of ion pairs, with preferential solvation in mixed solvents. The ultrasonic absorption of $MnSO_4$ in water, dioxane–water, and methanol–water solutions confirmed[427] a three-step association process where only the first step can be adequately described using the classical continuum model. The second and third steps most probably involve the successive removal of solvent molecules from between the ions, giving a contact ion pair in the final state. The basic difference between the association constant in various systems was confirmed for Bu_4NBr.[428]

Additional association data may be found in Refs. 291, 293, 294, 298, 299, 326, 327, 331, 338, 350–52, 356, 411.

It may be pointed out at this stage that the model proposed earlier (Section VI.2) should be quite satisfactory for multistep association processes, and moreover, that no physical significance can be assigned to the various $\mathring{a}_i$ parameters.

In another approach, similar to that proposed some years ago by Feakins and French,[429] it has recently been suggested [355] that in the plot of K_A versus $1/\varepsilon$, the solvent activity should enter as an active factor in the thermodynamic dissociation expression. It is claimed[347,355] that the inclusion of the molar concentration of the solvent makes isothermal values of K_A° independent of changes in the dielectric constant of solvent mixtures. It was reported to apply to electrolytes in water, water–organic, and organic solvent systems over a wide range of temperatures and pressures. Neglecting the activity coefficient of the solvent did not seem to affect the results obtained, but the very assumption of selective solvation may be responsible for the overall behavior. However, the views of Quist and Marshall have recently been questioned.

REFERENCES

[1]B. E. Conway, *Ann. Rev. Phys. Chem.* **17** (1966) 481.

[2]J. E. Desnoyers and C. Jolicoeur, in *Modern Aspects of Electrochemistry*, Vol. 5, Ed., J. O'M. Bockris, Butterworths, London, 1969.

[3]R. Fuoss, *Rev. Pure and Appl. Chem.* **18** (1968) 125, and references therein.
[4]J. Padova, *J. Phys. Chem.* **72** (1968) 692.
[5]A. J. Parker, *Quart. Rev.* **16** (1962) 163.
[6]A. J. Parker, *Chem. Rev.*, **69**(1), (1969) 1.
[7]F. Franks, in *Hydrogen Bonded Sytems*, p. 48, Ed., A. K. Covington and P. Jones, Taylor and Francis Ltd., London, 1968.
[8]R. G. Pearson, *J. Chem. Educ.* **45** (1967) 581, 643.
[9]F. Franks, in *Hydrogen Bonded Solvent Systems*, p. 31, Ed., A. K. Covington and P. Jones, Taylor and Francis Ltd., London, 1968.
[10]B. E. Conway and J. O'M. Bockris, in *Modern Aspects of Electrochemistry*, Chapter 2, Ed., J. O'M. Bockris, Butterworths, London, 1954; see also B. E. Conway in *Physical Chemistry*, Volume IXa, Chapter 1, Ed., H. Eyring, Academic Press, New York, 1970.
[11]P. Kebarle, M. Arshadi, and J. Scarborough, *J. Chem. Phys.* **49** (1968) 817; P. Kebarle, *Advan. Chem. Ser.* **72** (1968) 24, and references therein.
[12]K. P. Mischenko, *Acta Physicochim. URSS* **3** (1935) 693.
[13a]G. P. Kotlyarova and E. F. Ivanova, *Russ. J. Phys. Chem.* **40** (1966) 537.
[13b]G. P. Kotlyarova and E. F. Ivanova, *Russ. J. Phys. Chem.* **38** (1964) 221.
[14]V. V. Kuschenko and K. P. Mischenko, *Zh. Teor. Exp. Khim.* **4** (1968) 403.
[15]K. P. Mischenko and V. P. Tungusov, *Zh. Teor. Exp. Khim.* **1** (1965) 55.
[16]K. P. Mischenko and V. V. Sokolov, *Zh. Strukt. Khim.* **5** (1964) 819.
[17]G. Somsen, *Rec. Trav. Chim.* **85** (1966) 517.
[18]L. Weeda and G. Somsen, *Rec. Trav. Chim.* **86** (1967) 263.
[19]R. P. Held and C. M. Criss, *J. Phys. Chem.* **69** (1965) 2611.
[20]E. M. Arnett and D. R. McKelvey, *J. Am. Chem. Soc.* **88** (1966) 2598.
[21]Y. C. Wu and H. L. Friedman, *J. Phys. Chem.* **70** (1966) 2020.
[22]R. F. Rodewald, K. Mahendran, J. L. Bear, and R. Fuchs, *J. Am. Chem. Soc.* **90** (1968) 6698.
[23]R. P. Held and C. M. Criss, *J. Phys. Chem.* **71** (1967) 2487.
[24]L. Weeda and G. Somsen, *Rec. Trav. Chim.* **86** (1967) 893.
[25]G. A. Krestov and V. I. Klopov, *Zh. Strukt. Khim.* **5** (1964) 829.
[26]E. J. W. Verwey, *Rec. Trav. Chim.* **61** (1942) 564.
[27]L. H. Ahrens, *Geochim. Cosmochim. Acta* **2** (1952) 155.
[28]Y. C. Wu and H. L. Friedman, *J. Phys. Chem.* **70** (1966) 501.
[29]W. M. Latimer, K. S. Pitzer, and C. M. Slansky, *J. Chem. Phys.* **7** (1939) 108.
[30]H. L. Friedman, *J. Phys. Chem.* **71** (1967) 1723.
[31]C. M. Criss and E. Luksha, *J. Phys. Chem.* **72** (1968) 2966.
[32]W. M. Latimer, *The Oxidation States of the Elements and their Potentials in Aqueous Solutions*, 2nd Ed., Prentice-Hall, Englewood Cliffs, N.J., 1952.
[33]N. A. Izmailov, *Russ. J. Phys. Chem.* **34** (1960) 1142.
[34]R. G. Bates, in *The Chemistry of Nonaqueous Solvents*, Chapter 3, Ed., J. J. Lagowski, Academic Press, New York and London, 1966.
[35]R. G. Bates, in *Proc. Sympos. 10–12 Jan. 1968*, p. 49, Eds., A. K. Covington and P. Jones, Taylor and Francis Ltd., London, 1968.
[36]H. Strehlow, in Ref. 34, Chapter 4.
[37]V. F. Coetze, J. M. Simon, and R. J. Bertozzi, *Anal. Chem.* **41** (1969) 766.
[38]R. A. Robinson and R. H. Stokes, *Electrolyte Solutions*, 2nd ed., Butterworths, London, 1959.
[39]H. M. Koepp, H. Wendt, and H. Strehlow, *Z. Elektrochem.* **64** (1960) 483.
[40]A. Lauer, *Electrochim. Acta* **9** (1964) 1617.
[41]J. F. Coetzee and J. J. Campion, *J. Am. Chem. Soc.* **89** (1967) 2513.
[42]V. A. Pleskov, *Uspekhi Khim.* **16** (1947) 254.
[43]N. A. Izmailov, *Doklady Akad. Nauk* **149** (1963) 884.

[44]D. Feakins and P. Watson, *J. Chem. Soc.* (*London*) **1963**, 4734.
[45]S. Andrews, H. P. Bennetto, D. Feakins, K. G. Lawrence, and J. Tomkins, *J. Chem. Soc.* (*London*) **A1968**, 1486.
[46]M. Kh. Karapet'yants, *Khim. Prom.* **1961**(1), 33; N. E. Khomutov, *Russ. J. Phys. Chem.* **39** (1965) 336.
[47]C. L. De Ligny and M. Alfenaar, *Rec. Trav. Chim.* **84** (1965) 81.
[48]M. Alfenaar and C. L. De Ligny, *Rec. Trav. Chim.* **86** (1967) 929.
[49]O. Popovych, *Anal. Chem.* **38** (1966) 558.
[50]R. Alexander and A. J. Parker, *J. Am. Chem. Soc.* **89** (1967) 5549.
[51]E. Grunwald, G. Baughman, and G. Kohnstam, *J. Am. Chem. Soc.* **82** (1960) 5801.
[52]O. Popovych and H. J. Dill, *Anal. Chem.* **41** (1969) 456.
[53]M. E. Coplan and R. M. Fuoss, *J. Phys. Chem.* **68** (1964).
[54]J. F. Coetzee and J. J. Campion, *J. Am. Chem. Soc.* **89** (1967) 2513.
[55]J. F. Coetzee and G. P. Cunningham, *J. Am. Chem. Soc.* **87** (1965) 2529.
[56]B. Case and R. Parsons, *Trans. Faraday Soc.* **63** (1967) 1224.
[57]W. M. Latimer and W. L. Jolly, *J. Am. Chem. Soc.* **75** (1953) 4147.
[58]B. Jakuszewski and S. Taniewska–Osinska, *Lodz. Towarz. Nauk Wydzial III, Acta Chem.* **4** (1959) 17.
[59]B. Jakuszewski and S. Taniewska–Osinska, *Lodz. Towarz. Nauk Wydzial III, Acta Chem.* **7** (1961) 32.
[60]B. Jakuszewski and S. Taniewska–Osinska, *Lodz. Towarz. Nauk Wydzial III, Acta Chem.* **8** (1962) 11.
[61]C. M. Criss, R. P. Held, and E. Luksha, *J. Phys. Chem.* **72** (1968) 2970.
[62]G. A. Krestov, *Zh. Strukt. Chim.* **3** (1962) 516.
[63]J. Greyson, *J. Phys. Chem.* **66** (1962) 2218; B. E. Conway and L. Laliberte, *Trans. Faraday Soc.* **66** (1970) 3032.
[64]J. E. B. Randles, *Trans. Faraday Soc.* **52** (1956) 1573.
[65]F. O. Koenig, in *Compt. Rend. C.I.T.C.E. 3rd Reunion Manfredi* (*Milan*) *1952*, p. 299.
[66]N. Izmailov and Yu. F. Rybkin, *Dopovidi Akad. Nauk Ukr.RSR* **1962**(1), 69.
[67]N. Izmailov and Yu. F. Rybkin, *Dopovidi Akad. Nauk Ukr.RSR* **1962**(1), 1071.
[68]Yu. F. Rybkin, cited in Ref. 56.
[69]B. Case, N. S. Hush, R. Parsons, and N. E. Peover, *J. Electroanal. Chem.* **10** (1965) 360.
[70]M. Born, *Z. Phys.* **1** (1920) 45.
[71]J. Padova, *Electrochim. Acta* **12** (1967) 1227.
[72]J. A. Stratton, *Electromagnetic Theory*, McGraw-Hill, pp. 149–151, New York, 1941.
[73]K. J. Laidler and C. Pegis, *Proc. Roy. Soc.* **A241** (1957) 80; see also K. J. Laidler and J. S. Muirhead-Gould, *Trans. Faraday Soc.* **63** (1967) 953.
[74]J. Padova, *J. Chem. Phys.* **39** (1963) 2599.
[75]J. Padova, *J. Chem. Phys.* **39** (1963) 1552; J. E. Desnoyers, R. E. Verrall, and B. E. Conway, *J. Chem. Phys.* **43** (1965) 243.
[76]J. Padova, *J. Chem. Phys.*, to be published.
[77]H. S. Frank, *J. Chem. Phys.* **23** (1955) 2023.
[78]S. Drakin and Yu-Min, *Russ. J. Phys. Chem.* **38** (1964) 1526.
[79]M. Woycieka, *Rocz. Chem.* **38** (1964) 1207.
[80]B. Jakuszewksi, S. Tantewska-Osinska, and R. Logwinienko, *Bull. Acad. Pol. Sci.* **9** (1961) 127.
[81]B. Jakuszewski and S. Taniewska, *Bull. Acad. Pol. Sci.* **9** (1961) 133.
[82]N. Izmailov, *Doklady Akad. Nauk SSSR* **149** (1963) 1103.

[83]N. Izmailov, *Doklady Akad. Nauk SSSR* **149** (1963) 1364.
[84]J. A. Plambeck, *Can. J. Chem.* **47** (1969) 1401.
[85]R. M. Noyes, *J. Am. Chem. Soc.* **84** (1962) 513.
[86]R. M. Noyes, *J. Am. Chem. Soc.* **86** (1964) 971.
[87]A. D. Buckingham, *Disc. Faraday Soc.* **24** (1957) 151.
[88]D. D. Eley and D. C. Pepper, *Trans. Faraday Soc.* **37** (1941) 581.
[89]D. D. Eley and Evans, *Trans. Faraday Soc.* **34** (1938) 1093.
[90]G. Somsen, Enthalpies of Solvation of Alkali Halides in Formamide, Ph.D. Thesis, Amsterdam (1964).
[91]J. O'M. Bockris, *Quart. Rev. Chem. Soc.* (*London*) **3** (1949) 173.
[92]J. H. Swinehart, T. E. Rogers, and H. Taube, *J. Chem. Phys.* **38** (1963), 398.
[93]T. E. Rogers, J. H. Swinehart, and H. Taube, *J. Phys. Chem.* **69** (1965) 134.
[94]J. H. Swinehart and H. Taube, *J. Chem. Phys.* **37** (1962) 1579.
[95]Z. Luz and S. Meiboom, *J. Chem. Phys.* **40** (1964) 1058, 1066.
[96]Z. Luz and S. Meiboom, *J. Chem. Phys.* **40** (1964) 2686.
[97]S. Thomas and W. L. Reynolds, *J. Chem. Phys.* **44** (1966) 3148.
[98]N. A. Matwiyoff, *Inorg. Chem.* **5** (1966) 788.
[99]A. Fratiello, D. Miller, and A. Schuster, *Mol. Phys.* **12** (1967) 111.
[100]N. A. Matwiyoff and G. Movius, *J. Am. Chem. Soc.* **89** (1967) 6077.
[101]S. A. Al-Baldawi and T. E. Gough, *Can. J. Chem.* **47** (1969) 1417.
[102]S. Nakamura and S. Meiboom, *J. Am. Chem. Soc.* **89** (1967) 1765.
[103]H. H. Glaeser, H. W. Dodgen, and J. P. Hunt, *J. Am. Chem. Soc.* **9** (1967) 3067.
[104]A. M. Chmelnick and D. Fiat, *J. Chem. Phys.* **49** (1968) 2101.
[105]T. D. Alger, *J. Am. Chem. Soc.* **91** (1969) 2220.
[106]A. Jackson, J. Lemons, and H. Taube, *J. Chem. Phys.* **32** (1960) 533.
[107]T. S. Swift and G. Y. O. Lo, *J. Am. Chem. Soc.* **89** (1967) 3988.
[108]A. Passynski, *Acta Physicochim. URSS* **8** (1938) 385.
[109]J. Padova, *J. Chem. Phys.* **40** (1964) 691.
[110]J. G. Mikhailov, M. V. Rozina, and V. A. Shutilov, *Akust. Zh.* **10** (1964) 213.
[111]D. S. Allam and W. H. Lee, *J. Chem. Soc.* (*London*)**A1966**, 5.
[112]A. S. Kaurova and G. P. Roshchina, *Akust. Zh.* **12** (1966) 118.
[113]A. S. Kaurova and G. P. Roshchina, *Akust. Zh.* **12** (1966) 319.
[114]G. P. Roshchina, H. S. Kaurova, and S. Sharapova, *Ukr. Fiz. Zh.* **12** (1967) 93.
[115]G. P. Roshchina, A. S. Kaurova, and J. D. Kosheleva, *Zh. Strukt. Khim.* **9** (1968) 3.
[116]J. D. Hefley and E. S. Amis, *J. Phys. Chem.* **69** (1965) 2082.
[117]J. D. Macdougall, *Thermodynamics and Chemistry*, 3rd ed., John Wiley and Sons, New York, 1939.
[118]D. C. Grahame, *J. Chem. Phys.* **21** (1953) 1054.
[119]J. Padova and J. Abrahamer, *J. Phys. Chem.* **71** (1967) 2112.
[120]H. S. Frank and W. Y. Wen, *Disc. Faraday Soc.* **24** (1957) 141.
[121]B. E. Conway and R. E. Verrall, *J. Phys. Chem.* **70** (1966) 3952.
[122]G. Jones and M. Dole, *J. Am. Chem. Soc.* **51** (1929) 2050.
[123]H. Falkenhagen and E. L. Vernon, *Physik. Z.* **33** (1932) 140.
[124]R. W. Gurney, *Ionic Processes in Solution*, McGraw-Hill, New York, 1954.
[125]J. Padova, *J. Chem. Phys.* **38** (1963) 2635.
[126]D. Feakins and K. G. Lawrence, *J. Chem. Soc.* (*London*) **A 1966**, 212.
[127]R. H. Stokes, in *The Structure of Electrolyte Solutions*, Ed., J. Hamer, John Wiley and Sons, New York, 1959.
[128]R. H. Stokes and R. Mills, *Viscosity of Electrolytes and Related Properties*, Pergamon Press, New York, 1965.
[129]A. Einstein, *Ann. Physik* **19** (1906) 289; **34** (1911) 591.
[130]N. Izmailov, S. Alexandrov, and I. F. Ivanova, *Tr. Khim. Khark. Univ.* **18** (1957) 5.

[131]R. A. Robinson and R. H. Stokes, *Electrolyte Solutions*, p. 240, Butterworths, London, 1959.
[132]D. F. T. Tuan and R. M. Fuoss, *J. Phys. Chem.* **67** (1963) 1343.
[133]N. P. Yao and D. N. Bennion, UCLA Report 69–30, June 1969.
[134]B. B. Owen, *J. Am. Chem. Soc.* **54** (1932) 1758.
[135]E. Luksha and C. M. Criss, *J. Phys. Chem.* **70** (1966) 1496.
[136]J. Padova, *Israel J. Chem.* **4** (1966) 41.
[137]H. Uhlich, *Trans. Faraday Soc.* **23** (1927) 388.
[138]H. S. Frank, in *Chemical Physics of Ionic Solutions*, Eds., B. E. Conway and R. J. Barradas, John Wiley and Sons, New York, 1966.
[139]R. H. Stokes and R. A. Robinson, *Trans. Faraday Soc.* **53** (1957) 301.
[140]M. Born, *Z. Physik* **1** (1920) 221.
[141]R. M. Fuoss, *Proc. Nate. Sci.* **45** (1959) 807.
[142]R. H. Boyd, *J. Chem. Phys.* **35** (1961) 1281.
[143]R. Zwanzig, *J. Chem. Phys.* **38** (1963) 1603, 1605.
[144]E. J. Passeron, *J. Phys. Chem.* **68** (1964) 2728.
[145]R. A. Robinson and R. H. Stokes, *Electrolyte Solutions*, p. 125, Butterworths, London, 1959.
[146]J. A. V. Butler, *J. Phys. Chem.* **33** (1929) 1015.
[147]J. O'M. Bockris, J. Bowler-Reed and J. A. Kitchener, *Trans. Faraday Soc.* **47** (1951) 184.
[148]B. E. Conway, J. E. Desnoyers, and A. C. Smith, *Phil. Trans. Roy. Soc.* **A256** (1964) 389.
[149]E. Mooney, *J. Colloid Chem.* **6** (1951) 163.
[150]M. St.J. Arnold and K. J. Packer, *Mol. Phys.* **10** (1966) 141.
[151]J. Padova, *Can. J. Chem.* **43** (1965) 458.
[152]P. Debye, *Z. Physik Chem.* **130** (1926) 56.
[153]G. Scatchard, *J. Chem. Phys.* **9** (1941) 34.
[154]J. E. Ricci and G. J. Ness, *J. Am. Chem. Soc.* **64** (1942) 2305.
[155]H. L. Clever and F. H. Verhoek, *J. Phys. Chem.* **62** (1958) 1961.
[156]I. F. Efremov, T. A. Probof'eva, and Yu. P. Syrnikov, *Zh. Fiz. Khim.* **38** (1964) 2258.
[157]E. Grunwald, G. Baughman, and G. Kohnstam, *J. Am. Chem. Soc.* **82** (1960) 5802.
[158]F. Franks and D. J. C. Ives, *Quart. Rev.* **20** (1966) 1.
[159]D. Feakins, in *Physico-Chemical Processes in Mixed Aqueous Solvents*, p. 71, Ed., F. Franks, Heinemann, London, 1967.
[160]D. Feakins and C. M. French, *J. Chem. Soc.* (*London*) **1957**, 2581.
[161]D. Feakins, B. S. Smith, and L. Thakur, *J. Chem. Soc.* (*London*) **1966**, 714.
[162]H. P. Bennetto, D. Feakins, and K. G. Lawrence, *J. Chem. Soc.* (*London*) **A1968**, 1493.
[163]H. P. Bennetto and D. Feakins, in *Hydrogen-Bonded Systems*, p. 235, Ed., A. K. Covington and P. Jones, Taylor and Francis Ltd., London, 1968.
[164]A. J. Dill, Ph.D. Thesis, New York (1967).
[165]A. J. Dill, L. M. Itzkovitch, and O. Oopovych, *J. Phys. Chem.* **72** (1968) 4580.
[166]R. H. Stokes, *J. Am. Chem. Soc.* **86** (1964) 979.
[167]H. P. Bennetto and D. Feakins, in *Hydrogen-Bonded Systems*, p. 245, Ed., A. K. Covington and P. Jones, Taylor and Francis Ltd., London, 1968.
[168]A. Fratiello and D. C. Douglas, *J. Chem. Phys.* **39** (1963) 2017.
[169]C. M. Slansky, *J. Am. Chem. Soc.* **62** (1940) 2430.
[170]R. I. Moss and J. H. Wolfenden, *J. Chem. Soc.* (*London*) **1939**, 118.
[171]G. V. Karpenko, K. P. Mishchenko, and G. M. Poltoratski, *Zh. Strukt. Khim.* **8** (1967) 413.

[172]G. A. Krestov and I. V. Egorova, *Izv. Vyshikh. Uch. Zav. Khim., Khim. Tekhn.* **1967**, 750.
[173]D. Feakins, in *Physico-Chemical Processes in Mixed Aqueous Solvents*, p. 148, Ed., F. Franks, Heinemann, London, 1967.
[174]E. M. Arnett, in *Physico-Chemical Processes in Mixed Aqueous Systems*, p. 105, Ed., F. Franks, Heinemann, London, 1966.
[175]K. P. Mishchenko and S. V. Shadskii, *Teor. Eksp. Khim.* **1** (1965) 1.
[176]K. P. Mishchenko and S. V. Shadskii, *Doklady Akad. Nauk. SSSR* **167** (1966) 621.
[177]J. H. Stern and J. Nobilione, *J. Phys. Chem.* **42** (1968) 3937.
[178]G. A. Krestov and I. V. Egorova, *Teor. Eksp. Khim.* **3** (1967) 128.
[179]E. M. Arnett, W. G. Bentrude, J. J. Burke, and P. M. Duggleby, *J. Am. Chem. Soc.* **87** (1965) 1541.
[180]J. A. Pople, W. G. Schneider, and H. J. Bernstein, *High-Resolution Nuclear Magnetic Resonance*, Chapter 15, McGraw-Hill, New York, 1959.
[181]A. Fratiello and D. Miller, *Mol. Phys.* **11** (1966) 37.
[182]A. Fratiello, R. E. Lee, D. Miller, and V. M. Nishida, **13** (1967) 349.
[183]A. Fratiello, R. E. Lee, V. M. Nishida, and R. Schuster, *J. Chem. Phys.* **47** (1967) 4951.
[184]J. F. Hinton, M. McDowell, and E. S. Amis, *Chem. Comm.* **1966**, 776.
[185]A. Fratiello and D. Miller, *J. Chem. Phys.* **42** (1965) 796.
[186]J. F. Hinton and E. S. Amis, *Chem. Comm.* **1967**, 100.
[187]J. F. Hinton, E. S. Amis, and W. Mettetal, *Spectrochim. Acta* **25A** (1969) 119.
[188]A. Fratiello and E. G. Christie, *Trans. Faraday Soc.* **61** (1965) 306.
[189]A. Fratiello and R. Schuster, *J. Phys. Chem.* **71** (1967) 1948.
[190]G. E. Maciel, J. K. Hancock, L. F. Lafferty, P. A. Mueller, and W. K. Musker, *Inorg. Chem.* **5** (1966) 554.
[191]E. G. Bloor and R. G. Kidd, *Can. J. Chem.* **46** (1968) 22.
[192]A. Fratiello and R. Schuster, *Tetrahedron Lett.* **40** (1967) 4641.
[193]L. D. Supran and S. N. Sheppard, *Chem. Comm.* **1967**, 832.
[194]N. A. Matwiyoff and H. Taube, *J. Am. Chem. Soc.* **90** (1968) 2796.
[195]A. Fratiello and B. Schuster, *J. Chem. Educ.* **45** (1968) 91.
[196]A. Fratiello, R. E. Lee, V. M. Nishida, and R. Schuster, *J. Chem. Phys.* **48** (1968) 3705.
[197]A. Fratiello, R. E. Lee, V. M. Nishida, and R. Schuster, *J. Chem. Comm.* (1968) 173.
[198]A. Fratiello, R. E. Lee, and R. Schuster, *Chem. Comm.* **1969**, 37.
[199]A. Fratiello, R. E. Lee, V. M. Nishida, and R. Schuster, *Inorg. Chem.* **8** (1969) 69.
[200]L. S. Frankel, T. R. Stengle, and C. H. Langford, *Chem. Comm.* **1965**, 373.
[201]L. S. Frankel, T. R. Stengle, and C. H. Langford, *Can. J. Chem.* **46** (1968) 3183.
[202]L. S. Frankel, Ph.D. Thesis, Massachusetts, 1966.
[203]R. T. Iwamasa, Ph.D. Thesis, California, 1967.
[204]D. Geshke and H. Pfeiffer, *Zh. Strukt. Khim.* **5** (1964) 201.
[205]J. F. Hinton and E. S. Amis, *Chem. Rev.* **67** (1967) 367.
[206]J. Burgess and M. C. Symons, *Quart. Rev.* **22** (1968) 768.
[207]D. P. Olander, R. Marianelli, and R. C. Larson, *Anal. Chem.* **41** (1969) 1097.
[208]W. M. Cox and J. H. Wolfenden, *Proc. Roy. Soc.* **A145** (1934) 475.
[209]L. Onsager and R. M. Fuoss, *J. Phys. Chem.* **36** (1932) 2689.
[210]N. P. Yao and D. N. Bennion, UCLA Report No. 69-30-1, June 1969.
[211]K. Crickard and J. F. Skinner, *J. Phys. Chem.* **73** (1969) 2060.
[212]G. Jones and H. J. Fornwalt, *J. Am. Chem. Soc.* **57** (1935) 2041.
[213]H. T. Briscoe and W. Rinehart, *J. Phys. Chem.* **46** (1942) 387.
[214]J. C. Lafanechere, Ph.D. Thesis, Paris, June 1969.

[215]G. Delesalle, P. Devraine, and J. Heubel, *Compt. Rend.* **267C** (1968) 1464.
[216]D. Feakins, D. G. Freemantle, and K. G. Lawrence, *Chem. Comm.* (1968) 870.
[217]D. Feakins and K. G. Lawrence, *J. Chem. Soc.* **A1966**, 212.
[218]H. Mohapatra and D. B. Das, *J. Ind. Chem. Soc.* **44** (1967) 573.
[219]J. C. Lafanechere and J. P. Morel, **268C** (1969) 1222.
[220]R. Gillespie, in *Chemical Physics of Ionic Solutions*, p. 599, Eds., B. E. Conway and R. J. Barradas, John Wiley and Sons, New York, 1966.
[221]C. H. Langford and T. R. Stengle, *J. Am. Chem. Soc.* **91** (1969) 4014.
[222]A. Z. Golik, A. V. Oritschenko, and O. G. Artemchenko, *Dopovidi Akad. Nauk Ukr. RSR* **6** (1954) 457.
[223]F. H. Getman, *J. Am. Chem. Soc.* **30** (1908) 1077.
[224]R. Davis and H. C. Jones, *Z. Phys. Chem.* **81** (1913) 68.
[225]H. S. Frank and M. W. Evans, *J. Chem. Phys.* **13** (1945) 507.
[226]O. Ya. Samoilov, *Disc. Faraday Soc.* **24** (1957) 141.
[227]G. Engel and H. G. Hertz, *Ber. Bunsenges* **72** (1968) 808.
[228]E. von Goldamer and M. D. Zeidler, *Ber. Bunsenges* **73** (1969) 4.
[229]P. B. Das, P. R. Das, and D. Patnaik, *J. Ind. Chem. Soc.* **42** (1965) 166.
[230]P. B. Das and P. K. Das, *J. Ind. Chem. Soc.* **43** (1966) 58.
[231]G. R. Hood and L. P. Hohlfelder, *J. Phys. Chem.* **38** (1934) 979.
[232]T. B. Hoover, *J. Phys. Chem.* **68** (1964) 876.
[233]J. Padova, unpublished results.
[234]E. Asmus, *Z. Naturforsch.* **4A** (1949) 589.
[235]E. R. Nightingale, *J. Phys. Chem.* **63** (1959) 1381.
[236]V. Vand, *J. Phys. Chem.* **52** (1948) 277.
[237]W. L. Reynolds, Ph.D. Thesis, Yale University, 1966.
[238]W. D. Kraeft and J. Einfeldt, *Z. Physik Chem.* **237** (1968) 267.
[239]C. H. Springer, J. F. Coetzee, and R. L. Kay, *J. Phys. Chem.* **73** (1969) 471.
[240]G. J. Janz, A. E. Marcinkowsky, and I. Ahmad, *Electrochim. Acta* **9** (1964) 1687.
[241]G. A. Forcier and J. W. Olver, *Electrochim. Acta* **14** (1969) 135.
[242]D. F. Evans, C. Zawoyski, and R. L. Kay, *J. Phys. Chem.* **69** (1965) 3878.
[243]R. M. Fuoss, *J. Am. Chem. Soc.* **79** (1957) 3301.
[244]R. L. Kay, B. J. Hales, and G. P. Cunningham, *J. Phys. Chem.* **71** (1967) 3925.
[245]C. H. Springer, Ph.D. Thesis, University of Pittsburgh, 1968.
[246]D. K. McGuire, Ph.D. Thesis, University of Pittsburgh, 1964.
[247]J. S. Dunnet and R. P. H. Gasser, *Trans. Faraday Soc.* **61** (1965) 922.
[248]M. D. Archer and R. P. H. Gasser, *Trans. Faraday Soc.* **62** (1966) 3451.
[249]J. M. Crawford and R. P. H. Gasser, *Trans. Faraday Soc.* **63** (1967) 2758.
[250]D. Atlani, A. C. Justice, M. Quintin, and R. Dubois, *J. Chim. Phys.* **66** (1969) 180.
[251]D. Arrington, Ph.D. Thesis, University of Kansas, 1968.
[252]N. P. Yao and D. N. Bennion, UCLA Report No. 69-30-3, June 1969.
[253]J. F. Skinner and R. M. Fuoss, *J. Phys. Chem.* **70** (1966) 1426.
[254]R. D. Singh and S. Mishra, *Ind. J. Chem.* **4** (1966) 308.
[255]R. L. Kay, C. Zawoyski, and D. F. Evans, *J. Phys. Chem.* **69** (1965) 4208.
[256]R. L. Kay and J. L. Hawes, *J. Phys. Chem.* **69** (1965) 2787.
[257]B. Sesta, *Ann. Chim.* (*Rome*) **57** (1967) 129.
[258]D. F. Evans and T. L. Broadwater, *J. Phys. Chem.* **72** (1968) 1037.
[259]H. Tsubota and G. Atkinson, *J. Phys. Chem.* **71** (1967) 1131.
[260]T. A. Gover and P. G. Sears, *J. Phys. Chem.* **60** (1956) 330.
[261]M. Goffredi and T. Shedlovsky, *J. Phys. Chem.* **71** (1967) 2176, 2182, 4436.
[262]H. V. Venkatasetty and G. H. Brown, *J. Phys. Chem.* **66** (1962) 2075; **67** (1963) 954.
[263]D. F. Evans and P. Gardam, *J. Phys. Chem.* **73** (1969) 158.

[264]R. H. Davies and E. G. Taylor, *J. Phys. Chem.* **68** (1964) 3901.
[265]C. Carvaial, K. J. Tölle, J. Smid, and M. Swarc, *J. Am. Chem. Soc.* **87** (1965) 5548.
[266]D. N. Bhattacharya, C. L. Lee, J. Smid, and M. Swarc, *J. Phys. Chem.* **69** (1965) 608.
[267]J. Comyn, F. S. Dainton, and K. J. Ivin, *Electrochim. Acta* **13** (1968) 1851.
[268]J. C. Justice, R. Bury, and C. Treiner, *J. Chim. Phys.* **65** (1968) 1708.
[269]D. Nicholls, C. Sutphen, and M. Swarc, *J. Phys. Chem.* **72** (1968) 1021.
[270]R. Bury and C. Treiner, *J. Chim. Phys.* **65** (1968) 1410.
[271]M. E. Coplan, J. C. Justice, and M. Quintin, *J. Chim. Phys.* **65** (1968) 1152.
[272]A. K. R. Unni, L. Elias, and H. I. Schiff, *J. Phys. Chem.* **67** (1963) 1216.
[273]S. C. Blum and H. I. Schiff, *J. Phys. Chem.* **67** (1963) 1220.
[274]R. L. Kay, S. C. Blum, and H. I. Schiff, *J. Phys. Chem.* **67** (1963) 1222.
[275]A. M. Shkodin, L. P. Sadovnichaya, and V. A. Podolyanko, *Elektrokhimyia* **4** (1968) 718.
[276]H. O. Spivey and T. Shedlowsky, *J. Phys. Chem.* **71** (1967) 2165.
[277]G. D. Parfitt and A. L. Smith, *Trans. Faraday Soc.* **59** (1963) 257.
[278]A. Fratiello, Ph.D. Thesis, Brown University, 1962.
[279]J. L. Hawes and R. L. Kay, *J. Phys. Chem.* **69** (1965) 2420.
[280]J. O. Wear, Ph.D. Thesis, University of Kansas, 1962.
[281]F. Barbulescu, A. Greff, I. S. Popescu, and I. Sass, *Rev. Roum. Chim.* **11** (1966) 903.
[282]A. d'Aprano, *Ric. Sci.* **34**, *Rend. Cont. II* **7** (1964) 433.
[283]F. Accascina, L. Gardona, A. d'Aprano, and M. Goffredi, *Ric. Sci.* **34**, *Rend. Cont. II* **6** (1964) 63.
[284]F. Accascina, A. d'Aprano, and M. Goffredi, *Ric. Sci.* **34**, *Rend. Cont. II* **4** (1964) 443.
[285]F. Accascina and S. Petrucci, *Ric. Sci.* **29** (1959) 1640.
[286]R. Bury, *J. Chim. Phys.* **64** (1967) 83.
[287]S. Devi and P. B. Das, *J. Ind. Chem. Soc.* **42** (1965) 500.
[288]S. Devi and P. B. Das, *J. Ind. Chem. Soc.* **42** (1965) 501.
[289]G. Pistoia, *Ric. Sci.* **37** (1967) 731.
[290]R. Bury and J. C. Justice, *J. Chim. Phys.* **64** (1967) 1492; *Compt. Rend.* **260** (1965) 6089.
[291]M. Quintin and J. C. Justice, *Compt. Rend.* **260** (1965) 5255.
[292]G. Atkinson and Y. Mori, *J. Chem. Phys.* **45** (1966) 4716.
[293]B. R. Staples and G. Atkinson, *J. Phys. Chem.* **73** (1969) 520.
[294]R. D. Singh and S. Mishra, *Ind. J. Chem.* **7** (1969) 86.
[295]M. Goffredi and R. Triolo, *Ric. Sci.* **37** (1967) 1137.
[296]F. Accascina and A. d'Aprano, *Gazz.* **95** (1965) 1420.
[297]F. Accascina, A. d'Aprano, and R. Triolo, *J. Phys. Chem.* **71** (1967) 3469; F. Accascina and R. Triolo, *J. Phys. Chem.* **71** (1967) 3474.
[298]G. Pistoia, G. Pecci, and B. Scrosati, *Ric. Sci.* **37** (1967) 1167.
[299]G. Pistoia and B. Scrosati, *Ric. Sci.* **37** (1967) 1173.
[300]M. D. Monica, U. Lamanna, and L. Senatore, *J. Phys. Chem.* **72** (1968) 2124.
[301]M. D. Monica, U. Lamanna, and S. Janelli, *Gazz.* **97** (1967) 367.
[302]R. M. Gopal and O. N. Bhatnagar, *J. Phys. Chem.* **70** (1966) 4070.
[303]R. M. Gopal and O. N. Bhatnagar, *J. Phys. Chem.* **68** (1964) 3892.
[304]R. M. Gopal and O. N. Bhatnagar, *J. Phys. Chem.* **70** (1966) 3007.
[305]T. B. Hoover, *J. Phys. Chem.* **68** (1964) 3003.
[306]J. M. Notley and M. Spiro, *J. Phys. Chem.* **70** (1966) 1502.
[307]T. C. Wehman and A. I. Popov, *J. Phys. Chem.* **72** (1968) 4031.
[308]R. D. Singh, P. P. Rastogi, and R. M. Gopal, *Can. J. Chem.* **46** (1968) 3525.

[309]G. P. Johari and P. H. Tewari, *J. Phys. Chem.* **69** (1965) 696.
[310]P. H. Tewari and G. P. Johari, *J. Phys. Chem.* **67** (1963) 512.
[311]R. M. Gopal and O. N. Bhatnagar, *J. Phys. Chem.* **69** (1965) 2382.
[312]G. P. Johari and P. H. Tewari, *J. Phys. Chem.* **70** (1966) 197.
[313]W. R. Gilkerson, *J. Chem. Phys.* **25** (1956) 1199.
[314]F. Accascina and M. Goffredi, *Ric. Sci.* **37** (1967) 1126.
[315]F. Accascina, A. d'Aprano, and M. Goffredi, *Ric. Sci.* **34** *IIA*(6) (1964) 151.
[316]F. Accascina and S. Petrucci, *Ric. Sci.* **30** (1960) 808.
[317]S. Petrucci, G. C. Hemmes, and G. Battistini, *J. Am. Chem. Soc.* **89** (1967) 5552.
[318]A. d'Aprano and R. Triolo, *Ric. Sci.* **34** (IIA, 7) (1964) 443.
[319]A. Accascina and A. d'Aprano, *Ric. Sci.* **36** (1966) 257.
[320]G. C. Hemmes and S. Petrucci, *J. Am. Chem. Soc.* **91** (1969) 275.
[321]W. A. Adams and K. J. Laidler, *Can. J. Chem.* **46** (1968) 1977.
[322]W. A. Adams and K. J. Laidler, *Can. J. Chem.* **46** (1968) 1989.
[323]W. A. Adams and K. J. Laidler, *Can. J. Chem.* **46** (1968) 1999.
[324]L. Savedoff, *J. Am. Chem. Soc.* **88** (1966) 664.
[325]F. Accascina and S. Schiavo, *Ann. Chim.* (*Rome*) **43** (1953) 47.
[326]S. R. Hughes and S. H. White, *J. Chem. Soc.* (*London*) **A1966**, 1216.
[327]J. Barthel, *Inorg. Chem.* **7** (1968) 260.
[328]M. E. Coplan and R. M. Fuoss, *J. Phys. Chem.* **68** (1964) 1177.
[329]R. L. Kay, *J. Am. Chem. Soc.* **82** (1960) 2099; in *Electrolytes*, Ed., B. Pesce, p. 119, Pergamon Press, New York, 1962.
[330]G. P. Cunningham, Ph.D. Thesis, University of Pittsburgh, 1964.
[331]W. H. McMahan, Ph.D. Thesis, University of Kansas, 1965.
[332]F. Pushpanaden, Ph.D. Thesis, Indiana University, 1966.
[333]G. W. A. Fowles and W. R. McGregor, *J. Phys. Chem.* **68** (1964) 1342.
[334]I. R. Bellobono and G. Favini, *Gazz.* **1964**, 32.
[335]R. L. Kay, in *Trace Inorganics in Water*, p. 1, Am. Chem. Soc., Wash. D.C., 1968.
[336]K. H. Stern and E. S. Amis, *Chem. Rev.* **59** (1959) 1.
[337]G. J. Janz and M. J. Tait, *Can. J. Chem.* **45** (1967) 1101.
[338]H. Sadek, E. Hirsch, and R. M. Fuoss, in *Electrolytes*, p. 132, Pergamon Press, 1962.
[339]R. C. Paul, J. P. Singla, and S. P. Narula, *J. Phys. Chem.* **73** (1969) 741.
[340]R. M. Gopal and M. M. Husain, *J. Ind. Chem. Soc.* **40** (1963) 981.
[341]G. Atkinson and Y. Mori, *J. Phys. Chem.* **71** (1967) 3523.
[342]J. E. Prue and P. J. Sherrington, *Trans. Faraday Soc.* **57** (1961) 1795.
[343]L. G. Longsworth, *J. Phys. Chem.* **67** (1963) 689.
[344]H. K. Bodenseh and J. B. Ramsey, *J. Phys. Chem.* **69** (1965) 543.
[345]G. Atkinson and S. Petrucci, *J. Am. Chem. Soc.* **86** (1964) 7.
[346]E. Darmois, La Solvatation des Ions, *Mem. Sci. Phys.* **48** (1946).
[347]A. S. Quist and W. L. Marshall, *J. Phys. Chem.* **72** (1968) 1536.
[348]A. d'Aprano and R. M. Fuoss, *J. Am. Chem. Soc.* **91** (1969) 211.
[349]F. Conti, P. Delogu, and G. Pistoia, *J. Phys. Chem.* **72** (1968) 1396.
[350]F. Accascina, G. Pistoia, and S. Schiavo, *Ric. Sci.* **36** (1966) 560.
[351]F. Accascina, G. Pistoia, and S. Schiavo, *Ric. Sci.* **34** (IIA) (1964) 141.
[352]G. Pistoia, A. M. Polcaro, and S. Schiavo, *Ric. Sci.* **37** (1967) 227, 300, 309.
[353]L. D. Pettit and S. Bruckenstein, *J. Am. Chem. Soc.* **88** (1966) 4783.
[354]J. F. Skinner and R. M. Fuoss, *J. Phys. Chem.* **69** (1965) 1437.
[355]W. L. Marshall and A. S. Quist, *Proc. Natl. Acad. Sci. U.S.* **58** (1967) 901.
[356]J. C. Justice, *J. Chim. Phys.* **64** (1967) 353.

[357]E. Pitts, R. F. Tabor, and J. Daly, *Trans. Faraday Soc.* **65** (1969) 849.
[358]R. M. Fuoss and F. Accascina, *Electrolytic Conductance*, Interscience Publishers, New York, 1959.
[359]C. Evers and R. L. Kay, *Ann. Rev. Phys. Chem.* **11** (1960) 21.
[360]R. M. Fuoss, *J. Am. Chem. Soc.* **80** (1958) 5059.
[361]R. M. Fuoss, in *Chemical Physics of Ionic Solutions*, p. 463, Eds. B. E. Conway and R. J. Barradas, John Wiley and Sons, New York, 1966.
[362]R. M. Fuoss and L. Onsager, *J. Phys. Chem.* **61** (1957) 668.
[363]D. A. MacInnes, *The Principles of Electrochemistry*, p. 91, Dover, New York, 1961.
[364]L. G. Longsworth, *J. Am. Chem. Soc.* **69** (1947) 1288.
[365]W. V. Child and E. S. Amis, *J. Inorg. Nucl. Chem.* **16** (1960) 114.
[366]J. O. Wear, C. V. McNully, and E. S. Amis, *J. Inorg. Nucl. Chem.* **18** (1961) 48.
[367]J. O. Wear, C. V. McNully, and E. S. Amis, *J. Inorg. Nucl. Chem.* **19** (1961) 278.
[368]J. O. Wear, C. V. McNully, and E. S. Amis, *J. Inorg. Nucl. Chem.* **20** (1961) 100.
[369]J. O. Wear, T. J. Curtis, Jr., and E. S. Amis, *J. Inorg. Nucl. Chem.* **24** (1962) 93.
[370]J. O. Wear and E. S. Amis, *J. Inorg. Nucl. Chem.* **24** (1962) 903.
[371]R. G. Griffin, E. S. Amis, and J. O. Wear, *J. Inorg. Nucl. Chem.* **28** (1966) 543.
[372]H. Strehlow and H. Koepp, *Z. Elektrochem.* **62** (1958) 373.
[373]H. Schneider and H. Strehlow, *Ber. Bunsenges* **69** (1965) 674.
[374]H. Schneider and H. Strehlow, *Ber. Bunsenges* **66** (1962) 309.
[375]E. F. Ivanova, *Russ. J. Phys. Chem.* **39** (1965) 766.
[376]H. Fowler and C. A. Kraus, *J. Am. Chem. Soc.* **62** (1940) 2237.
[377]R. M. Fuoss and E. Hirsch, *J. Am. Chem. Soc.* **82** (1960) 1013.
[378]G. G. Stokes, *Cambridge Phil. Soc. Trans.* **9** (1856) 5.
[379]H. S. Harned and B. B. Owen, *The Physical Chemistry of Electrolytic Solutions*, 3rd ed., Reinhold Publishing Co., New York, 1958.
[380]P. Walden, H. Uhlich, and G. Busch, *Z. Physik Chem.* **123** (1926) 429.
[381]D. F. Evans and R. L. Kay, *J. Phys. Chem.* **70** (1966) 366.
[382]R. L. Kay and D. F. Evans, *J. Phys. Chem.* **70** (1966) 2325.
[383]S. Schmick, *Z. Phys.* **24** (1924) 56.
[384]R. L. Kay, G. P. Cunningham, and D. F. Evans, in *Hydrogen-Bonded Solvent Systems*, p. 249. Eds, A. K. Covington and P. Jones, Taylor and Francis Ltd., London, 1968.
[385]D. F. Evans and P. Gardam, *J. Phys. Chem.* **72** (1968) 3281.
[386]R. L. Kay, G. P. Cunningham, and G. A. Vidulich, unpublished data, cited in Ref. 384.
[387]T. Shedlovsky and R. L. Kay, *J. Phys. Chem.* **60** (1956) 151; T. Shedlovsky, in *The Structure of Electrolytic Solutions*, Ed., J. Hamer, John Wiley and Sons, New York, 1959.
[388]S. Petrucci, *Acta Chem. Scand.* **16** (1962) 760.
[389]F. Accascina, and S. Petrucci, *Ric. Sci.* **30** (1960) 1164.
[390]J. E. Lind and R. M. Fuoss, *J. Phys. Chem.* **65** (1961) 999.
[391]J. E. Lind and R. M. Fuoss, *J. Phys. Chem.* **65** (1961) 1414.
[392]J. E. Lind and R. M. Fuoss, *J. Phys. Chem.* **66** (1962) 1727.
[393]R. W. Kunze and R. M. Fuoss, *J. Phys. Chem.* **67** (1963) 911.
[394]R. W. Kunze and R. M. Fuoss, *J. Phys. Chem.* **67** (1963) 914.
[395]J. C. Justice and R. M. Fuoss, *J. Phys. Chem.* **67** (1963) 1707.
[396]T. L. Fabry and R. M. Fuoss, *J. Phys. Chem.* **68** (1964) 971.
[397]T. L. Fabry and R. M. Fuoss, *J. Phys. Chem.* **68** (1964) 974.
[398]R. M. Fuoss and C. A. Kraus, *J. Am. Chem. Soc.* **79** (1957) 3301.
[399]A. d'Aprano and R. M. Fuoss, *J. Phys. Chem.* **72** (1968) 4710.

[400] A. d'Aprano and R. M. Fuoss, *J. Am. Chem. Soc.* **91** (1969) 279.
[401] I. D. McKenzie and R. M. Fuoss, *J. Phys. Chem.* **73** (1969) 1501.
[402] J. R. Bard, J. O. Wear, R. G. Griffin, and E. S. Amis, *J. Electroanal. Chem.* **8** (1964) 419.
[403] C. Treiner and R. M. Fuoss, *Z. Physik Chem.* **228** (1965) 343.
[404] L. N. Erbanova, S. I. Drakin, and M. Kh. Karapet'yants, *Russ. J. Phys. Chem.* **38** (1964) 1450.
[405] L. N. Erbanova, M. Kh. Karapet'yants, and S. I. Drakin, *Russ. J. Phys. Chem.* **39** (1965) 1467.
[406] M. E. Clark and J. L. Bear, *J. Inorg. Nucl. Chem.* **31** (1969) 2619.
[407] H. Schneider and H. Strehlow, *Z. Physik Chem. N.F.* **49** (1966) 44.
[408] J. Padova, Ph.D. Thesis, London, 1954.
[409] H. L. Yeager and B. Kratochvil, *J. Phys. Chem.* **73** (1969) 1963.
[410] H. Brusset and T. Kikindal, *Bull. Soc. Chim. France* (**1962**), 1150.
[411] A. Than and E. S. Amis, *Z. Physik Chem. N.F.* **58** (1968) 196.
[412] U. Sen, K. R. Kundu, and M. N. Das, *J. Phys. Chem.* **71** (1967) 3665.
[413] G. R. Haugen and H. L. Friedman, *J. Phys. Chem.* **72** (1968) 4549.
[414] D. W. Ebdon, Ph.D. Thesis, University of Maryland, 1967.
[415] T. R. Griffith and M. C. R. Symmons, *Mol. Phys.* **3** (1960) 90.
[416] N. Bjerrum, *Kge. Danske Videnskab. Selskab.* **7** (9) (1926).
[417] P. H. Flaherty and K. H. Stern, *J. Am. Chem. Soc.* **80** (1958) 1034.
[418] J. T. Denison and J. B. Ramsey, *J. Am. Chem. Soc.* **77** (1955) 2615.
[419] R. M. Fuoss, *J. Am. Chem. Soc.* **80** (1958) 5059.
[420] J. H. Delap, Ph.D. Thesis, Duke University, 1960.
[421] H. Reiss, *J. Chem. Phys.* **25** (1956) 400.
[422] F. H. Stillinger, Jr. and R. Lovett, *J. Chem. Phys.* **48** (1968) 3858, 3869.
[423] C. A. Kraus, *J. Phys. Chem.* **60** (1956) 120.
[424] E. A. Guggenheim, *Trans. Faraday Soc.* **56** (1960) 1159.
[425] J. E. Prue, in *Chemical Physics of Ionic Solutions*, p. 163, Eds., B. E. Conway and R. J. Barradas, John Wiley and Sons, New York, 1966.
[426] R. M. Fuoss and K. L. Hsia, *Proc. Natl. Acad. Sci. U.S.*, **57** (1967) 1550; **58** (1967) 1818.
[427] G. Atkinson and S. Kor, *J. Phys. Chem.***69** (1965) 128.
[428] S. Petrucci and G. Atkinson, *J. Phys. Chem.* **70** (1966) 2550.
[429] D. Feakins and C. M. French, *J. Chem. Eoc.* (1957) 2581.
[430] P. B. Das and D. Patnaik, *J. Ind. Chem. Soc.* **39** (1962) 13.
[431] E. Glueckauf, *Trans. Faraday Soc.* **51** (1955) 1255.

2

Solvated Electrons in Field- and Photo-Assisted Processes at Electrodes

B. E. Conway*

Universities of Southampton and Newcastle-on-Tyne England

I. INTRODUCTION

Solvated electrons can arise in electrode processes by photo-injection from the cathode under suitable conditions of illumination, by direct injection from the cathode into certain solvents by application of a sufficient potential difference at the cathode–solution interface, and by base-metal dissolution in special solvents such as ammonia and some amines. The possibility of photoassistance of electrode processes was recognized† in early work by Bowden[1] and the involvement of solvated electrons by direct cathodic injection into suitable solvents was demonstrated, e.g., in the case of liquid NH_3, by Birch[3] and by Laitinen and Nyman.[4] Early indications of such a process were found by Palmaer.[5] The involvement of solvated electrons in electrochemical processes has recently become of considerable fundamental interest in regard to (a) the nature of electrochemical photoeffects not involving electronically excited states in solution, such as arise in the case of dye molecules; (b) the

*Commonwealth Visiting Professor (1969–70) at the Universities of Southampton and Newcastle-on-Tyne; permanent address, Chemistry Department, University of Ottawa, Ottawa, Canada.

†The earliest report of an effect of light upon electrodes appears to be that of Becquerel.[2]

reactions which can be made to occur with solvated electron intermediates; (c) possible electrochemical engineering applications in cathodic reduction; and (d) in recent years, the hydrated electron has been proposed as a primary intermediate in cathodic hydrogen evolution[8] and in metal dissolution in aqueous media[9,10]; the former proposal is obviously of great interest in regard to the whole problem of the mechanisms of the h.e.r. under various conditions and the latter idea is pertinent to the question of the role of supposed low-valence metal ions, e.g., Be^{+}, Mg^{+}, in base-metal dissolution.[9a]

In the present chapter, we shall examine the evidence for solvated electron intermediates in electrode processes* and the experimental data which support their involvement in direct cathodic and photocathodic processes in various media. The electroorganic chemistry of reductions by solvated electrons will be treated only briefly since this is a separate field more connected with the chemistry of the Birch reduction reaction.[11] However, insofar as these reductions give evidence for the cathodic production of solvated electrons, e.g., in ammonia and amine solvents, they will receive brief examination from an electrochemical point of view.

Most of the reactions in which the production of solvated electrons can be studied involve examination of the kinetics of scavenging reactions in which the electrons, once injected into the solvent, undergo secondary reactions with electron acceptors. In the case of photoeffects, the relation between electrode potential and critical quantum energy for the release of electrons from the metal into the solvent is a central question. The extent to which various metals may dissolve in suitable solvents with the direct release of solvated electrons (to be denoted as e_s^- in the general case) is also of interest as a process parallel to the normal electrochemically controlled corrosion of metals. The problem of whether this is a significant process in the case of the water solvent will receive some attention in this chapter in relation to the possible role of e_{aq}^- as an intermediate in the hydrogen evolution reaction (h.e.r.).[8] We first examine some energetic conditions relevant to the electrochemical production of solvated electrons.

*Since this chapter was written (1969), another review on this topic, by Walker, has appeared (in *Electroanalytical Chemistry*, Ed. A. J. Bard, p. 1, Vol. 5, Marcel Dekker Inc., New York, 1970) but presents the topic from a rather different point of view.

II. ENERGETIC FACTORS

1. Effective Work Function for Production of Solvated Electrons

This question may be examined in energetic terms related to the probability of electron tunneling[13] or adiabatic electron transfer[110] in the process $e_M \rightarrow e_{aq}^-$. The situation at a *single* metal–solution interface will first be considered, to illustrate the elementary processes involved. However, the practical conditions for possible cathodic electron injection into a solvent must always involve *two* metal–solution interfaces and the measured p.d. between the metals in a cell. Such a situation will be examined in a later section.

Radiationless electron transfer from an electrode to solution to form hydrated electrons must obey Gurney's condition[13] that the so-called "neutralization energy" U is greater than or equal to $\phi_e - eV$, where ϕ_e is the work function* of the cathode. For electron injection into the solvent, no ionization potential or electron affinity is involved as is the case in electron transfer with cations or anions, and only the solvation energy S_e of electrons in a solvent cage determines U. That is, the energetic condition for formation of e_{aq}^- by electron transfer to solvation sites is simply $\phi_e - eV \leq S_e$; other kinetic conditions will determine the rate of such a process if the above energetic condition is fulfilled.

Frumkin[12] considered the evaluation of $\phi_{e,S}$, the work function relevant to the escape of an electron into solution. Representing this transfer by means of a cycle involving the following steps: electron in metal to electron *in vacuo*; electron *in vacuo* to state near the surface of the solution; transfer through the solution interface, and finally, return to the metal through the solution–metal interface after hydration, Frumkin[12] wrote

$$\phi_{e,S} = \phi_e + V_{M,soln} - G_{s,e} \tag{1}$$

where $V_{M,soln}$ is the metal solution Volta p.d. at a given measured electrode potential and $G_{s,e}$ is the "real" free energy of solvation of the electron from its state in a vacuum, i.e., including the surface potential work term at the vacuum–solution interface. Using

*Here, and subsequently, ϕ_e will refer to the work function for the metal/vacuum interface.

Baxendale's[13] value for $G_{s,e}$ of 1.75 V,* and taking $\phi_{e,Hg} = 4.52$ eV and $V_{M,soln}$ as -0.26 V at the potential of zero charge according to Randles,[14] $\phi_{e,S} = 2.5$ V. From this value, and assuming the principles of thermal† electron emission into vacuum to be applicable to the metal–solution interface, Frumkin estimated an electron emission current of $\sim 10^{-2}$ A cm^{-1} at a rational potential of -2.2 V. Such an effect may occur but would normally be masked by a cathodic current for direct electrolytic H_2 evolution arising from the overall reaction $H^+_{aq} + e_M \rightarrow \frac{1}{2}H_2$ assuming that the latter does not itself proceed (see below) by a mechanism involving[8–10] primary production of solvated electrons. At the p.z.c., $\phi_{e,S}$ will probably be about 3 V.

Conway and MacKinnon[15] performed a similar calculation in relation to estimates of the surface potential χ and the absolute metal–solution p.d. at the p.z.c. and concluded that direct electron emission could not, from an energetic point of view,‡ be appreciable until potentials were attained which would be comparable with the estimated standard potential of e^-_{aq}. Conway and MacKinnon[15] took S_e as 1.74 eV,[13] so that e^-_{aq} will tend to be formed only when $\phi - eV \leq 1.74$ eV. For Ag, ϕ is ~ 4.5 and for Pt, ~ 5.2 eV; hence, the absolute p.d. at the electrode interface must be at least ~ -2.76 V for Ag and ~ -3.46 V for Pt for electron emission to occur, forming e^-_{aq}. Conversion of such absolute p.d.'s to figures relative to a practical scale of measurement, e.g., with respect to the reversible hydrogen electrode, raises the usual and old problem of the absolute p.d.[72] at the interface of a hydrogen electrode at, e.g., Pt or Ag (since the absolute *single* p.d. will depend on the metal) and, in particular, the influence of the χ potential.

*A molecular orbital calculation of the hydration energy of a primitive positive charge of zero size has recently been made by Burton and Daly,[15a] who obtain a value of -150 kcal mole^{-1}. This is much higher than the existing estimates for e^-_{aq}. A substantial difference may, in fact, be expected due to the delocalization of e^- in the water cage and to the negative rather than positive charge. This would tend to diminish the charge density in the solvent cage and decrease the magnitude of the hydration energy.

†At an electrode, from which electron tunneling to sites in the solvent may occur, cold-emission conditions may be a more appropriate basis for this kind of approach; compare the situation of field-assisted, cold emission in field-emission microscopy.[28]

‡From a kinetic point of view, if the energetic conditions are satisfied, the probability of electron transfer to solvent-sequestered states in solution could be made by means of a quantum-mechanical calculation taking into account the vibration-libration states of H_2O in the liquid water lattice and the electronic state of e^-_s in the solvent cage.[116]

Recent estimates of the latter quantity indicate that it might be rather smaller[73,74] than hitherto assumed,[74] so that together with the recent reliable estimate[75] of the hydration energy of the proton, -261 ± 2.5 kcal mole^{-1}, and the known ionization potential of H and dissociation energy of $\frac{1}{2}H_2$, the single p.d. at, e.g., the standard H_2/Pt electrode can hardly be numerically in excess of -0.7 ± 0.3 V. Thus, the condition for $\phi - eV$ is only realized for Pt when the potential V is at least $-3.46 - (-0.7)$ V, i.e., ~ -2.8 V referred to the reversible hydrogen scale. This result seems at least not inconsistent with the thermodynamic argument (see p. 126) that e^-_{aq} could be produced cathodically only at very negative potentials, based on the calculated standard potential of the electron of -2.67 V. Under these conditions, the normal cathodic H_2 evolution mechanism involving H^+_{aq} and e in the metal would provide a current many orders of magnitude in excess of the current for direct production of hydrated electrons.

2. Comparison of Energetics of Ion/Electrode and Electron Injection Processes

(*i*) *Processes at a Single Metal/Solution Interface*

Although the experimental study of electrode processes always requires the combination of two electrodes, and hence two metal–solution interfaces, it is useful first to examine the conditions that determine the energy changes associated with the transfer of electron charges across the metal–solution boundary. Considerable confusion has arisen from treatments of this problem by Pyle and Roberts[9] and by Bass,[16,17] and their conclusions have been quoted uncritically in other papers. Here and in Section II.4, we shall therefore reconsider this question in some detail and attempt to clarify the situation.

In the normal case of an electron transfer process involving a metal and its cation M^{z+}, the energy ΔU of a single interface process such as

$$M \rightarrow M^{z+}_{aq} + ze_M \tag{2}$$

is determined by means of a Born–Haber cycle

$$\begin{array}{ccccc} M_g & \xrightarrow{I_z} & M^{z+}_{(g)} & + & ze_{(g)} \\ \uparrow U_{sub} & & \searrow S_{M^{z+}} & & \searrow -z\phi_e \\ M & \xrightarrow{\Delta U} & M^{z+}_{aq} & + & ze_M \end{array} \tag{3}$$

so that ΔU can be represented for zero metal-solution p.d. as

$$\Delta U = U_{\text{sub}} + I_z + S_{\text{M}^{z+}} - z\phi_{\text{e}} \quad (4)$$

where U_{sub} is the sublimation energy of M to give monomeric M species, I_z is their ionization energy to give cations of charge *ze*, $S_{\text{M}^{z+}}$ (a large negative quantity) is the hydration energy of M^{z+}, and ϕ_{e} is, as usual, the electron work function of the metal M. For an ionic redox reaction involving species M^{z+} and $\text{M}^{(z+1)+}$, the corresponding energy is

$$\Delta U = S_{\text{M}^{(z+1)+}} - S_{\text{M}^{z+}} + I_{z/z+1} - \phi_{\text{e}} \quad (5)$$

Analogous free-energy expressions, which determine the reversible electrode potentials for these reactions, can obviously be written.[28]

The case of a metal dissolution reaction involving e_{aq}^- is analogous to reaction (2) but with the electron left in the solvent instead of the metal:

$$\text{M} \rightarrow \text{M}_{\text{aq}}^{z+} + z\text{e}_{\text{aq}}^- \quad (6)$$

In this case, the reaction is not an electrochemical one; reaction (6) can hence proceed with no electrical limitations, in principle, concerning charge separation across an interface, while the single-interface process (2) *cannot* continuously occur without being coupled to a suitable cathodic process (e.g., discharge of $\text{H}^+ \rightarrow \text{H}_2$ or reduction of O_2), as in corrosion, or with a cathodic process at another connected electrode, as in a battery.

The energy ΔU for this process is determined by the relation

$$\Delta U = U_{\text{sub}} + I_z + S_{\text{M}^{z+}} + zS_{\text{e,aq}} \quad (7)$$

where $S_{\text{e,aq}}$ is the solvation energy of the electron.

Clearly, the energy for process (6) differs from that for (2), the normally considered electron-transfer equilibrium at an electrode interface, simply by the quantity $-z\phi_{\text{e}} - zS_{\text{e,aq}}$. It is evident that in comparison with equation (4), ΔU for the e_{aq}^- process does not directly involve ϕ_{e}. This may seem an unexpected result, since an electron is physically injected from the metal into the solution. It arises, however, from the manner in which the process and its component energy terms have been represented. The direct field-assisted electron injection (at sufficiently cathodic potentials) or the photoassisted process does, however, involve the work function when the single-interface process is considered since in that case, no formation of a metal cation is involved. An alternative *formal* representation of reaction (6) would be reaction (2),

$$M \rightarrow M_{aq}^{z+} + ze_M$$

coupled with

$$ze_M \rightarrow ze_{aq}^{-} \tag{8}$$

from which it is seen that $z\phi_e$ would not directly enter into determining the energetics of a metal dissolution process (considered at a single interface) which resulted in formation of hydrated (or, in the general case, solvated) electrons, since the $z\phi_e$ quantity cancels out. U_{sub} and $I_z - z\phi_e$ can, in fact, be regarded as an "energy of ionization of the metal," independent of how the electron itself ends up in the physical charge-separation process.

By analogy with the quantities discussed by Pyle and Roberts[9] and by Bass,[16,17] the difference $-z\phi_e - zS_{e,aq}$ defines an energetic condition for preferred dissolution of the metal as ions and electrons in the solvent or as ions + electrons in the metal [in the latter case, of course, an electrochemical equilibrium is involved, because charge separation has occurred *across* a single interface, so that the reaction *cannot continue unless the process is coupled with a cathodic reaction*[16] at another electrode or a cathodic reaction at another site on the same electrode (corrosion conditions)]. This requirement for a mechanism producing ions in solution plus electrons left in the metal must always apply where conditions are considered for metal dissolution in terms of a partial anodic process. The *thermodynamic* conditions for a metal dissolution process or "metal stability" of the type $M \rightarrow M^{z+} + ze_M$ must always take into account the coupled cathodic process. Since $S_{e,aq}$ is ~ -40 kcal mole^{-1}, ϕ_e must evidently be $< \sim 1.74$ eV for metal dissolution as ions + solvated electrons. In practice, further *kinetic* limitations, e.g., the rate of H_2 evolution, will usually apply. The above energetic condition incidentally eliminates the possibility of electron injection from alkali metal amalgams unless they are quite concentrated with respect to base metal content; this condition will not only determine the relevant ϕ_e for the nonelectrochemical metal dissolution condition, but also the mixed potential for dissolution (see p. 96) in the case of the usually considered electrochemical mechanism involving coupled anodic and cathodic processes.

Conditions for photoassisted production of solvated electrons at electrodes will be considered in Section III.2.

The relation between chemical and electrochemical metal dissolution was also considered by Laitinen and Nyman,[4] who regarded the dissolution to form ions and solvated electrons as a mixed cathodic and anodic process for which two expressions for the potential could be written:

For the dissolution of the metal as ions,

$$E_M = E_M^\circ + (RT/zF) \ln a_{M^{z+}} \tag{9}$$

and for the z electrons per atom of M,

$$E_e = E_e^\circ - (RT/zF) \ln[a_{e_s}]^z \tag{10}$$

For a common potential to arise, it is therefore evident that

$$E_M^\circ - E_e^\circ = -(RT/zF) \ln a_{M^{z+}}[a_{e_s}]^z \tag{11}$$

It is evident also that the term in the ln function of equation (11) is equivalent to a solubility product for metal dissolution as $M^{z+} + ze_s$. This equation leads to the expectation that univalent metals whose standard potentials are more negative than that for the electron electrode will have a solubility greater than 1 mole liter^{-1}. Theoretically, all metals will have a finite solubility determined by equation (11), but in practice, it may be infinitesimally small. The relationships shown above were considered by Makishima[18] and used by him as a basis for theoretical evaluation of the potential of the electron electrode in liquid NH_3 (cf. Ref. 4).

(*ii*) *Processes at an Electrode in a Practical Cell*

The discussion above defines the conditions under which the work function enters into the energetics of a metal dissolution process at a *single* electrode interface; it must be emphasized, however, that when the process is considered as part of an overall electrochemical reaction in a practical cell involving a pair of electrodes between which a measured p.d. is set up or applied, the work function cancels out since the *measured* p.d., E, determined by means of a potentiometer in the circuit shown in Figure 1, is given by $E + (\phi_{M_2} - \phi_{M_1}) + (\phi_S - \phi_{M_2}) + (\phi'_{M_1} - \phi_S) = 0$ where the ϕ terms are the Galvani potentials of the indicated phases. The work functions ϕ_{e,M_1} and ϕ_{e,M_2} of the metal electrodes M_1 and M_2 enter into the determination of the single interface potential differences $\phi'_{M_1} - \phi_S$ and $\phi_{M_2} - \phi_S$ at each of the two electrode interfaces, as in cycle (3). They also determine the contact potential $\phi_{M_2} - \phi_{M_1}$ in the measuring circuit at the junction M_1/M_2; ϕ_{e,M_1} and ϕ_{e,M_2} hence cancel out when the

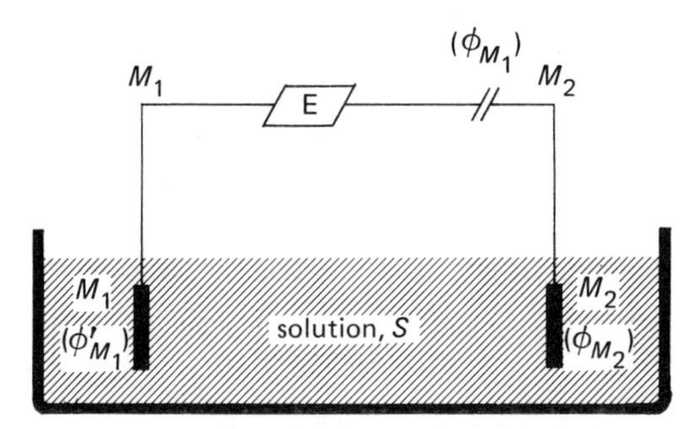

Figure 1. Potentiometer circuit.

measured potential E is considered in relation to the coupled processes at M_1 or M_2 electrodes in a *cell.* This situation has not always been well understood in electrochemistry and a more detailed discussion of this cancellation of ϕ_e terms may be found in the treatment given elsewhere by Conway.[28]

At a particular measured cell potential E, solvated electron generation at a cathode as a net cathodic process $e_M \rightarrow e_S$ [in distinction to the open-circuit dissolution process (6)] must therefore always be subject to the same conditions as apply to other cathodic or anodic reactions and, in particular, will not directly involve the electronic work function of the metal M. The *kinetics* of electron injection from the metal M into the solvent may, however, be indirectly determined by the properties of that metal through the influence of the metal surface on solvent structure[42] and solvent dipole orientation at the interface, which determine the energy of the transition state.

3. Standard Electrode Potentials for e_s^- : Liquid Ammonia and Water

In the case of liquid NH_3 solutions, Laitinen and Nyman[4] made two estimates of the $E°$ value referred to an H_2 electrode standard in the same solvent.

In the first, the current–potential curve for sodium metal solution was considered. The potential at which the current–voltage curve crosses the axis of zero current is a function of the (surface) concentration of the e_s^-, and at this potential, $i = 0$. Hence, the surface concentration is equal to that in the bulk. Laitinen and Nyman[4] found the zero-current potential at -2.388 V for a concentration of 0.006 mole liter^{-1}. This gave an $E°$, referred to a mercury pool reference electrode in the same solution, of -2.49 V. Correction to the hydrogen standard electrode potential in liquid ammonia, using data of Pleskov and Monosson,[19] gave an $E°$ value of -1.86 V.

From polarographic experiments, the current–voltage curve can be represented[4] by

$$E_e = A - (RT/F) \ln i \qquad (12)$$

for production of solvated electrons in liquid NH_3. The quantity A is given by

$$A = E_e^\circ + (RT/F) \ln 6.05 \times 10^5 D_e^{1/2} M^{2/3} t^{1/6} \qquad (13)$$

where D_e is the diffusion constant of the electron in NH_3, M is the flow rate of Hg in mg sec^{-1}, and t is the drop time in sec.

Using equation (12) for A, a second method for estimation of $E°$ can be proposed. Evaluation[4] of the diffusion constant for $e^-_{NH_3}$ at -36°C gives a value of 1.93×10^{-4} cm^2 sec^{-1}. With the experimental data from polarograms of saturated solutions of n-Bu_4NI in liquid NH_3, $A = -2.36$ V w.r.t. the Hg pool electrode in the same solution. Substituting these data in equation (13) gave $E° = -2.55$ V w.r.t. Hg pool or -1.92 V w.r.t. standard hydrogen electrode in liquid ammonia at -36°C. The value estimated theoretically by Makishima[18] was -2.04 V.

The problem of evaluation of the $E°$ value for electrons in *aqueous* media is much more difficult and no direct experimental method exists, owing to the short lifetime of e^-_{aq} in the liquid water phase (in ice at low temperature, the lifetime is greater). However, by an indirect procedure involving determination of relative values of rate constants, a value of -2.6 to -2.75 V w.r.t. standard aq. H_2 has been derived.[13,21] Marcus[21,22] has pointed out that a correction factor should be applied to this value to take into account the translational entropy of the electron, and on this basis a more negative value of -2.9 V is derived which determines, in any thermodynamic calculations, the true relative reducing power of the hydrated electron.

Baxendale[13] proposed that the standard potential for e^-_{aq} should be used as a basic standard for the scale of electrochemical potentials of the elements in an aqueous medium. Dainton[20] considered a similar process involving e^-_{aq} for the calomel half-cell reaction. Baxendale regarded the energy change for the process $e_M \rightarrow e^-_{aq}$ as a more realistic basis for an electrode potential scale than that commonly used, namely $H^+_{aq} + e_M \rightarrow \frac{1}{2}H_2$. That is, the free energy of a redox reaction would be considered in relation to a process such as*

$$M^{z+}_{aq} + e^-_{aq} \rightleftharpoons M^{(z-1)+}_{aq} \quad (14)$$

rather than

$$M^{z+} + \tfrac{1}{2}H_2 \rightleftharpoons M^{(z-1)+} + H^+_{aq} \quad (15)$$

He assumed there are difficulties with the H_2 electrode reaction on account of uncertainty of the work function (see below).

A number of other objections to this proposal also arise, not least of which is the uncertainty of the reversible electrochemical behavior of e^-_{aq} and the difficulty of directly setting up an electrode at which its potential could be established experimentally in relation to that of another electrode. Other difficulties arise, as pointed out by Conway and MacKinnon[14] (see Section VI) if it is supposed[8,9] that most electrode processes in aqueous solution proceed by an e^-_{aq} mechanism such as (14). Second, in any practical measurement of e.m.f. of cells, the electrons in the half-cell processes must surely be associated with the metal, otherwise e.m.f. measurements "by definition" would be neither feasible nor indeed unambiguous and reliable thermodynamically. Although Baxendale[13] regarded processes such as (14) as so-called "real" reactions, they are in fact not to be included in such a category since they are not susceptible to experimental measurements with regard to a cell p.d.,† nor are they "electrochemical" in nature. In an experimental measurement in electrochemical thermodynamics, there is of course never any practical question of the absolute state of the electrons involved: the half-reactions for the cell must be written so as to add up to an overall *chemical* reaction with no net electrons left over, e.g.,

$$Fe^{3+}_{aq} + e_{Pt} \rightleftharpoons Fe^{2+}_{aq}$$

*Such a reaction as this is not strictly an "electrochemical" one, since no electron charge transfer occurs in the M redox reaction itself across the metal–solution interface.

†For example, the reference electrode standard, whatever it is chosen to be, must be capable of being used in a cell and of giving a reliable and reproducible potential. It must also correspond to a thermodynamically unambiguous reaction; the electrons must enter or leave the metal for the reaction to be used electrochemically as a half-cell process.

coupled with $H^+_{aq} + e_{Pt} \rightleftharpoons \frac{1}{2}H_2$ is obviously equivalent to

$$Fe^{3+}_{aq} + \tfrac{1}{2}H_2 \overset{Pt}{\rightleftharpoons} Fe^{++}_{aq} + H^+_{aq}$$

Here, it is also useful to note that it is not the potential for the isolated half-cell reaction $H^+_{aq} + e_{Pt} \rightleftharpoons \frac{1}{2}H_2$ that is assumed conventionally to be zero, but rather the potential associated with that reaction at a metal M (not necessarily Pt) *in combination in a measuring circuit* with possibly, but not necessarily, another metal M′ at which some other electrode process is occurring reversibly. Such circuits always involve the contact potential between M and M′ in addition to the metal/solution interfacial p.d.'s at the two electrode metals and it is the algebraic sum of such p.d.'s that is the *measured e.m.f.* In order to avoid this practical complication, it may, however, be useful to suggest[15] that what can be assumed to be zero on the conventional hydrogen electrode scale is the p.d. at the interface of a metal M at which the reversible hydrogen electrode process is occurring when that metal M is chosen to be *the same* as that at which the other electrode process in the cell is taking place. Contact p.d.'s are thus avoided in the discussion of the e.m.f. and in principle (but not always in practice), any metal can be used for establishing the reversible hydrogen electrode reaction (the only practical limiting factor is, of course, whether the reaction has a sufficiently high exchange current to be useful as a reversible electrode reaction and whether side reactions, e.g., metal dissolution, that may be potential-determining, are insignificant or not). Baxendale also indicated,[13] erroneously, that a primary difficulty with the scale of potentials associated with the usual half-cell reaction of hydrogen involving electrons in the metal is the question of the value to be assigned to the work function ϕ_e of the metal at which the H_2/H^+ equilibrium is set up. This, however, is not a practical difficulty since in both cases considered above (i.e., where M and M′ are different, or where hypothetically M also constitutes the metal at which the hydrogen electrode reaction is established reversibly), the work functions cancel out; in one case, because the metal at each of the two interfaces is the same; in the second case (metal M and reference electrode process established at M′), because the different work function terms involved at each metal solution interface are almost cancelled by the difference of work functions (and hence the contact potential) at the M/M′ interface in the external circuit. ϕ_e only enters into *a priori* kinetic calculations.

4. Energy Conditions for Production of Electrons in Water

The conditions under which the reaction

$$e_{aq}^- + e_{aq}^- \xrightarrow{k_1} H_2 + 2OH^- \tag{16}$$

could be responsible for hydrogen evolution in cathodic electrolysis of aqueous solutions and in metal dissolution have been considered.[8–10]. The other relevant reactions with e_{aq}^- are

$$H_2O + e_{aq}^- \xrightarrow{k_2} H + OH^-; \qquad k_2 = 16.1 \quad \text{mole}^{-1}\,\text{sec}^{-1} \tag{17}$$

$$H_3O^+ + e_{aq}^- \rightarrow H + H_2O \quad \text{(acidic solution)} \tag{18}$$

If the e_{aq}^- concentration near the surface of the metal exceeds $\sim 10^{-7}$ mole liter^{-1}, reaction (16) will be predominant since $k_1/k_2 > 10^7$. The following energetic factors have been examined.

Bass[16] proposed a relation, analogous to Gurney's relation[23] for electron transfer and ion neutralization that determined, energetically, the condition for spontaneous formation of H atoms in metal dissolution:

$$\phi_e \leq (I - S_{H+}) + 6rekT/\mu \tag{19}$$

where ϕ_e is the work function of the metal, I is its ionization potential, S_{H^+} is the solvation energy of the proton, and $2r$ is the diameter of a water molecule and μ its dipole moment. Using equation (19), Bass predicted that there would be an upper limit to ϕ_e of ~ 2.3 eV above which the metal would apparently be stable in an aqueous medium (cf. Ref. 12). Pyle and Roberts[9] argued that while this (unconventional) treatment gives an idea of why the alkali metals are *unstable*, it did not account for the *p*H dependence of the stability of base metals such as Zn or Mg. They seemed to be unaware that the latter effect, and the question of instability of the alkali metals in aqueous media, could, however, be well accounted for by the classical theory of corrosion processes in which an anodic partial process $M \rightarrow M^{z+} + ze_M$ (with electrons left in the metal) is coupled with a cathodic process, e.g., $ze + zH^+ \rightarrow \frac{1}{2}zH_2$ (in the absence of O_2) and the corrosion rate is usually determined by the kinetics of both steps (Figure 2). Since the H_2 evolution rate is dependent on H_3O^+ concentration (particularly at concentrations of $H^+ > 0.1$ m) due to the finite positive reaction order for the h.e.r. in $[H^+]$ and due to double-layer effects, the cathodic partial reaction will always tend to become faster as *p*H is lowered and the corrosion or dissolution

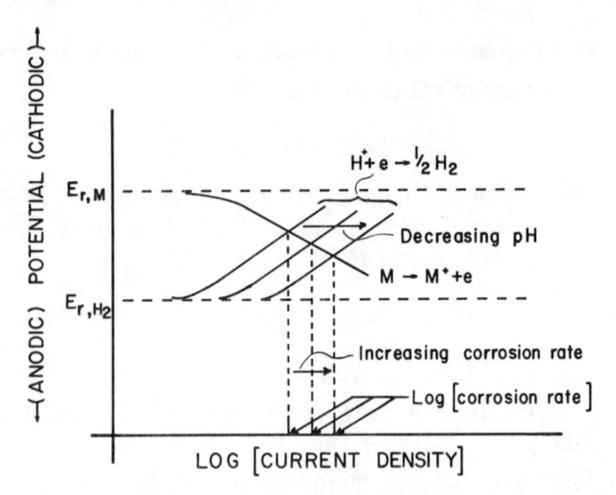

Figure 2. Schematic current–potential relations for corrosion of a metal at different *p*H values.

rate will become greater, depending on the relative slopes of the polarization lines for the coupled cathodic and anodic partial reactions. For metals which dissolve in alkaline solutions, e.g., Zn, as complex anions, the reverse is the case above *p*H 7, namely the Zn will dissolve *more* rapidly the higher is the *p*H. In this *p*H region, apart from double-layer effects, the h.e.r. component current will be independent of *p*H above *p*H ~7 since the reaction order is zero in $[H^+]$ $(H_2O + 2e_M \rightarrow \frac{1}{2}H_2 + 2OH^-)$[24]; the hydrogen reversible potential, of course, will become more negative with increasing *p*H.

Pyle and Roberts continued their application of Bass's criteria by considering the conditions under which a metal could, supposedly, dissolve dissociatively into metal ions + hydrated electrons. They proposed that if the work done to remove an electron from a metal depends, as it presumably must, on the distance to which it is removed, a metal will be "unstable" if the work done to remove the electron through the double layer to "infinity" is less than the energy gained in solvating that electron at an "infinite" distance from the metal (in practice, probably ~10–20 Å).[25] This corresponds to the reaction $M \rightarrow M^+ + e^-_{aq(\infty)}$. Pyle and Roberts, using Bass's criterion, concluded that a metal would be "unstable" with respect to dissolution as ions + solvated electrons if

$$\phi_e \leq S_e + 3kTed/\mu \qquad (20)$$

where d is a thickness of the double layer and S_e is the solvation energy of e^-. For liquid ammonia, at $-70°C$, the limit of ϕ_e was estimated as $2.5 \pm 0{\cdot}4$ eV, with d taken as 5 Å, the radius of the metal ion + the diameter of NH_3. S_e was taken as -1.7 ± 0.4 eV.[26] For water, ϕ_e was evaluated as also ≤ 2.5 eV, with S_e taken as -1.75 eV,[27] which leads to the same conclusions as the relation of Bass[16] regarding the relative stability of metals for this type of reaction. According to Gurney's criteria,[23] however, for an *electrochemical* metal dissolution reaction $M \rightarrow M^{z+} + ze_M$, the condition

$$z\Delta\phi_{s-M} + z\phi_e \geq I - S_{M^{z+}} + U_{sub} \tag{21}$$

must obtain for dissolution to occur.* In fact, with all metals there will, of course, always be some tendency for dissolution and the real thermodynamic condition for electrochemical equilibrium in the absence of a coupled depolarization reaction is[28]

$$\begin{aligned} \mu_M &= \bar{\mu}_{M^{z+}} - z\bar{\mu}_{eM} \\ &= \mu^\circ_{M^{z+}} + RT \ln a_{M^{z+}} - z\mu_{eM} + z\phi_s F - z\phi_M F \end{aligned} \tag{22}$$

where ϕ_s and ϕ_M are the inner potentials of the solution and metal phases, respectively. Since $\mu_M \equiv \mu^\circ_M$ and $\mu_{eM} \equiv \mu^\circ_{eM}$, the above relation gives the *standard* chemical potential change for metal dissolution in terms of the absolute metal–solution p.d., $\phi_s - \phi_M$:

$$\mu^\circ_{M^{z+}} - z\mu^\circ_{eM} - \mu^\circ_M = \Delta\mu^\circ = -zF(\phi_s - \phi_M) - RT \ln a_{M^{z+}} \tag{23}$$

For the *standard* state conditions, the energy terms ϕ_e, I, and $S_{M^{z+}}$ determine $\Delta\mu^\circ$ and the standard p.d. $\Delta\phi^\circ_{s-M}$.

For metals with ϕ_e greater than the critical value estimated above, Pyle and Roberts regarded solvated electrons as being formed very close to the electrode surface (contrast the photoeffect[25]) and unable to diffuse away into the bulk. In the case of water, these electrons were regarded as being able to react to form hydrogen at an integrated rate given by[9]

$$v_{H_2} = 10^{-3} \sum_{x=0}^{x=\infty} c^2_{e,x} k_2 \, \delta x \tag{24}$$

where $c_{e,x}$ is the concentration of e^-_{aq} at a distance x from the metal surface. The summation requires a knowledge of the distribution of

*Gurney's original treatment did not include lattice or adsorption energy terms.

electrons in successive solvent laminae near the surface. The concentration of e_{aq}^- at the inner Helmholtz plane (i.H.p.) was estimated assuming that the electron is solvated within one molecular layer of water from the interface. This seems to be inconsistent with the estimates of the partial molar volume of electrons in ammonia at $-33°C$, where a large value of 84 ml mole^{-1} was derived.[29] The value can hardly be less in water, where hydrogen bonding and structure effects usually lead to smaller electrostrictions, than in other solvents of lower cohesive energy densities. The above figure implies a spherical cage of 140 Å^3 in volume, i.e., with a diameter of ~6.4 Å (cf. Ref. 20). Electrons would thus not be completed solvated until well beyond the i.H.p.

It was concluded[9] that in the summation for v_{H_2}, only the electrons in the first plane layer of water molecules could contribute significantly to the rate of H_2 evolution, if the latter, in fact, occurred by that mechanism.

Pyle and Roberts expressed the view that the production of OH^- ions in reaction (16) would lead to a film of metal hydroxide which eventually would inhibit continuing H_2 evolution associated with metal dissolution, which was supposedly proceeding by e_{aq}^- formation [reaction (6)]. They extended this idea to cathodic hydrogen evolution at Hg in order to attempt to account for the (incorrectly interpreted,[80a,b] see Section VI) observations of Hills and Kinnibrugh[30] that the volume of activation for the h.e.r. was negative. The formation of $Hg(OH)_2$ and $Hg(OH)^+$ species (in a cathodic reaction!) was postulated.[9] This is contrary to the known electrochemical thermodynamic conditions required for formation of such compounds, which can only arise by *oxidation* of Hg, and indeed, no evidence of an oxidized surface species is indicated by cathodic transients or e.m.f. decay[31] at a Hg cathode. The views of Pyle and Roberts are evidently untenable on elementary grounds and illustrate the difficulties into which they and other authors have been led in postulating a general role of e_{aq}^- intermediates in electrochemical hydrogen evolution (cf. Refs. 8, 10). The negative $\Delta V^\ddagger$ could, of course, as Hills considered, be the result simply of production of a charge in the transition state "$(e_{aq}^-)^\ddagger$" rather than its partial removal in the neutralization of H_3O^+. This viewpoint is, however, difficult to accept for other reasons[15] which are considered later in this chapter (p. 125).

5. State of Solvated Electrons in Polar Media

As in the case of electron transfer to reducible ions in solution, the state of the electron after transfer to sites in the solvent will be of importance in considering any e_s production process. The redox process in the case of solvated electrons is "water cage" $+ e_M \rightarrow e_{aq}^-$ for aqueous media instead of $M_{aq}^{z+} + e_M \rightarrow M_{aq}^{(z-1)+}$ for aqueous ion reduction at a metal M.

In polar solvents, the energy levels of a solvated electron have been calculated by Landau[114] and treated by Jortner[115] on the basis of a continuum model in which trapping of the electron is considered in terms of the polarization it induces in the dielectric medium and with which it interacts. The charge distribution and energy levels depend on the interaction of electrons with the polarized dielectric. The electron–medium interaction is treated as a long-range interaction and this is a serious limitation in the model. In the liquid lattice, motions occur at infrared frequencies and the properties of the electrons are determined by these frequencies; the electron motion follows adiabatically the fluctuations of the polarization field in the molecular medium, as in the theory of adiabatic reactions.

Calculations based on the continuum dielectric model have been performed for e_{aq}^- in the limit of zero cavity size.[115] A variational calculation is made for hydrogenic-type wave functions for the ground and first excited states, and is based on a Hartree–Fock scheme where the Coulombic and exchange interactions of electrons with the medium are replaced by the dielectric polarization energy.

The electronic energies W_{1s} and W_{2p} for 1s and 2p states are obtained using the wave functions

$$\psi_{1s} = (\gamma^3/\pi)^{1/2} e^{-\lambda r} \tag{25a}$$

$$\psi_{2p} = (\alpha^5/\pi)^{1/2} r e^{-\alpha r} \cos\theta \tag{25b}$$

as

$$W_{1s} = \frac{h^2\gamma}{8\pi^2 m} - \frac{\beta e^2}{R_0} + \frac{\beta e^2}{R_0}(1 + \gamma R_0)e^{-2\gamma R_0} \tag{26a}$$

and

$$W_{2p} = \frac{h^2\alpha^2}{8\pi^2 m} - \frac{\beta e^2}{R_0} + \frac{\beta e^2}{R_0}\left(1 + \frac{3}{2}\alpha R_0 + \alpha^2 R_0^2 + \frac{1}{3}\alpha^3 R_0^3\right)e^{-2\alpha R_0} \tag{26b}$$

where R_0 is the cutoff cavity radius, $\beta = \varepsilon_\infty^{-1} - \varepsilon_0^{-1}$, the difference of reciprocals of the optical and static dielectric constants, and α and γ are parameters of the wave functions, determined by the variational procedure. The electronic polarization energy U is introduced as a perturbation by writing the total energy $E = W + U$, where

$$U = -\tfrac{1}{2}\int_{R_0}^{\infty} (ep/r^2)4\pi r^2\, dr$$

and p is the electronic polarization. The energy W is maximized using

$$U_{1s} = -(e^2\gamma/3)[1 - (1/\varepsilon_\infty)] \tag{27a}$$

$$U_{2p} = -(e^2\alpha/5)[1 - (1/\varepsilon_\infty)] \tag{27b}$$

and the values of α and γ so obtained are employed to determine the energies W, U, and, finally, E. For the electron in NH_3, Figure 3 shows the energy levels and potential functions for $e^-_{NH_3}$. Similar principles apply to e^-_{aq}, since the electronic spectrum is similar to that in NH_3, except that the cavity size is expected to be smaller owing to the larger local surface free-energy change for cavity formation in H_2O in comparison with that for liquid NH_3. The optical 1s → 2p transition is shown as $h\nu$ in Figure 3, together with the potential-well corresponding to the cavity diameter.

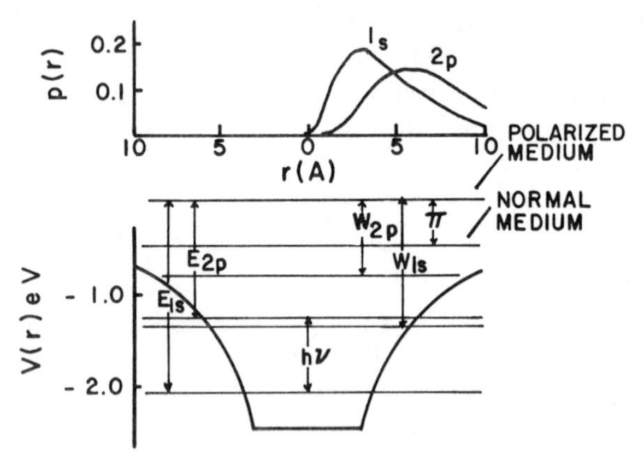

Figure 3. Energy levels and potential functions (schematic) for the solvated electron in NH_3 (after Jortner[116]).

Since only fair agreement with experiment is found (heat of solution: theoretical, 1.32 eV; experimental, 1.6 eV; first optical excitation energy: theoretical, 1.35 eV; experimental, 1.7 eV), the assumption that the main contribution to electron binding in polar media arises from long-range (polaron) interactions evidently provides only an approximate basis for evaluation of the properties of e_{aq}^-. The short-range interactions, particularly in a structured solvent such as water, require more detailed examination. For example, the short-range anisotropies and perturbations due to the arrangement of the first solvation layer, will modify the otherwise spherically symmetric potential seen by the electron on account of the polarization interactions. These effects will be most important in considering the kinetics of an adiabatic transition $e_M + \square \rightarrow e_{aq}^-$ at high negative potentials at an electrode.

III. PHOTOASSISTED ELECTRON INJECTION FROM CATHODES

1. Introduction

The effects of ultraviolet or visible light on electrode reactions and electrochemical potentials have been studied for some years,* in particular with regard to photoelectrode potentials associated with redox couples involving dye molecules[6,7] and the electrochemistry of molecules in electronically excited states.[38–40] More direct photoeffects were also studied by Hillson and Rideal[7] in their work on the hydrogen evolution reaction (h.e.r.) but the magnitude of the change of overpotential in the h.e.r. at a given current was small. Recently, photoelectrochemical effects involving solvated electrons have been thoroughly investigated by Barker and co-workers,[25,32] Delahay and Srinivasan,[33] and Heyrovsky.[34–36] The first basic equations for electrochemical kinetic photoeffects were given by Bockris.[6]

2. Energetic Conditions for Photoelectrochemical Emission

The restrictions on direct cathodic injection of electrons into a solvent, which have been discussed in Section II from the point of view of the minimum energies required, are changed when the

*See Refs. 1, 6, and 7, as well as various papers by Audubert referred to in Ref. 6.

process is assisted by absorption of a quantum of energy $h\nu$. We consider now the results of such photoassistance in electrochemical electron injection.

In the well-known photoelectric effect at a metal–vacuum interface, the condition for electron emission is simply

$$\phi_e + \tfrac{1}{2}mv^2 = h\nu \tag{28}$$

for irradiation at a frequency ν at a metal having a work function ϕ_e; $\frac{1}{2}mv^2$ is the classical kinetic energy of the emitted electron. The important result corresponding to the above equation is that electron emission can only occur when $h\nu \geq \phi_e$. Practically, the photocurrent is measured as a function of a retarding electric potential V, so that values of V_c, the critical potential for zero photocurrent, are evaluated for various ν and extrapolated to zero V corresponding to electrons emitted just with zero kinetic energy. Then, $h\nu_c = \phi_e$ at a potential V_c.

A similar situation can be established at an electrode–*solution* interface. Normally, electrons will only be transferred to states in the solution if the condition $\phi_e - eV \leq S_e$ obtains and limitingly, for a radiationless transition, when $\phi_e - eV = S_e$, where V is the absolute metal–solution p.d. and S_e has been estimated as -40 kcal mole^{-1}.[13,20] Since for most metals (excluding those of the alkali series), $\phi_e > 4$ eV, it is evident that direct field-assisted electron injection into the solution can only arise under normal conditions when V is appreciably negative, $> \sim 2$ V. Provision of photoassistance, however, in effect lowers the work function or, in other terms, diminishes the value of eV required to establish the equality required by equation (28); that is,

$$\phi_e - eV - h\nu = S_e \tag{29}$$

for photoassisted electron emission into solution. These effects, associated with direct electron emission, must be distinguished from the much more extensively studied effects associated with photoexcitation[38–40] of species in solution (e.g., dyes) which then undergo electrochemical reactions at the metal interface. The case of *direct* electron injection has been termed the "electrochemical photoeffect" by Heyrovsky,[35] who first studied it quantitatively (cf. Barker *et al.*,[25] who investigated the effects of varying wavelength of the exciting light).

3. Experimental Characterization of the Electrochemical Photoeffect

Apart from early qualitative observations of photoeffects on electrode potentials, photoeffects in the absence of electronically excitable, electrochemically active molecules were first characterised by Bowden[1] in his investigations on the h.e.r. Further work was carried out by Price[37] and Hillson and Rideal.[7] The most satisfactory reproducibility of results on the electrochemical photoeffect is obtained at renewed Hg surfaces in a polarographic technique. This was employed by Berg,[38–40] who found residual photocurrents at Hg electrodes in the absence of light-absorbing substances in solution. That electrons are the photoemitted species was indicated by the important experiments of Barker *et al.*,[25,32] who examined the effects of scavengers for hydrated electrons on the photo current behavior of Hg electrodes and showed that the rate constants for the scavenging processes were the same as those evaluated (see below) in pulse radiolysis experiments where hydrated electrons are well-characterized. Similar conclusions were reached by Delahay and Srinivasan[33] from Coulostatic experiments (see below). Photocurrents of the order of 0.2 μA are observed at Hg electrodes illuminated with a 1-kW Hg arc source and the currents are usually proportional to intensity at a given v. Of greater interest is the dependence of the photo current i_p on wavelength and electrode potential. Price[37] found i_p was experimentally dependent on potential in aq. H_2SO_4 while the polarographic results of Heyrovsky[35] in aq. H_2SO_4 indicate a corresponding logarithmic (Tafel) relation between i_p and V with two linear regions (Figure 4), the one at lower V having a smaller Tafel slope. Price [37] and Hillson and Rideal[7] also found i_p was an exponential function of the light frequency. (The results of Barker *et al.*[25] give a different relation in the presence of scavengers; $i_p^{1/2}$ is proportional to potential.)

Connecting these two results[35] leads to the important conclusion that for a given photocurrent, V and hv are *linearly* related as in the analogous case of the photoelectric effect at the metal–vacuum interface. Heyrovsky verified this relation by examining (a) the "threshold potential" V_c for appearance of significant i_p at various frequencies of the illumination (cf. Ref. 25) and (b) for given electrode potentials, the values of hv_0, the critical energy of the light, at which significant i_p first appears (cf. the evaluation of hv_c in the case of the

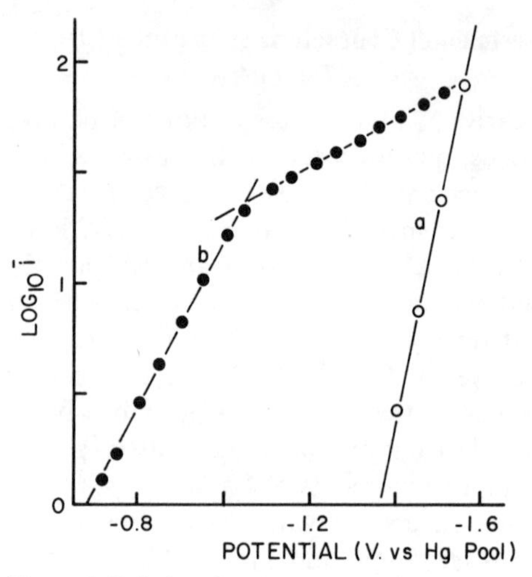

Figure 4. Relations between i_p and potential for the electrochemical photoeffect at Hg in 0.1 *M* H_2SO_4. Units of i scale are 4×10^{-9} A. (a) current for h.e.r., (b) photocurrent (after Heyrovsky[35]).

metal–vacuum interface). V_c and $h\nu_0$ are linearly related as shown typically in Figure 5.[35] The critical condition was referred to as the "red-limit" wavelength, $\lambda_0 = c/\nu_0$, when c is the velocity of light. The linear relation of Figure 5 characterizes the electrochemical photoeffect. The intercept of the line on the energy axis gives the minimum quantum required to produce a photocurrent at the potential of zero charge, i.e., in the absence of a local field at the surface due to net charge on the metal and ionic charge in the double layer. This does not correspond of course to zero absolute metal–solution p.d. or field in the electrode interphase, owing to the operation of the metal–solution surface p.d. χ_0 due to electron asymmetry and solvent dipole orientation, etc., at the p.z.c.

The magnitudes of the i_p are approximately equal for various 0.1 *M* solutions of simple salts and are *cathodic* (see below), but the intercepts of the plots of potential versus $h\nu_0$ (red-limit energies) vary between 3.4 (Figure 5) and 4.4 eV, while the slopes of the lines are between 0.9 and 1.4 eV V^{-1}. Dilution of uni-univalent supporting electrolytes usually resulted in a marked increase of the photocurrent at a given wavelength, but with 2:1 salts, and more so with a

3:1 salt, the effect of dilution was much smaller. In ethanolic solutions, i_p is always much smaller than in aqueous solutions. Various workers have found that the photocurrents are larger in acid than in neutral or alkaline solutions for otherwise similar conditions of illumination and potential. This is due to scavenging effects arising with the proton[25,32] ($e^-_{aq} + H_3O^+ \rightarrow H + H_2O$).

Anodic photocurrents are also found[35] under certain conditions at Hg, e.g., in the presence of various carboxylic acids, amino acids, and with glyoxal and diacetyl, but are not exhibited in solutions of, e.g., formate, acetate, fumarate, maleate, and some other ketones and aldehydes. The anodic photocurrents are not connected with dissolution of the Hg electrode and they are independent of the nature of the solvent (EtOH and H_2O). The anodic photocurrents were interpreted in terms of charge-transfer adsorption complexes[41] of groupings such as

$$-\underset{\underset{\text{O}}{\|}}{\text{C}}-\underset{\underset{\text{X}}{|}}{\text{C}}=$$

with the Hg surface and also having optical absorption dependent on the electrode potential. This effect may arise from polarization of the molecule, interaction with π orbitals, and orientation at the interface, e.g., as indicated by direct thermodynamic studies of adsorption of ketones, pyridine, and aromatic hydrocarbons[44] at the Hg–H_2O and Hg–alcohol interfaces.[42,43] This type of charge

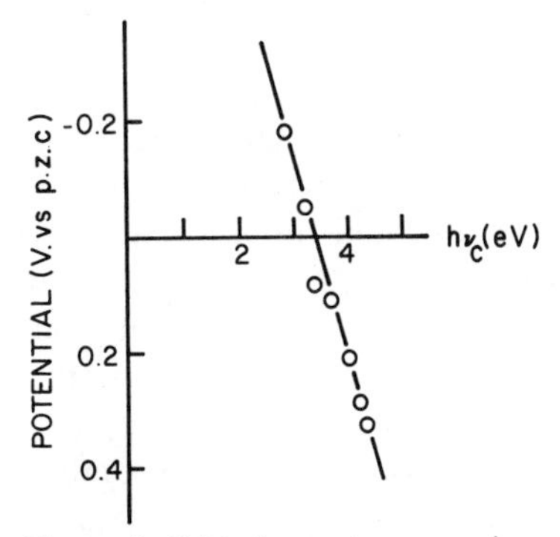

Figure 5. Critical quantum energies $h\nu_0$ and electrode potentials V_c for photoelectron emission at a Hg electrode (after Heyrovsky[35]).

transfer effect may be particularly strong in the case of adsorption of thiourea at Hg, where the behavior of the adsorbate is known to approach that of specifically adsorbed anions.

In the cathodic photocurrent observations, several cases were distinguished, depending on the intercept on the plot of V versus $h\nu_0$. For a number of inorganic substances, this intercept is the same, 3.4 eV, and appears with acids of $pK_a < 6$ and with other efficient electron scavengers. This situation implies that at the potential of zero charge, the electron is transferred to the solution with a common energy (the behavior in EtOH is qualitatively different, but here the solvent itself can participate in succeeding radical reactions) and it is suggested that the electron transfer takes place to an *adsorbed* water molecule (cf. Ref. 9), producing, temporarily, H_2O^-. This entity may then either return the electron to the metal, dissociate into $OH^- + H$ (or H adsorbed as in the normal course of the h.e.r.), or transfer the electron to neighboring water molecules to form a normal hydrated electron in a solvent cage. It is interesting to note that the value of 3.4 eV for $h\nu_0$ at the potential of zero charge should correspond to the value of the effective work function referring to the direct escape of an electron into aqueous solution, which Frumkin[12] estimated (see Section II.1) to be higher than 3 eV. In this connection, the question whether this work function value should include the surface potential χ_d due to dipoles is interesting. If the electron was injected through the surface layer to some distance in the bulk (as argued by Barker *et al.*[25,32]), the relevant work function would be $\phi_e + \chi_d$ (with proper recognition of the sign of χ_d). If, however, an adsorbed water molecule initially traps the electron,[9] only ϕ_e itself (cf. p. 85) may be involved. More generally, since the electron will normally have to pass through *part* of the layer of oriented dipoles, the relevant work function should probably be $\phi_e + \gamma\chi_d$, where γ is a fraction of χ_d. A similar situation arises in Boudart's[45] representation of chemisorption arising from partial electron transfer within a dipole ad-layer ("induced heterogeneity"). If a value for $\phi_{e,Hg}$ is assumed, and the experimental intercept is taken corresponding to the red-limit energy $h\nu_0$ at the potential of zero charge, it is evident that equation (29) enables, in principle, the absolute single p.d. at the electrode surface to be estimated for this condition. It is, of course, equal to the *total* surface potential at the metal–solution interface. Its value depends obviously on the reliability with which the solvation energy of the electron S_e in the relevant situation near the interface can be assigned. For electron transfer to cavities in the bulk, the value of S_e is ~ -40 kcal mole^{-1}, as mentioned above, and $\gamma = 1$.

4. Effects of Double-Layer Structure

The structure of the double layer at an electrode surface will obviously have an effect on the field-assisted and photoassisted emission of electrons into the solution, particularly if the electron becomes solvated or scavenged near the electrode interface.[9] Barker *et al.*[25] calculated the effect of the diffuse layer on emitted electron distribution and also pointed out the possible complications arising from discreteness-of-charge effects in the Helmholtz layer.

Heyrovsky[35] found small specific differences in the photocurrents at Hg for aqueous solutions of alkali and alkaline earth cations with anions of halogens, OH^-, ClO_4^-, ClO_3^-, N_3^-, CNO^-, SO_3^{--}, SO_4^{--}, PO_4^{3-}, OAc^-, and ϕCOO^-. Of greater interest was the observation of a substantial effect of dilution of the electrolyte on the i_p behavior, which depended also on whether the cation was monovalent or polyvalent. Solutions of alkaline earth salts require more than 100-fold dilution in comparison with 1:1 salts in order to give i_p values of the same magnitude (Figure 6). The potential red limits are somewhat modified by the type of salt and its concentration, and the slopes of the V versus $h\nu_0$ relations are also changed. The above "double-layer" effects were attributed by Heyrovsky to a photoreaction running in parallel with the production of e_{aq}^-, e.g., $H_2O^- \rightarrow H + OH^-$; $H \rightarrow \frac{1}{2}H_2$. The cations were regarded as interfering with this reaction due to electrostatic polarization of water molecules in the electrode interphase. Barker *et al.*[25] also observed the salt dilution effect on i_p even in the presence of scavengers, and i_p can increase *above* the diffusion current for the scavenger present. The effect does not appear to be due to the influence of the diffuse layer on the movement or distribution of electrons (discussed by Barker *et al.*[25]), as in diluted 1:1 electrolytes, the current already appears at potentials corresponding to positive surface charge on Hg and increases through the potential of zero charge.

Taking into account discreteness-of-charge effects in the double layer, Barker *et al.*[25] suggested that discreteness of the

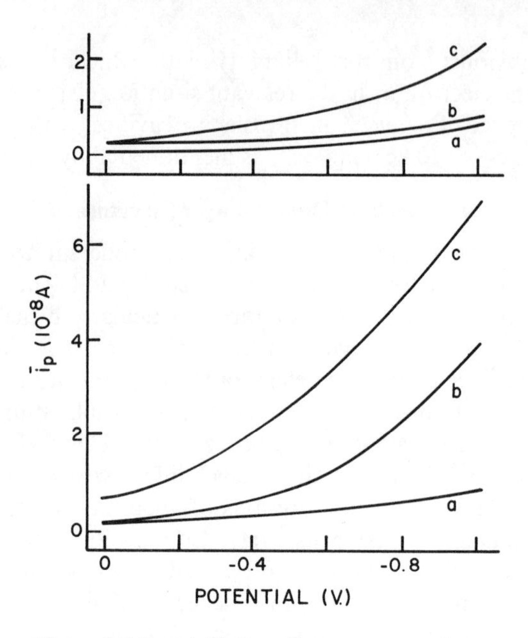

Figure 6. Salt and dilution effects on mean primary photocurrents i_p at Hg. Upper diagram: $CaCl_2$; lower, LiCl at concentrations (a) 10^{-1} *M*; (b) 10^{-3} *M*; (c) 10^{-5} *M*. Potentials versus p.z.c. (rational scale).

negative charge at adsorbed anions would only lead to a small local diminution of electron injection and the average rate would be much the same as if the charge had been smeared out over the inner Helmholtz plane. Localized charge at cations, however, tends to produce a substantial enhancement of emission over a small fraction of the surface, i.e., more than if the charge were smoothed out in an outer Helmholtz layer. On this basis, a qualitative explanation of the change of slope of the i_p versus potential relation near the p.z.c. was sought in terms of the local lowering of ϕ_e for the mercury surface.

IV. SCAVENGER BEHAVIOR IN THE ELECTROCHEMICAL PHOTOEFFECT

Barker *et al.*[25] considered various possibilities for the origin of the photoeffect, including homogeneous or heterogeneous photolysis of components in the solution, formation of excited Hg ions at the surface that react with solutes in solution, H atoms formed by dissociation of adsorbed H_2O (cf. Heyrovsky[35]), and production of solvated electrons both near the electrode (to which some return and become "reoxidized") and over a range of distances into the solution. The observed dependence of i_p on the square root of reducible solute concentration c_s when the latter is small, together with the independence* of i_p on supporting electrolyte concentration (between 0.2 and 1 *M*) and on the type of cation at negative electrode charges, support the involvement of e_{aq}^-. The influence of scavengers is basic to the understanding of the photoeffect since, in the absence of reactions which remove the injected electronic charge, reoxidation by the electrode would always occur as it does, to a small extent, under most conditions. The residual or primary photoeffect in the absence of conventional scavengers was referred to in the discussion of Heyrovsky's work[35] and the investigations by Berg.[38–40]

In the work of Delahay and Srinivasan,[33] a differential procedure was adopted for the study of photocurrents in which one of a pair of electrodes was flash irradiated (compare Ref. 32) so that effects due to potential drifts are virtually eliminated. Potential–time variations (after an initial adjustment of potential) at a Hg electrode were followed on open circuit following the flash and resulting e_{aq}^- formation. The procedure is analogous to the Coulostatic method for charge injection across the double layer, and in fact, the photocharge Q produced at the electrode was determined by various Coulostatic discharges from a condenser until the Coulostatic potential–time transient was the same as that due to flash irradiation.

The value of Q depends, of course, on electrode potential and on proton concentration due to the scavenging effects of H_3O^+ ion (Figure 7). The latter removes e_{aq}^- irreversibly [see reaction (18), Section 11.4]; if this were not the case, i_p would be much smaller due to an anodic partial current arising from reoxidation of e_{aq}^- in an *anodic* process (cf. the standard potential for e_{aq}^- discussed in Section

*Heyrovsky[35] found, however, considerable dependence of the photocurrent on the cation in going from monovalent to divalent species and appreciable variation of i_p with electrolyte concentration, but over a wider range of dilutions than those used by Barker *et al.*[25]

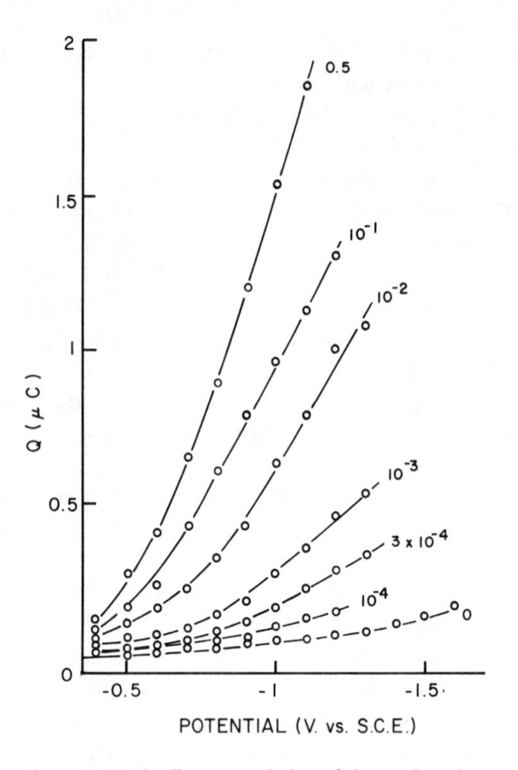

Figure 7. Flash effect on variation of charge Q at the electrode interface with potential for different molar concentrations of HCl indicated. Total anion concentration is 0.5 M in aq. KCl + HCl mixtures (after Delahay and Srinivasan[33]).

II.3). e^-_{aq} is also removed by the "bimeolecular" mechanism (cf. Refs. 9, 27), (16):

$$e^-_{aq} + e^-_{aq} \xrightarrow{k_2} H_2 + 2OH^-, \qquad k_2 = 0.5 \times 10^{10} \text{ liter mole}^{-1} \text{ sec}^{-1}$$

In relation to Barker's calculations[32] of the steady-state photocurrent associated with thermalization at a mean distance δ from the electrode, Q as a function of the saturation value of Q (written as Q_s) that would be observed if all emitted electrons were scavenged ($c_s \to \infty$), can be obtained as

$$Q/(Q_s - Q) = \delta(k_s c_s/D)^{1/2} \tag{30}$$

where k_s is the scavenging rate constant (in this case, for H_3O^+) and other terms have been defined. Of course, thermalization of emitted electrons occurs over a range of distances from the electrode[15,25] and δ is only an effective mean transmission distance.

As discussed by Delahay and Srinivasan,[33] the relation for Q given above arises from Barker's treatment[32] as follows: Fick's law is written with the additional term $-kc_s c_e$ where c_e is the concentration of e^-_{aq} and k and c_s are as defined previously. It is assumed that e^-_{aq} diffuses toward the solution from a plane at the (mean) fixed distance δ from the electrode, and the steady-state i_p is obtained as

$$i_p = Fc_e^\delta (k_s c_s D)^{1/2} \tag{31}$$

The condition for a saturation current $i_{p,s}$ is then introduced, i.e., for $k_s c_s \to \infty$, and c_e is calculated as a function of $i_{p,s}$ by polarographic principles, giving at distance δ:

$$c_e^\delta = \delta/FD(i_{p,s} - i_p) \tag{32}$$

These equations apply to steady-state conditions, so that Q and Q_s in equation (30) are proportional to i_p and $i_{p,s}$, respectively, whence the origin of equation (30) becomes evident.

The predictions of equation (30) for Q as a function of c_s are confirmed experimentally for acid solutions, where H_3O^+ is the electron scavenger, since $Q \propto c_s^{1/2}$, as indicated in Figure 8 for low c_s. The relations depend, as expected, on potential, but the initial

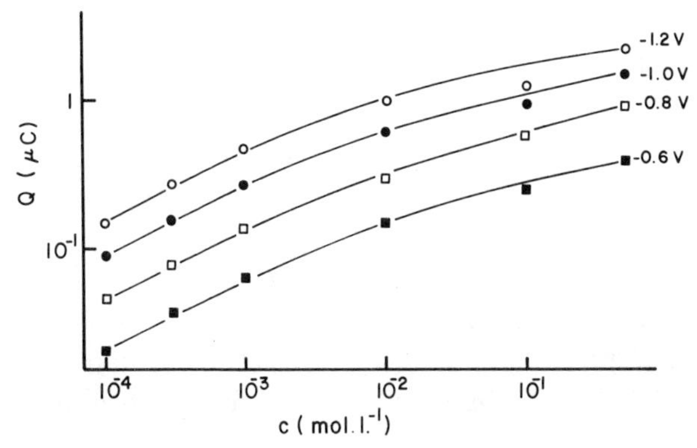

Figure 8. Dependence of Q on acid concentration c_s for different potential values (versus s.c.e.). Here, H_3O^+ is acting as the scavenger (after Delahay and Srinivasan[33]).

slopes are 0.5, and it is $Q^{1/2}$ which is approximately proportional to potential at various acid (HCl) concentrations. The origin of this form of the dependence of Q on potential is, at present, obscure.*

The most detailed studies of the kinetics and mechanisms of scavenging and succeeding free radical reactions have been made by Barker *et al.*[25,32] using N_2O, H_3O^+, and ethanol in their work on the electrochemical photoeffect.

Employment of N_2O as the electron scavenger has the advantage that it is specific, on a relative rate-constant basis, for e_{aq}^- (as found in pulse radiolysis work) and gives rise to photocurrents over a wide potential range if[35] light of sufficiently high energy ($\lambda = 2537$ Å) is employed in the irradiation of the electrode. For low N_2O concentrations, i_p at constant potential is proportional to $c_{N_2O}^{1/2}$ indicating outward diffusion and homogeneous scavenging of an entity produced at[9] or near[25] the illuminated electrode interface. The photocurrents are similar, with a solution of fixed ionic strength, if either H_3O^+ or N_2O is present at comparable low concentrations. This not only indicates that scavenging near the electrode can proceed equally well by H_3O^+ ions, but also indicates that a mechanism involving homogeneous scavenging of H atoms produced by photodissociation of adsorbed H_2O molecules at the electrode surface need not be considered. Also, since OH radicals do not react with either H_3O^+ or N_2O, the photocurrents cannot be due to scavenging of OH radicals formed heterogeneously by light absorption. All these observations, as in the work of Heyrovsky,[35] lead to the conclusion that e_{aq}^- is the primary species responsible for the electrochemical photoeffect.

The N_2O electron capture reaction is

$$e_{aq}^- + N_2O \rightarrow N_2O^- \tag{33}$$

followed by

$$N_2O + H_2O \rightarrow N_2 + OH + OH^- \tag{34}$$

The half-life of N_2O is believed to be of the order of nsec, so that the production of OH radicals occurs very near (~ 20 Å) the site of the primary electron capture. The OH radicals become heterogeneously

*The corresponding variation of photocurrent i_p with potential is a square-root relation, with $i_p^{1/2}$ proportional to V, as found *in vacuo*. The theory of Gurevich *et al.*[108] predicts a linear relation between $i_p^{0.4}$ and V for monochromatic light.

reduced at the electrode in a diffusion-controlled process with a further loss of an electron from the electrode. Thus, the photocurrent associated with the production of e_{aq}^- involves actually the passage of two electrons across the interface in this case. In the presence of scavengers which react with OH, obviously this electron number will be less than two and i_p correspondingly smaller. The homogeneous rate of capture of e_{aq}^- is often somewhat smaller than the rate of photoelectrochemical injection, so that any uncaptured e_{aq}^- rapidly diffuse back to the Hg electrode, which, over the normally accessible range of potentials, acts as an *anode* sink for the electrons (cf. p. 109):

$$e_{aq}^- + Hg(-) \rightarrow e_{Hg} \tag{35}$$

This matter will be referred to below in another connection with regard to the question of e_{aq}^- formation in the *absence* of photo-assistance. The mean thickness of the electron capture boundary layer may be estimated as $\sim 1.4 \times 10^{-6}$ cm.[32] At neutral or slightly alkaline pH, the OH radicals will be heterogeneously reduced at the electrode and in the thin boundary layer, the rate of reduction will rapidly approach their rate of formation, so that for each e_{aq}^- produced, a further electron is consumed from the solution as mentioned above. At potentials more negative than -0.9 V (s.c.e.), and for acidic pH's, the H^+ ion is the scavenger for e_{aq}^-:

$$H_3O^+ + e_{aq}^- \rightarrow H + H_2O \tag{36}$$

Under these conditions, i_p varies with V in a similar way to that in the case of N_2O photocurrents, up to potentials where the normal heterogeneous* h.e.r. becomes important.

Barker[32] considered in detail the following types of radical and scavenger (S) reactions, following the initial photoproduction of e_{aq}^-. The first step is a primary scavenging reaction:

$$e_{aq}^- + S \rightarrow \dot{S}^- \qquad (\text{for } S = H_3O^+, N_2O;\ S^- = H \text{ or } OH) \tag{37}$$

$$\dot{S}^- \rightarrow S + e_M \qquad (\text{case of } S = H_3O^+ \text{ only}) \tag{38}$$

$$\dot{S}^- + e_M \rightarrow S^{2-} \qquad (S^{2-} = H_2, OH^-) \tag{39}$$

*The question of whether the cathodic h.e.r. itself proceeds by a mechanism involving e_{aq}^- was considered by Walker[8] and will be examined in a later section of this chapter.

The following is a secondary scavenging process:

$$\dot{S}^- + A \rightarrow \dot{A}R \qquad (AR = CH_3 \cdot \dot{C}HOH \text{ or } \dot{C}H_2OH) \qquad (40)$$

with succeeding electrode oxidation or reduction reactions involving the intermediate $\dot{A}R$:

$$\dot{A}R + e_M \rightarrow AR^- \qquad (A = \text{MeOH, EtOH}) \qquad (41)$$

$$\dot{A}R \rightarrow e_M + \text{aldehyde} \qquad (42)$$

The simplest case arises when scavenging proceeds by the first reaction and the photocurrent is influenced by the rate of production of e_{aq}^- in the solution and by the boundary layer reciprocal thickness parameter $(k_s c_s / D_{e_{aq}^-})^{1/2}$, where k_s and c_s are the rate constant for the primary scavenging reaction and the concentration of the scavenger S, respectively, and $D_{e_{aq}^-}$ is the diffusion constant for e_{aq}^- in water estimated by Barker *et al.*[25] as 5×10^{-5} cm sec^{-1} from pulse radiolysis experiments; the value is similar to that for the OH^- ion in water.

Experimentally, Barker[32] conducted two types of investigation, one employing a quasi-steady-state procedure involving modulation of light intensity at 2537 Å over 45 msec periods, and the other with a flash arrangement operating at ~3600 Å. The dependence of the transient i_p on time (200 μsec/division in the figures below) and electrode potential (versus 0.2 *M* calomel electrode) is shown in Fig. 9a for saturated 10% N_2O/N_2 containing 0.2 *M* aq. KCl, while Figs. 9b and 9c show the effects of 0.2 *M* EtOH and MeOH, respectively, in the 0.2 *M* KCl supporting electrolyte. The relative sizes of the initial photocurrent pulses at −1.4 and −1.6 V with and without EtOH present give an electron number of two, as discussed above for conditions where homogeneous scavengers for the S^- intermediate (following reactions in the scheme above) are absent.

Results at low acidity (*p*H 4), obtained near the potential at which H_3O^+ ion depletion near the surface becomes significant due to discharge in the h.e.r. process, indicate that i_p falls abruptly. This is a diffusion-layer effect which arises because, for each molecule of N_2O consumed by e_{aq}^- capture, two OH^- ions are ultimately

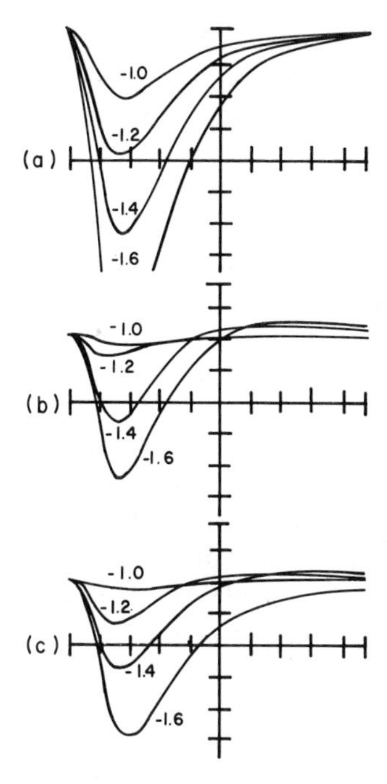

Figure 9. i_p transients for flash electrochemical photoeffects in saturated 10% N_2O/H_2 solutions containing: (a) 0.2 *M* KCl; (b) 0.2 *M* EtOH + 0.2 *M* KCl; (c) 0.2 *M* MeOH + 0.2 *M* KCl (after Barker[32]).

formed. These neutralize H_3O^+ in the diffusion layer and enhance the proton depletion. Thus, although a photocurrent may still be detected, it is almost exactly neutralized by a decrease in the relatively large current contribution for H_3O^+ reduction in the h.e.r. Apart from being consistent with the mechanism suggested by Barker for the N_2O photocurrent, it provides important evidence (cf. Ref. 15) for the "cathodic" h.e.r. being a process which is electrochemically and kinetically *independent* (see below) of e^-_{aq} formation, by photoelectron injection from the cathode.

Barker suggested that photoemission should commence close to the e.c.m. at Hg with light of $\lambda = 2537$ Å, i.e., for $h\nu$ approximately equal to $\phi_{e,Hg}$. He also considered the effect of transferring a fast-moving electronic charge into a polarizable medium and estimated that with $\varepsilon \doteqdot 2$ (the high-frequency value of the dielectric constant), the change of energy level of the electron would be about 40 kcal mole^{-1}. The conversion of a thermalized electron to an hydrated electron, he considered, would involve the further energy change associated with orientation polarization and related effects. The energy of hydration from an *in vacuo* reference state has been variously estimated as between 40 and 50 kcal mole^{-1}[46] and an energy change for hydration of a thermal electron was estimated as no larger than 10–15 kcal mole^{-1}.

Barker *et al.*[25] envisaged indirect formation of e_{aq}^- some distance from the electrode in three distinct steps: (a) ejection of the electron from the electrode (cf. the case of the ejection from a surface *in vacuo* when $h\nu_c \geqslant \phi_e$), (b) thermalization of the electron, and finally (c) its hydration; the latter is a relatively slow process. An appreciable fraction of ejected electrons will return to the electrode before hydration is achieved. That the electrons are ejected to some distance from the surface is indicated by the departure from linearity of the relation between current and square root of scavenger concentration.

For ejection to a mean distance δ from the electrode, Barker *et al.* showed [compare equations (30) and (32) discussed by Delahay and Srinivasan[33]] that

$$i_p/i_{p,\text{sat}} = Q\delta/(1 + Q\delta) \tag{43}$$

where $i_{p,\text{sat}}$ is the saturation photocurrent for $c_s \to \infty$, $Q = (k_s c_s/D)^{1/2}$, k_s is the scavenging rate constant, and c_s is the scavenger concentration.

Studies with light of varying wavelength (but not perfectly monochromatic) showed that $i_p^{1/2}$ for N_2O solutions in KCl was linear in potential over certain ranges of potential. The threshold potential changed with ν as expected, with a proportionality of almost 96%. On the other hand, initial results due to Heyrovsky[34] with N_2O and other scavengers indicated that the proportionality constant between rate of change of threshold potential and quantum energy in eV was 0.5.

In scavenging experiments designed to measure comparative effects of two scavengers, and in separate competitive scavenging experiments (with solutions containing two electron scavengers, e.g., NO_3^- and N_2O; NO_2^- and N_2O; NO_3^- and H_3O^+), the rate constants for scavenging processes were evaluated from the photocurrent data and compared with data obtained from pulse radiolysis work. The various pairs of data for a variety of scavenger systems were sufficiently in agreement for it to be concluded that the photo effects, e.g., with N_2O are virtually entirely due to capture processes involving solvated electrons as they are in pulse radiolysis experiments. The solvated electron reactions were simulated by transmission-line equivalent circuits. The effects of the diffuse double layer at the electrode were also considered. It was shown that the diffuse double-layer field increases the stationary concentration of e_{aq}^- at the plane of mean penetration distant δ from the electrode, and hence i_p, by a factor

$$f = \delta^{-1} \int_0^\delta \exp(-\psi F/RT)\, dx \tag{44}$$

where x is the general distance coordinate. For a 1:1 supporting electrolyte and for $2\delta \gg 1/\kappa$, where κ is the Debye–Hückel reciprocal thickness of the double layer,

$$f = 1 + \delta^{-1} \int_0^\delta \exp[(-\psi F/RT) - 1]\, dx \tag{45}$$

$$= 1 + \Gamma_+/c_\pm \tag{46}$$

where Γ_+ is the surface excess of univalent cations and $c_\pm$ is the electrolyte concentration in mole cm^{-3}. For $\delta = 40$ Å, f is of the order of 40 at $c_\pm = 10^{-5}$ mole cm^{-3}, when the rational electrode potential is very negative. In practice, i_p is increased to a smaller extent than the theory predicts, partly on account of discreteness-of-charge effects.

V. CHEMICAL EVIDENCE FOR DIRECT CATHODIC INJECTION OF ELECTRONS INTO SOLVENTS

Although direct production of e^- in aqueous solution is unlikely, as has been discussed above (see also Section VI), in other solvents, from which H_2 cannot be competitively generated so easily, e_s^- production is a feasible reaction* at sufficiently cathodic potentials

*Electrochemical engineering applications are hence attractive, due to the possibility of conducting homogeneous reactions with the electrogenerated e_s^-.

and is supported by the types of reaction which can be made to occur with suitable organic substrate solutes.

In the case of liquid NH_3, the general requirements for direct electrolytic production of e_s^- were considered by Makishima.[18] For Pt electrodes, metal deposition from the electrolyte, reduction of valence state of a metal ion in solution, hydrogen discharge from the solvent (e.g., $NH_3 + e_M \rightarrow NH_2^- + \frac{1}{2}H_2$; $NH_4^+ + e_M \rightarrow NH_3 + \frac{1}{2}H_2$), or direct electron dissolution are possible processes. It is the later type of process (considered theoretically in Section II) that is of special interest here and requires, of course, an electrolyte the ions of which are "nonreducible." The tetraalkylammonium ions are an obvious choice to meet such a requirement and could, in principle, give e_s^- or lead to decomposition of the cation. Various experiments have been carried out in liquid NH_3. In very early work, Palmaer[5] obtained blue solutions in this solvent by electrolysis of Me_4NOH, Me_4NCl, and Et_4NCl with Pt electrodes at $-34°C$ and attributed the color to the free electron or to "free alkylammonium radicals." Other early investigations were made by Emmert[47] and similar results were found by Schlubach,[48] Kraus,[49] and Forbes and Norton,[50] who electrolyzed a variety of R_4N^+ iodides. The fact that blue solutions are obtained suggests that it is the species e_s^- that is produced rather than radicals formed from the cations, though these are evidently generated as intermediates in other solvents.[51] Southworth *et al.*[52] showed more recently that reduction of substituted ammonium salts with alkyl groups containing less than 6 C atoms occurs with difficulty due to formation of an amalgam which eventually decomposes to NH_4OH (in water). Various other electrolytic studies[53–55] have been carried out in H_2O, MeOH, dimethylacetamide, and CH_3CN. A thorough study of the decomposition of R_4N^+ salts and the resulting formation of radicals was carried out by Dubois *et al.*,[51] who treated the reaction in terms of normal cathodic neutralization of the cation by a transferred electron:

$$R_4N^+ + e_M \rightarrow R_4N^{\cdot} \rightarrow R_3N + R^{\cdot} \tag{47}$$

When R is a saturated alkyl function, no dimers R_2 were found. Dimeric products only arose in the case of R = allyl or benzyl. Alkanes and alkenes corresponding to R are the usual products arising from H abstraction from residual water contained in the solvent hexamethylphosphorotriamide (HMPA).[64,65] The possibility that homogeneous reduction by e_s^- occurred was not considered, but if it did, it is unlikely that this pathway could be distinguished in the cases investigated from that corresponding to direct electrochemical electron transfer. Formation of colored solutions was not reported. The reduction potentials employed (up to -3.3 V versus Ag/Ag^+ in the same solvent for the aliphatic R_4N^+ salts) were sufficiently high, however, to allow the possibility of direct injection of electrons into the solvent.

This possibility, in the HMPA solvent used, is rendered more likely by the later observation of Dubois and Dodin[66] that alkali metal solutions (prepared by dissolution of the alkali metal in the HMPA solvent or by electrolysis of an alkali metal salt at a Pt cathode) can lead to reduction of benzene. In weak solutions of e_s^- in HMPA prepared by electrolysis of 0.2 *M* $LiClO_4$ or $NaClO_4$ at Pt, 0.1 *M* water is unreactive but 0.5 *M* H_2O gives H_2. Benzene and toluene are reduced to 1,3-cyclohexadiene and 1-methyl,1,4-cyclohexadiene, respectively. In more concentrated metal solutions, produced by direct dissolution of Li in HMPA, the radical anions of naphthalene and biphenyl are formed from benzene. The latter reaction is regarded as occurring via metal monomers in a metallation reaction while the former, in more dilute solutions, occurs via a Birch reduction[11] involving e_s^-.[58]

A direct electrolytic reduction of *m*-tolylmethylether in liquid ammonia was reported by Birch[3] in relation to the well-known Birch reduction method employing alkali metals dissolved in liquid ammonia. Benkeser *et al.*[68] demonstrated similar selective reductions of aromatic hydrocarbons by electrolysis of Li salts in methylamine. The solutions in the cathode compartment become blue due to e_s^- formation either by dissolution of electrodeposited Li or by direct electron injection (see below). These reductions are generally believed to occur homogeneously, but the possibility of direct heterogeneous electrochemical reduction was recognized.[68] The similarity of products obtained, in such cases, from the electrolytic method and from homogeneous reaction with metal–amine or metal–ammonia solutions lends support to the first view.

Laitinen and Nyman[4] made a careful study of the polarographic behavior of alkali and R_4N^+ salts in liquid ammonia at the Hg electrode and distinguished the possible cathodic processes of (a) metal discharge and amalgamation; (b) hydrogen evolution; and (c) electron injection into the solvent. *n*-Bu_4NI was used as an electrolyte to avoid process (a). Process (b) in liquid NH_3 should depend on NH_4^+ concentration, i.e., the "pH." The polarographic current which appears beyond -2.3 V (versus 0.001 M NaI/Hg pool electrode) in a 0.001 M NaI solution (i.e., beyond the observed limiting current for Na^+ deposition at $E_{1/2} \cong -1.65$ V) was therefore attributed to direct dissolution of electrons in ammonia. Reduction of R_4N^+ ion to form an organic amalgam was considered but no limiting current for such a process could be detected up to -4.0 V (with Me_4N^+, a limiting current *is* found, confirming, in this case, that an organic amalgam can be formed; cf. the work of McCoy *et al.*[56,57])

It was also shown[4] that for Bu_4N^+ solutions, the electron dissolution process was essentially the same at a Pt or a Hg surface, i.e., the value of A in equation (12) and the slope of the current–voltage relation were essentially the same.

Estimates of the standard potential for $e^-_{NH_3}$ were made and have been discussed in Section II.3.

Direct electrolytic generation of solvated electrons in HMPA was recently demonstrated by Sternberg *et al.*[58] This solvent, having the structure $[(CH_3)_2N]_3PO$, is capable of dissolving alkali metals.[59] Attempts to generate e_s^- in ethanol are unsuccessful[58] owing to the relatively strong proton-donor properties of this solvent since hydrogen evolution is found to occur[58,60] well below the potential for release of e_s^- into the solvent. However, in binary solutions of ethanol with HMPA, cathodic release of e_s^- can be achieved and the solvated electrons can add to a benzene ring. When a 0.3 M LiCl solution in HMPA is electrolyzed, dark blue globules said to be characteristic of "solvated lithium" form at the cathode surface at a potential of -2.3 V (versus Ag wire in the same solution) and the catholyte becomes dark blue. The half-life of e_s^- is ~38 min in this solvent, a figure which was determined by measuring the decay of peak height of the ESR signal at room temperature.[58] In the presence of ethanol, the color is less intense, depending on the ethanol concentration.* When both ethanol and benzene are present, however, the solution remains colorless. Electrolysis of

*The lifetime of the e_s^- in the ethanol-containing solutions is presumably decreased as it is in H_2O–NH_3 solutions, where an approximately linear dependence on composition is found.[61] e_s^- can be distinguished from organic free radicals which may also be formed in various cathodic reactions by reference to the ESR spectrum,[58] which shows a characteristic fine or hyperfine structure depending on the radical carrying the unpaired electron.

tetralin in 62 mole % HMPA–ethanol mixture at −2.5 V (Ag) gives 2 (vol.) % hexalin, 1 % octalin, and 17 % decalin with 80 % tetralin remaining. The current efficiency for tetralin hydrogenation was 54 %. In the absence of HMPA, i.e., with only ethanol present, copious H_2 evolution occurs and tetralin reduction is negligible. Ethanol acts as a proton donor, since in pure HMPA solution, no reduction of tetralin occurs.

In the presence of ethanol, as in the case of reduction of benzene in ethylene diamine,[62] benzene is reduced initially to the benzene radical anion, which becomes protonated in the presence of a suitable donor. Further electronation and protonation then completes the reduction of a double bond, as discussed by Krapcho and Bothner-By.[63]

Electrochemical reduction of the benzene ring (in benzene and tetralin) by electrolysis of ethylene diamine solution saturated with LiCl was also demonstrated earlier by Sternberg *et al.*[67] The same solution gave a blue color characteristic of the solvated electron upon electrolysis in the absence of reducible substrates. Since it was shown that similar reductions could be performed in the absence of Li ion using Bu_4NI as the electrolyte, this work supports the view[3,4] that solvated electrons can be produced directly by cathodic electron injection into the solvent provided that the appropriate conditions of electrode potential (see Section II.3) and solvation energy are satisfied, and fast annihilation reactions are insignificant.

Sternberg *et al.*[62] showed that in the case of anthracene, naphthalene, and biphenyl, the half-wave potentials for reduction fell well below that for Li^+ ion reduction, so that reduction occurred by direct electron transfer and not via Li cation as a charge transfer agent. In the case of benzene, however, it was concluded that the charge transfer step was the reduction of solvated Li^+ to "solvated Li," which then reacts with benzene in solution to give the benzene radical anion:

$$Li_s^+ \cdots e_s^- + B_s \rightarrow Li_s^+ + B_s^-$$

The B_s^- anion is then protonated from the ethylene diamine solvent.

That the electron transfer does not take place directly from the cathode to the reducible substrate was deduced[67] by reference to the following observations: (a) electrolysis of LiCl in ethylene diamine produces the blue color characteristic of $Li_s^+ \cdots e_s^-$ (solvated electron); in the presence of tetralin or benzene, this color is not observed; (b) no reduction occurs when NH_4Cl is used; (c) no wave attributable to reduction of tetralin is observed at any potential less negative than the decomposition potential of the ethylene diamine solvent, -2.4 V, while naphthalene and Li^+ cations give rise to easily recognized waves. If direct electron transfer to tetralin occurs, it must do so at ~ -2.9 V, i.e., about 1 V more negative than in the reduction of naphthalene and considerably more negative than the decomposition potential of the solvent. Hence, with benzene, little direct cathodic electron transfer occurs and the reduction proceeds predominantly by uptake of electrons from solvated Li, i.e., from solvated electrons. That similar reduction is found using Bu_4N^+ cation suggests that it is the "free" solvated electron that is involved rather than a solvated $Li_s^+ \cdots e_s^-$ complex.

Similar reductions of benzene and toluene were achieved by Misono *et al.*[69] in a diglyme–H_2O mixture $\sim$0.5 *M* in Bu_4NBr.

In CH_3CN–H_2O mixtures, Misono *et al.*[70] investigated conditions for selective reduction of naphthalene to the 1,4-dihydro product. In anhydrous CH_3CN, green solutions resulted but the overall current efficiency for reduction of naphthalene was low. In the presence of Et_4N^+ salts, gaseous products including C_2H_4 and C_2H_6 were formed, presumably on account of decomposition of Et_4N radicals. In this case, the role of solvated electrons is not clear.

Using alcohol–HMPA solvent mixtures with 0.5 *M* LiCl as electrolyte, Asahara *et al.*[71] demonstrated the electrolytic reduction of naphthalene to 1,4- and 1,2-dihydronaphthalenes, the relative yields depending on the HMPA content. No reduction occurred in the absence of the HMPA component. The solutions containing HMPA became blue, as in the work of Sternberg *et al.*,[58] and the reductions occurred below the normal half-wave potential (cf. Ref. 62). Again, the ESR spectrum must be used to provide a definitive indication of e_s^- in distinction to colored organic free radicals.

VI. HYDRATED ELECTRONS IN AQUEOUS ELECTROCHEMICAL REACTIONS

1. Some General Problems

In the preceding sections, it has been shown how solvated electrons can arise under photoelectrochemical conditions and directly, in the case of certain special nonaqueous solvents, by cathodic injection.

The possibility that a number of cathodic reactions, including electrolytic H_2 evolution, proceed in *aqueous* media with the direct formation of hydrated electrons as the primary intermediates has been proposed in various recent papers.[8,9,20,30] The idea has also been extended to the case of corrosion-type reactions[9] (cf. p. 95) where the anodic partial process is metal oxidation and the cathodic one, evolution of H_2 or reduction of oxygen, both occurring at a common potential on open-circuit. Various experiments have been devised[15,30,76,77] to test these proposals. A number of electrochemical kinetic and thermodynamic difficulties arise if such mechanisms of cathodic processes are accepted, and in this section, we examine the consequences of such a point of view. The question is clearly one of quite general importance in electrode processes and in aqueous organic electrochemistry where reductions are involved.

The hydrated electron has been well-characterized[78a,b] in photochemical and radiolytic processes in water and the kinetics and mechanisms of its homogeneous reactions have been extensively investigated.

In the case of the hydrogen evolution reaction (h.e.r.), mechanisms of the reaction involving hydrated electrons are regarded[8,9] as the following:

$$e_M(\text{cathode}) \rightarrow e_{aq}^- \qquad \text{(primary ``discharge'' step)} \tag{48}$$

with (17) or (18), i.e.,

$$e_{aq}^- + H_2O \rightarrow H + OH^- \qquad \text{or} \qquad e_{aq}^- + H_3O^+ \rightarrow H + H_2O$$

$$2H \rightarrow H_2$$

Alternatively, we have according to (16),

$$e_{aq}^- + e_{aq}^- \rightarrow H_2 + 2OH^-$$

In the case of metal dissolution in the anodic partial process of corrosion, e.g., of Na amalgam, the step involving e_{aq}^- is regarded[9] as

$$M \rightarrow M^{z+} + ze_{aq}^- \tag{49}$$

With regard to the h.e.r., these mechanisms differ in obvious ways from pathways previously considered in the large body of work which has been published since Tafel's primal contributions. While the question of the basic nature of the cathodic discharge act had not until recently been raised, a number of the recent papers on the h.e.r. provide certain critical lines of evidence indicating involvement of protons directly discharged at the metal.

The view that hydrated electrons are the precursors of cathodically evolved hydrogen in electrochemical hydrogen production and metal dissolution in *aqueous* solutions seems to have originated in the papers of Pyle and Roberts[9] and Walker,[8] who proposed this mechanism as a general basis for cathodic hydrogen evolution. Experimental indications of this view were obtained[8,76] by him under certain conditions by means of an ingenious multiple-reflectance light-absorption experiment, and by chemical experiments involving the use of scavengers.[76] An interesting observation[77] by Yurkov that deposition of Cu occurred *between* an anode and cathode of Cu was attributed to neutralization of Cu ions by electrons in solution, but it is difficult to accept this explanation, and also one that could be based on diffusion of H atoms from the cathode,[79] owing to the macroscopic distances involved.[77]

Hills and Kinnibrugh[30] also made tentative suggestions that e_{aq}^- entities might be involved as intermediates in the h.e.r.; they based this view on the observation of an (apparent) negative volume of activation in the kinetics of the h.e.r. studied at high pressures. However, it may be shown that this negative value becomes a (normal) positive value when the data, which were derived for constant *overpotential*, are corrected* for pressure effects on the potential of the reversible reference electrode. Hence, the $\Delta V^\ddagger$ results do not support participation of e_{aq}^- in the h.e.r.

The apparent *negative* value of $\Delta V^\ddagger$ for the h.e.r. at Hg seemed[30] to indicate e_{aq}^- as a primary intermediate since, in such a mechanism for the h.e.r., a charge would be produced rather than removed in the cathodic process. In the case of ammonia, the electron has a large positive partial molal volume[29] and a similar situation might be expected for the more structured solvent water. An estimate of the volume by Barr and Allen[81] gave,† however, the figures -5.5 to -1.1 ml mole^{-1} for e_{aq}^-. This would be the right order of magnitude to account for the (apparent) negative $\Delta V^\ddagger$ were it not for the fact

*See Ref. 80a and, for later discussions of this point, Refs. in 80b.
†A more recent paper [81a] gives a value between $+2.7$ and -1.7 ml mole^{-1}.

that in measurements of pressure effects on the kinetics of an electrochemical reaction, the change of potential of the reference electrode with pressure must be allowed for.[80a] Hills and Kinnibrugh made a careful experimental evaluation of the pressure dependence of the e.m.f. of the cell

$$H_2Pt \left| 0.1\ M\ HCl \right\vdots M\ KCl\ Hg_2Cl_2 \left| Hg \right.$$

but expressed their experimental kinetic results as the derivative

$$-RT[\partial(\ln i)/\partial p]_{T,\eta} = -3.40 \pm 0.15 \quad \text{ml mole}^{-1}$$

i.e., for constant overpotential η assuming that this condition gave $\Delta V^{\ddagger}$. Conway[80a] pointed out that a correction for the pressure dependence of the reversible H_2 electrode should be made and in a later publication this matter was considered in more detail by various contributors in discussion.[80b] At that time,[80b] there seemed to be some ambiguity regarding how the experimental pressure correction referred to by Conway[80a] should be made and this matter was further considered by Krishtalik.[80b] It may be shown that the derivative $[\partial(\ln i)/\partial(\ln p)]_{\eta,T}$ does not, in fact, give $\Delta V^{\ddagger}/RT$. Thus, if the rate of the cathodic reaction is written as

$$\begin{aligned} i &= zFkC_{H^+} \exp(-\beta VF/RT) \\ &= zFkC_{H^+} \exp(-\beta\eta F/RT) \exp(-\beta V_r F/RT) \end{aligned} \tag{50}$$

where $\eta = V - V_r$ and V_r is the reversible potential for the electrode process at equilibrium,

$$\frac{d(\ln i)}{dp} = \frac{d(\ln k)}{dp} - \frac{\beta F}{RT}\frac{dV_r}{dp} \tag{51}$$

at constant η and T. Since $zFV_r = -\Delta G_r^{\circ}$,

$$\left[\frac{\partial(\ln i)}{\partial p}\right]_{\eta,T} = -\frac{\Delta V^{\ddagger}}{RT} + \frac{\beta}{z}\frac{\Delta V_r^{\circ}}{RT} \tag{52}$$

It is evident that the derivative $RT[\partial(\ln i)/\partial p]_{\eta,T}$ taken at constant *overpotential* gives $-\Delta V^{\ddagger}$ modified by the term $(\beta/z)\Delta V_r^{\circ}$, involving the volume change ΔV_r° in the reversible electrode reaction. This result, with regard to pressure, is thus similar[80a,b] to that arising in evaluations of *heats* of activation for electrode reactions where only an apparent value is obtained from $-R[\partial(\ln i_0)/\partial(1/T)]_{\eta=0}$.

Krishtalik[80b] estimated the true $\Delta V^\ddagger$ corresponding to the results of Hills and Kinnibrugh[30] as $+5.6$ ml mole^{-1}. This value, now being positive, is consistent with a normal H^+ neutralization in the h.e.r. mechanism, with release of electrostricted water. An alternative explanation for negative $\Delta V^\ddagger$ to that involving production of e^-_{aq} was suggested by Conway[80a] in terms of the effects of penetration of H^+ into the double layer as the transition state becomes formed. This idea was developed further by Heusler and Gaiser,[107] who showed how it could account for results of new measurements[107] of the pressure effect on the kinetics of the h.e.r.

Apart from the question of the significance of $\Delta V^\ddagger$, a number of other thermodynamic and kinetic difficulties present themselves if reaction (48) is considered to be the primary electrode process in the h.e.r.:

(i) The first arises from thermodynamic considerations of the conditions for electron emission into solution. The standard potential for the hydrated electron e^-_{aq} on the hydrogen scale has been estimated from kinetic data to be ~ -2.67 V,* yet hydrogen can be easily evolved on the more catalytic metals at appreciable rates ($>10^{-4}$ A cm^{-2}) already at 0.1–0.2 V cathodic to the reversible H_2 potential E_H (even in alkaline solutions, e.g., 10^{-4} A cm^{-2} in 0.1 N NaOH at 20°C at Ni[13]). If the cathodic process were (cf. Ref. 2)

$$e_M + H_2O \rightarrow e^-_{aq} \quad (\rightarrow H \rightarrow \tfrac{1}{2}H_2)$$

rather than direct discharge of a proton out of H_2O or H_3O^+, it would be necessary to suppose that the e^-_{aq} intermediate was produced at these potentials in a quasiequilibrium† concentration C of the order of 10^{-40} g-electrons liter^{-1}. Diffusion effects also operate[9,76] and hence render the above figure an upper limit. Since hydrogen evolution occurs already at 10^{-5}–10^{-3} A cm^{-2} at the more catalytic metals (Ni, Pt) at an overpotential $\eta = -0.2$ V, say, it is kinetically difficult to envisage such a rate arising by the e^-_{aq} mechanism. Even at dissolving dilute amalgams, where $\eta = \sim -1.5$ V, C would still be unrealistically small.

*This also corresponds to the energy of the absorption maximum at 720 nm as discussed by Matheson on p. 45 in Ref. 78. However, there is no simple relation between the solvation energy of e^-, taken from the gas phase, and the electronic spectrum (cf. Fig. 2). For the value of E°_e for NH_3, see Ref. 82. Also see Ref. 13.

†Scavenging processes cause a departure from any quasiequilibrium which may exist on a short time scale and the consequences of this effect are considered below.

In terms of normal electrode kinetic principles, adsorbed H and molecular H_2 can be produced at appreciable rates at any significant overpotential from alkaline or acid solutions since such potentials are (by definition) negative w.r.t. E_H. The sources of protons are H_2O or H_3O^+ under these two respective conditions.[83,84] It is easy to see that the opposite is the case with e_{aq}^- with regard to "overpotential"; it must be produced at potentials very positive to its standard potential $E^\circ_{e_{aq}^-}$ yet at a rate equal to twice the overall Faradaic rate of H_2 evolution. In kinetic terms, this would imply an unreasonably high exchange current for the process $e_M \rightarrow e_{aq}^-$ occurring at $E^\circ_{e_{aq}^-}$ ($= -2.67$ V). Even if it is recognized that a more realistic potential for this calculation should perhaps be $2.67 - (0.06 \times 6) = 2.30$ V, i.e., the potential corresponding to the practical concentration limit for detection of e_{aq}^-, namely $\sim 10^{-6}$ mole liter^{-1} by sensitive absorption spectrophotometry.[8,76]

(ii) Application of the principle of microscopic reversibility illustrates other difficulties. At the reversible potential (but not necessarily at other potentials where the electrochemical reaction is not at equilibrium, and kinetic pathways may not be identical[85] with those at equilibrium when $\eta = 0$), a commonly considered mechanism, e.g., for catalytic metals is

$$H^+ + e + M \rightarrow MH \tag{II}$$

with

$$2MH \rightarrow H_2 \tag{III}$$

or with the alternative desorption step $MH + H^+ + e$ giving H_2. With the e_{aq}^- path, on the other hand, the mechanism at the reversible potential would have to be written

$$e_M + H_2O \rightleftharpoons M + e_{aq}^- \tag{IV}$$

$$e_{aq}^- + H_2O \rightleftharpoons H + OH^- \tag{V}$$

$$H \rightarrow \tfrac{1}{2}H_2 \tag{III}$$

Pyle and Roberts[9] also considered the step $e_{aq}^- + e_{aq}^- \rightarrow H_2 + 2OH^-$.

In terms of this e_{aq}^- scheme, the back-reaction pathway would therefore have to be written in terms of H giving e_{aq}^- spontaneously, with the latter being the entity in charge-transfer equilibrium with the electrode. Similar processes should presumably occur in metal deposition (cf. the calomel half-cell reaction discussed by Dainton[20] and written $Hg_2^{++}2Cl^- + 2e_{aq}^- \rightarrow 2Hg + 2Cl_{aq}^-$) if this type of mechanism applied. Such a mechanism would require spontaneous e_{aq}^- formation (which would have to occur with very low concentrations of e_{aq}^-) at a rate sufficient to account for the kinetics of reactions, including the h.e.r., which can proceed at quite high, non-diffusion-controlled exchange current densities of 10^{-3}–10^{-1} A cm^{-2}, e.g., metal ion reduction or deposition at certain metals.[86]

(iii) The principle of microscopic reversibility enters into the discussion in another way, too. If we can consider an anodic electrolysis of a metal very near to its reversible potential (so that the process is subject to the requirements of microscopic reversibility), i.e., in the back reaction of a metal deposition reaction, e.g., $Hg + 2Cl_{aq}^- \rightarrow Hg_2Cl_2 + 2e_{Hg}$ or $Na/Hg \rightarrow Na_{aq}^+ + e_{Hg}$, it is clear that electrons must pass along the metal electrode and conducting wires. It is contrary to the observed electrolytic behavior and also intuitively unlikely that electrons are first produced anodically in solution as considered, for example, by Pyle and Roberts,[9] and then return to the metal in steps such as

$$M \rightarrow M^+ + e_{aq}^- \tag{VI}$$

$$e_{aq}^- + M \rightarrow e_M \quad \text{(anodic current in the electrolytic experiment)} \tag{VII}$$

e.g., in anodic amalgam dissolution* near the reversible potential for Na/Hg. In any final electrolytic process such as (VII), the electron in the metal M(e) must pass down the external circuit. Similar considerations apply to the corrosion of a metal on open-circuit, e.g., amalgam dissolution.

A suggestion[20] has been made that the absolute potential of the calomel half-cell reaction, written as [20]

$$2Hg + 2Cl^- \rightleftharpoons Hg_2Cl_2 + 2e_{aq}^- \tag{VIII}$$

can be calculated if the hydration energy of e_{aq}^- is known. For *electrochemical* occurrence of the calomel reaction, it seems necessary,[87] however, to regard the electron released in reaction (VIII)

*Under open-circuit conditions of corrosion of the amalgam, the anodic reaction (VI), if it occurred, would be coupled with cathodic steps in the h.e.r. such as those considered above. It is unlikely that anodic (electrolytic) dissolution near the reversible potential would go by a mechanism qualitatively different from that occurring on open-circuit corrosion (spontaneous dissolution of the amalgam).

as entering the Fermi level in the metal, otherwise no current passes into the external circuit when the calomel potential is experimentally measured (a small-signal current is, of course, always required in the measuring instrument; also, a charge separation *across* the interface is required to set up the p.d. or when the electrode is slightly polarized). The work function of Hg must hence be involved instead of the solvation energy of e^- in determining the electrode potential. Other problems of a different kind regarding the calculation of absolute electrode interfacial p.d.'s have been discussed by Frumkin[72] in relation to the role of the surface potential at metal–solution interfaces.

(iv) Finally, it has been shown[88,89] in recent years that electrochemically adsorbed H can be controllably produced and quantitatively determined at the noble metals such as Pt, Rh, or Ir by means of electrochemical transient procedures. Also, at Ag, the coverage by atomic H can be followed[111] as a function of cathode potential. The coverage–potential relations (Fig. 10) for H adsorbed at various metals are also independent of stirring so that the H cannot arise from a diffusion-controlled reaction involving $e^-_{aq} \rightarrow H + OH^-$. Also, it is well known that atomic H can be electrolytically generated and diffused *through* various metals such as Fe or Pd at rates which can be controlled by the discharge of protons at the metal–aqueous solution interface. Also, the potential dependence of the coverage by H follows relations that can be derived from an electrochemical quasiequilibrium of the type $H_2O + M(e) \rightarrow MH_{ads} + OH^-$ occurring at various types of sites on the metal. Processes of formation and removal of adsorbed H atoms at the surface involving e^-_{aq} as a transient intermediate therefore seem unlikely. They would have to be written, for example, as

$$M(e) + H_2O \rightarrow e^-_{aq} \qquad \text{(I)}$$

$$e^-_{aq} + H_2O \rightarrow H + OH^- \quad \text{in solution} \qquad \text{(IX)}$$

with the H which is detected in a chemisorbed condition arising adventitiously from readsorption of H atoms from solution, produced in step (IX), i.e. (cf. Ref. 9),

$$H_{soln} + M \rightarrow MH_{ads} \qquad \text{(X)}$$

Similar difficulties must apply to the corresponding back reactions, since electrochemical adsorption of H (prior to H_2 evolution

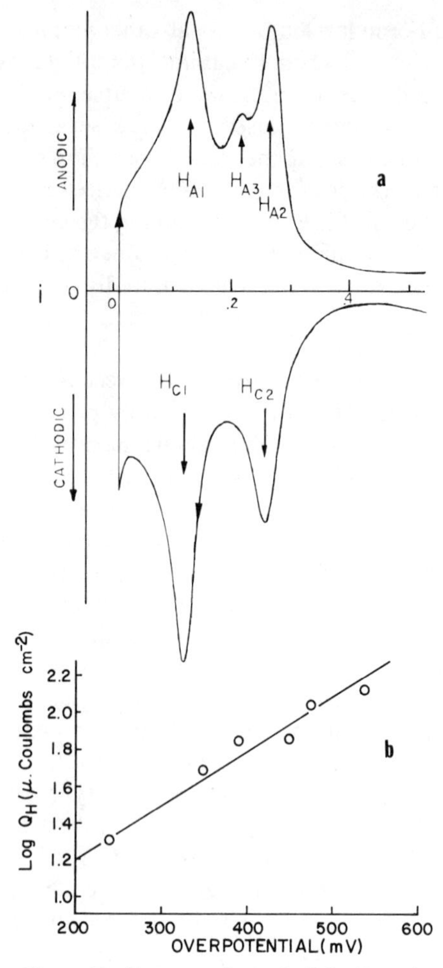

Figure 10. Coverage of atomic H by Pt and Ag as a function of electrode potential (after Bockris *et al.*[111] and Will and Knorr[89]). (a) Pt differential charging curves for electrochemisorption of H at Pt. (Integral under the curves between base line and peaks gives charge for progressive coverage of surface by H. Note distinguishable regions H_{c1}, H_{c2}, H_{A1}, H_{A2}, H_{A3} for cathodic deposition and anodic stripping, respectively, of the chemisorbed H). (b) Integral charge Q_H for electrochemisorption of H at Ag.[111]

at slightly more cathodic potentials) is known to be highly reversible,[89] so all the steps considered above would have to be (a) reversible and (b) capable of proceeding at appreciable rates if an e_{aq}^{-} pathway for H adsorption (and hence H_2 evolution) were involved. With regard to the relative stability of e_{aq}^{-} in acid and alkaline solutions, it is important to note that the electrochemisorption behavior of H at Pt, Rh, and Ir is found to be almost the same,[90] independent of pH, so it is unlikely to arise from H homogeneously produced from e_{aq}^{-} as in reactions (I) and (IX). For example, e_{aq}^{-} is irreversibly converted[91] to H much faster in acid than in alkaline or neutral media, so that homogeneous recombination of H to H_2 would be favored over adsorption under the latter conditions of pH.

The current–potential relations for electrochemisorption of H obtained by the potentiodynamic method are best interpreted in terms of a *direct* discharge process. Thus, the charge associated with adsorbed H production is almost independent[89] of the rate of potential sweep and the pseudocapacitance charging currents at the peak maxima are proportional to sweep rate, so that only a direct *surface* discharge, and not a diffusion-controlled[92] process involving species in solution, can be taking place.

Conway and MacKinnon[15] pointed out that the closely related question of the interpretation of the fundamental kinetic coefficient

$$b[= dV/d(\ln i)]$$

in Tafel's equation[86] characterizing the dependence of rate of the electrochemical reaction on potential V gives supporting evidence for chemisorbed H rather than e_{aq}^{-} in solution as an intermediate in the h.e.r. For certain metals, e.g., Pd[84] and Pt,[93] and some alloys,[84] $b = [RT/(1 + \beta)]F$ or $RT/2F$, where β is the usual electrochemical Brønsted factor[94] (≈ 0.5). Experimentally, such values of b are well characterized over an appreciable range of potentials and theoretical deductions of the observed values have been made on the satisfactory and quantitative basis that the coverage θ_H by *adsorbed* atomic hydrogen is appreciable and potential-dependent according to a relation of the form[95]

$$\theta_H/(1 - \theta_H) = K \exp(-VF/RT) \exp(-r\theta) \qquad (53)$$

for the quasiequilibrium, potential-dependent reaction*

$$H^+ + e + M \rightleftharpoons MH_{ads}$$

where the desorption process is the rate-controlling one characterizing the kinetics (and in particular, b) in a step such as

$$MH_{ads} + H^+ + e \rightarrow H_2 \qquad b = RT/(1 + \beta)F$$

or

$$MH_{ads} + MH_{ads} \rightarrow H_2 \qquad b = RT/2F$$

following the primary act of H^+ ion discharge.

Electrochemical production of solvated electrons at various distances in the solution, followed by homogeneous reaction to give H_2, could not give values of $b < RT/\beta F$ such as are observed in the kinetics of the h.e.r. at a number of metals including platinum† studied by Walker[8,76] in regard to N_2O reduction in his scavenging experiments. If the e^-_{aq} path (reactions I, IX, X) were involved, a value of $b = RT/\beta F \doteqdot 120$ mV, would be found for *all* metals, corresponding to the electron tunneling process of Gurney.[23] Furthermore, one of the outstanding characteristics of the kinetics of the h.e.r. is that they are highly specific to the cathode material,[96] an effect which has been attributed‡ to the role of chemisorbed H. A common mechanism for all metals of electron emission into solution with subsequent homogeneous hydrogen production could therefore hardly give the observed specificities with regard to the marked dependence of values of i_0 and b on the metal (Figure 11).

The difficulties referred to above concern principally processes occurring near the reversible potential for the h.e.r. or that for Na/Hg dissolution. It is necessary to recognize, however, that if cathodic potentials relative to the hydrogen electrode are sufficiently high, say

*In point of fact, the H adsorption is characterized by two or three Langmuir constants K in the case of Pt, due to different binding energies at distinguishable types of sites on the surface.

†The competitive reduction studies at Pt using N_2O are rendered ambiguous[15] by the heterogeneous reduction of N_2O by adsorbed H and by the presence of dissociatively adsorbed methanol present as competitor.

‡Analysis of the factors connected with metal properties which determine, e.g., the exchange current for the h.e.r. at various metals, shows that such directly metal-dependent factors as the electronic work function ϕ_e do not enter in a primary way into the determination of the kinetics since ϕ_e also is involved in determining the potential of the reference electrode.[28]

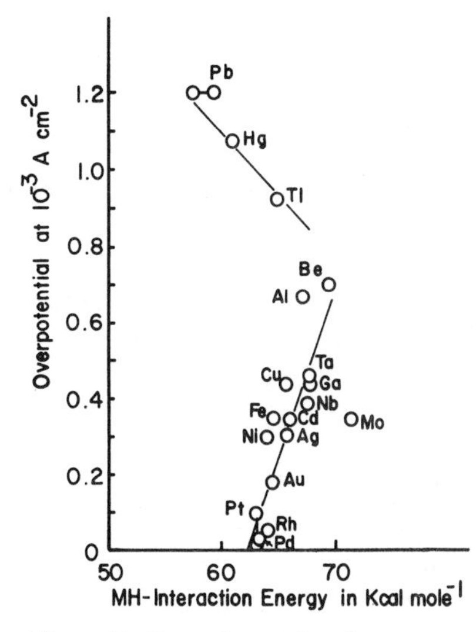

Figure 11. Dependence of exchange current for the h.e.r. on properties of the cathode metal such as metal–H adsorption energy. (after Conway and Bockris[96]).

>1.9–2.0 V, it would be kinetically feasible for the e^-_{aq} path to operate at such potentials. Only in the case of dissolving base metals or strong amalgams does it seem possible that a process[76] of direct electron injection into the water solvent to form e^-_{aq} might occur to a small extent as a *parallel* reaction with normal hydrogen evolution as the main cathodic partial process. This may be the explanation of Walker's observations at amalgams with scavengers. Such a path would not, however, constitute a general mechanism for cathodic processes as has been implied, particularly for the "lower overpotential" metals.

Bockris and Azzam,[112] in high-c.d., direct-current experiments, and Nürnberg,[113] with Faradaic rectification measurements, studied the h.e.r. at Hg and other metals up to high cathodic potentials but did not observe any indications of a change of mechanism consistent with e^-_{aq} production, e.g., in the case of Hg, the Tafel slope remained exactly constant up to the most negative potentials attainable.

The conclusions from the above arguments depend, in the main, on the value of the calculated standard redox potential[13,82] for e^-_{aq} being approximately correct. If e^-_{aq} entities were involved appreciably in the kinetics of the h.e.r. up to, say, -0.5 V E_H, it would be necessary to suppose that the $E°$ value for e^-_{aq} would have to be in error by at least 2 V. A discrepancy of such magnitude seems unlikely since it would require revision of one of the rate constants (on which the $E°$ value is based[13]) by many orders of magnitude. However, the use of equilibrium $E°$ values under conditions of irreversible scavenging by the main solution species H_2O or H_3O^+ must now be examined.

2. Standard Potential for e^-_{aq} and the Nonequilibrium Situation Arising from Annihilation Processes

The thermodynamic difficulties discussed above (but not, however, the kinetic ones based on observed Tafel slopes and electrochemisorption behavior of H) could be diminished if the $E°$ for e^-_{aq} were substantially in error and/or if the "equilibrium" potential were inapplicable. The latter situation should obviously receive further consideration because of the conditions imposed by the short lifetime of the electron in acid media. Ejection of electrons into the solution may arise under suitable conditions of potential and/or irradiation[25,35] (see Section III), but e^-_{aq} may not exist long enough for any reversible equilibrium with the electrode to be kinetically established. This is equivalent to requiring that in the following supposed equality of rates for quasiequilibrium,

$$\vec{v}_I = k_I \exp(-\beta VF/RT) = \overleftarrow{v}_I = k_{-1}C_{a\bar{aq}} - \exp[(1-\beta)VF/RT] \quad (54)$$

the $\overleftarrow{v}_1$ term is actually appreciably less than the $\vec{v}_1$ term. Under conditions of annihilation of e^-_{aq}, a steady-state condition for production and removal of e^-_{aq} by reaction (IX) at a given potential V must be considered, and an approximate* form of such a relation would be written

$$d\bar{C}_{e\bar{aq}}/dt = k_1 \exp(-\beta VF/RT) - k_{-1}\bar{C}_{e\bar{aq}} \exp[(1-\beta)VF/RT] - k_{IX}\bar{C}_{e\bar{aq}} = 0 \quad (55)$$

*The approximate nature of this relation arises (a) because a distribution of electron injection rates will be involved and (b) since the back reaction for electrons injected appreciable distances into the solution will be diffusion-controlled.

where $\bar{C}_{e_{aq}^-}$, for the approximation, expresses a mean concentration[25] of solvated electrons near the surface and the rate constants refer to the reactions designated by (I) and (IX) earlier. The difference between the first two terms in equation (55) is equivalent to the net cathodic current. Then,

$$\bar{C}_{e_{aq}^-} = [k_{\mathrm{I}} \exp(-VF/2RT)]/[k_{-1} \exp(VF/2RT) + k_{\mathrm{IX}}] \quad \text{for} \quad \beta = 0.5 \qquad (56)$$

If e_{aq}^- could be produced under true equilibrium conditions (e.g., as is almost the case[4] in liquid NH_3), the same concentration $\bar{C}_{e_{aq}^-}$ would be attained when the potential was equal to the reversible, but not the standard, value V^* given by

$$\bar{C}_{e_{aq}^-} = K_{\mathrm{I}} \exp(-V^*F/RT) \qquad (57)$$

We can express the difference between the potential V required to reach any given concentration in the steady state and that V^* for attainment of the same concentration under quasiequilibrium conditions as[15]

$$\exp[-(V - V^*)F/RT] = 1 + K_{\mathrm{I}}(k_{\mathrm{IX}}/k_{\mathrm{I}}) \exp(-VF/2RT) \qquad (58)$$

It will be noted that the r.h.s. is always >1, so that $V - V^*$ will be negative and since negative potentials have been considered throughout for the cathodic process, V is evidently a more cathodic potential than V^*. This, of course, is to be expected on general kinetic principles for an irreversible electrode process such as the sequence (I), (IX), where the "reagent" in the back reaction is being drained off in a chemical step.

As an alternative basis for the argument, it is possible to show that the concentration $\bar{C}^*_{e_{aq}^-}$ that would be produced under hypothetical equilibrium conditions will be greater rather than smaller than the concentration $\bar{C}_{e_{aq}^-}$ for steady-state conditions at any given potential V. The two concentrations may be obtained using equations (57) and (56), respectively, for a given value of V in each expression, and are:

$$C^*_{e_{aq}^-} = K_{\mathrm{I}} \exp(-VF/RT)$$

$$C_{e_{aq}^-} = \frac{K_{\mathrm{I}} \exp(-VF/RT)}{1 + (k_{\mathrm{IX}}/k_{-1}) \exp(-VF/2RT)} \qquad (59)$$

Then, since under all conditions, $1 + (k_{IX}/k_{-1})\exp(-VF/2RT) > 1$, $C_{e_{aq}^-} < C_{e_{aq}^-}^*$, a result which is consistent with the foregoing analysis in terms of V and V^*. It must hence be concluded from this calculation that the estimates of limiting concentration of e_{aq}^- which were made earlier in this chapter in terms of a quasithermodynamic equation will be an *upper* rather than a lower limit; the "thermodynamic" conclusions regarding electrochemical production of e_{aq}^- are hence not invalidated by the fact that nonequilibrium conditions generally obtain in water and acid solutions and in the presence of other scavengers.

3. Illumination Effects and Reflectance

In Walker's experiments[76] at Ag, an apparent direct indication of the presence of hydrated electrons near a Ag electrode surface was obtained by spectrophotometry. By a "lock-in" amplification and detection system, he was able to show that the absorbance, measured at one wavelength only (6330 Å), was associated only with the cathodic end of an ac polarization of amplitude ~1.1 V. This is an important and interesting result that requires further examination. From the experimental details that were recorded,[76] it does not seem possible that potentials sufficiently cathodic would have been reached for e_{aq}^- to be generated, although no reference electrode was employed with respect to which a definite potential scale can be stated. At Ag, for example, it is known[97] that H_2 will be cathodically evolved at extremely high rates (see below) at any potentials sufficiently close to $E_{e_{aq}^-}^\circ$ (~2.67 V) for it to be thermodynamically feasible that a detectable concentration (say 10^{-6} mole liter^{-1}) of e_{aq}^- could be generated in the "steady state." In fact, for such a concentration to be achieved in the electrode boundary layer,[9] the hydrogen overpotential would have to be at least (see below) ~2.3 V, at which potential hydrogen would normally be evolved at ~10^{16} A cm^{-2}! ($i_{0,Ag} \doteqdot 10^{-8}$ A cm^{-2}, $b = 0.09$ V for the h.e.r. at Ag.[44])

Under conditions of illumination at an electrode, it has been shown[35] that photoelectrochemical generation of e_{aq}^- becomes possible and, in effect, the reversible potential for e_{aq}^- production is then lowered by $300\,h\nu/e$ volts, where ν is the excitation frequency and e the electronic charge. It may therefore be proposed that the reason why Walker[76] was able apparently[98] to detect e_{aq}^- spectrophotometrically was because the laser illumination of the silver electrode surface at 6330 Å might already provide energy equivalent to ~1.96 V for excitation of electrons in the surface of the metal above the Fermi level, so that only a small extra potential would be required to reach the e_{aq}^- "reversible potential" and liberate electrons to a significant extent by photoassistance. The practical question remains whether the intensity of the illumination was sufficient to produce the observed effects.

An alternative explanation is that the absorption observed at 6330 Å in Walker's phase-sensitive detection experiments was due to potential-modulated reflectance[99] of the silver surface itself. In unpublished experiments, Schindewolf[98] carried out similar studies but at *various* wavelengths and found that the optical changes on cathodic polarization were more connected with the reflectance spectrum of Ag[99] than with absorption spectrum of cathodically ejected e_{aq}^-. Bewick and Conway[117] carried out similar electric

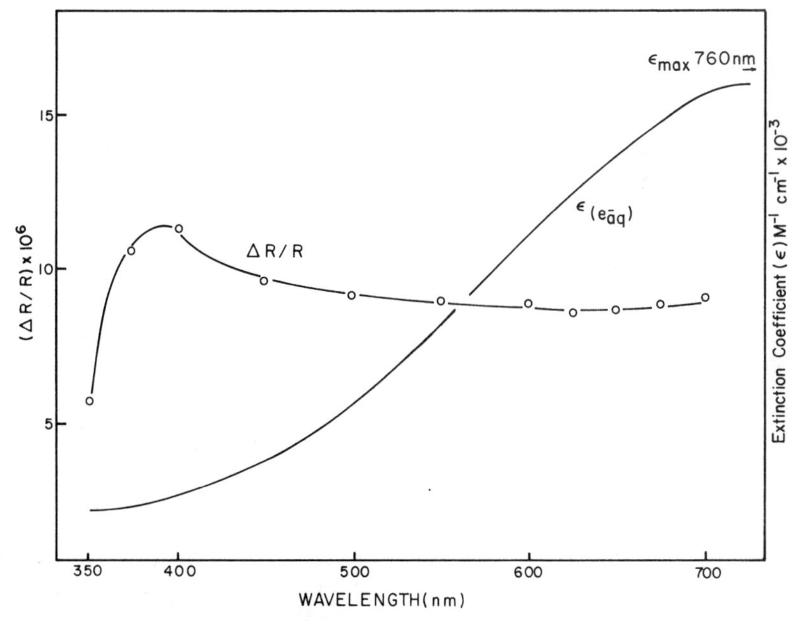

Figure 12. Comparison of the e_{aq}^- spectrum and the reflectance effects at Ag cathodes in 0.25 *M* Na_2SO_4 at 25°C as a function of wavelength (after Bewick and Conway[117]).

modulated reflectance studies under controlled-potential conditions* and their results are compared with the e_{aq}^- spectrum in Figure 12.

4. Solvated Electrons and Organic Electrochemical Reactions

In liquid NH_3, amine, and other solutions, there is good evidence, which has been discussed above, that cathodic reduction can proceed via solvated electrons[58,62,64–71] in homogeneous reactions within the boundary layer near an electrode. A final question remains of whether organic cathodic reductions can proceed in *water* by such a mechanism. In a number of experiments carried out with suitable compounds, it has been shown that aqueous electrochemical reductions of organic substances at the mercury cathode are stereo-selective, e.g., in the electrolysis of diphenylcyclopropyl bromides,[100] dimethyl maleic and fumaric acids,[101] and 2-chloro, 2-phenyl-propionic acid.[102] The chemistry of such reactions shows that it is difficult to account for the stereoselectivity except by direct electron transfer from the cathode to an adsorbed molecule, preferentially oriented in one direction at the electrode interface. The kinetics are usually also specific to the metal, e.g., in reduction of ketones and alkyl halides.[103] If such reductions occurred homogeneously by an *SN*2 type of mechanism through hydrated electrons (cf. the reaction of chloracetic acid with e_{aq}^-; Ref. 104), it is difficult to see how the observed stereoselectivity could arise. There are no difficulties, however, if the view hitherto held in electrode kinetics is retained, namely that cathodic reduction requires a suitable thermal re-organization of the oxidant molecule and/or its solvation shell in the activation step and this condition allows radiationless electron transfer directly from the metal.[23,105] This is a general mechanism of electroreduction and there does not seem to be any reason why it may not occur just as well with solvated electron acceptors such as N_2O or $ClCH_2COOH$ (which, in homogeneous process, are speci-fically more reactive with e_{aq}^- in distinction to H) as with many reagents in other cathodic processes. For example, at Hg or amalgam cathodes, there is no question electrochemically[106] of organic reductions occurring by an atomic hydrogen mechanism; this only

*J. D. McIntyre[118] points out that in Walker's experiments with "a.c." current conditions, the main changes of polarization will have occurred at the central wire electrode rather than at the large area cylinder from which the optical signal was collected.

arises at the more catalytic metals such as Ni, Pt, Rh. For example, N_2O itself can be catalytically and competitively reduced in the presence of CH_3OH at a Pt electrode by adsorbed atomic H or on open circuit by means of Pt powder and H_2.[15] Similar decomposition of N_2O can be effected by Pd/H.[109] In these cases, N_2O can be reduced to N_2 without participation of e^-_{aq}.

In conclusion, it must be remarked that direct transfer of an electron to the thermally reorganized transition state of a reactant must be distinguished from the e^-_{aq}, which is a true intermediate having an appreciable lifetime, in comparison with that of a kinetic transition state in the electrode double layer, which has a transitory existence comparable with the period of a moleculecular vibration or solvent lattice fluctuation.

ACKNOWLEDGMENT

The facilities of the Universities of Southampton and Newcastle on Tyne, where this article was written during the tenure of a Commonwealth Visiting Professorship, are gratefully acknowledged.

REFERENCES

[1] F. P. Bowden, *Trans. Faraday Soc.* **28** (1931) 505.
[2] E. Becquerel, *Compt. Rend.* **9** (1839) 145.
[3] A. J. Birch, *Nature* **158** (1946) 60.
[4] H. A. Laitinen and C. J. Nyman, *J. Am. Chem. Soc.* **70** (1948) 3002.
[5] W. Palmaer, *Z. Elektrochem.* **8** (1902) 729.
[6] J. O'M. Bockris, in *Modern Aspects of Electrochemistry*, Chapter 4, Eds., J. O'M. Bockris and B. E. Conway, Butterworths, London, 1954.
[7] P. J. Hillson and E. K. Rideal, *Proc. Roy. Soc. (London)*, **A216** (1953) 458.
[8] D. C. Walker, *Can. J. Chem.* **45** (1967) 807.
[9] T. Pyle and C. Roberts, *J. Electrochem. Soc.* **115** (1968) 247.
[9a] E. J. Casey, R. E. Bergeron, and G. D. Nagy, *Can. J. Chem.* **40** (1962) 463.
[10] D. C. Walker, *Quart. Rev.* **21** (1967) 79.
[11] A. J. Birch, *Quart. Rev.* **4** (1950) 69; **12** (1958) 17.
[12] A. N. Frumkin, *J. Electroanal. Chem.* **9** (1965) 173.
[13] J. Baxendale, *Rad. Research Suppl.* **4** (1964) 139.
[14] J. E. B. Randles, *Trans. Faraday Soc.* **52** (1956) 1573; cf. A. N. Frumkin, *Electrochim. Acta* **2** (1960) 351.
[15] B. E. Conway and D. J. MacKinnon, *J. Phys. Chem.* **74** (1970) 3663.
[15a] R. E. Burton and J. Daly. *Trans. Faraday Soc.* **66** (1970) 1281.
[16] L. Bass, *Proc. Roy. Soc. (London)* **A277** (1964) 129.
[17] L. Bass and T. Pyle, *Nature* **202** (1964) 1003.
[18] S. Makishima, *J. Fac. Eng., Tokyo Imp. Univ.* **21** (1938) 115.
[19] V. A. Pleskov and A. M. Monosson, *Acta Physicochim. URSS* **1** (1935) 871.
[20] F. S. Dainton, *Endeavour* **26** (1967) 115.
[21] R. A. Marcus, *J. Electrochem. Soc.* **113** (1966) 1199.

[22]R. A. Marcus, *Adv. Chem.* **50** (1965) 138; see also *J. Chem. Phys.* **43** (1965) 3477.
[23]R. W. Gurney, *Proc. Roy. Soc. London* **A134** (1932) 137.
[24]K. J. Vetter, in *Trans. Symp. on Electrode Processes, The Electrochemical Society, 1958*, Ed., E. Yeager, John Wiley and Sons, New York, 1960.
[25]G. C. Barker, A. W. Gardner, and D. C. Sammon, *J. Electrochem. Soc.* **113** (1966) 1183; *Trans. Faraday Soc.* **66** (1970) 1498, 1509.
[26]W. D. Jolly, in *Progress in Inorg. Chem.*, Vol. 1, p. 235, Interscience, New York, 1959.
[27]E. J. Hart, *Science* **146** (1964) 19.
[28]B. E. Conway, *Theory and Principles of Electrode Processes*, Ronald Press, New York, 1965.
[29]U. Schindewolf, R. Vogelgesang, and K. W. Boddeker, *Angew. Chem.* **6** (1967) 1076.
[30]G. J. Hills and D. R. Kinnibrugh, *J. Electrochem. Soc.* **113** (1966) 1111.
[31]J. A. V. Butler and G. Armstrong, *Trans. Faraday Soc.* **29** (1933) 1261; *Proc. Roy. Soc. (London)* **A137** 1932; 604; *J. Chem. Soc.* **1934**, 743.
[32]G. C. Barker, *Electrochim. Acta* **13** (1968) 1221; cf. paper presented at C.I.T.C.E. meeting Moscow, 1963; the latter communication does not seem to have been published at the time (see Ref. 33).
[33]P. Delahay and V. S. Srinivasan, *J. Phys. Chem.* **20** (1966) 420.
[34]M. Heyrovsky, *Nature* **206** (1965) 1356.
[35]M. Heyrovsky, *Proc. Roy. Soc. (London)* **A301** (1967) 411.
[36]M. Heyrovsky and R. G. W. Norrish, *Nature* **200** (1963) 880.
[37]L. E. Price, Ph.D. thesis, Cambridge, 1938; quoted by M. Heyrovsky, Ref. 35.
[38]H. Berg, *Rev. Polarography (Kyoto)* **11** (1963) 29.
[39]H. Berg and H. Schweiss, *Mber. Deutsch Akad. Wiss.* **2** (1960) 546.
[40]M. Berg and H. Schweiss, *Electrochim. Acta* **9** (1964) 425.
[41]R. S. Mulliken, *J. Am. Chem. Soc.* **74** (1952) 811.
[42]B. E. Conway and L. G. M. Gordon, *J. Phys. Chem.* **73** (1969) 3523.
[43]B. E. Conway and E. J. Rudd, *Disc. Faraday Soc.* **45** (1968) 87.
[44]B. E. Conway and L. G. M. Gordon, in course of publication (1971); see L. G. M. Gordon, Ph.D. thesis, Univ. of Ottawa (1969).
[45]M. Boudart, *J. Am. Chem. Soc.* **72** (1952) 1531, 3566.
[46]J. Jortner, *Rad. Research Suppl.* **4** (1964) 24.
[47]B. Emmert, *Z. Elektrochem.* **42** (1909) 1507; **45** (1912) 430.
[48]H. H. Schlubach, *Z. Elektrochem.* **53B** (1920) 1689; **54** (1921) 2811.
[49]C. A. Kraus, *J. Am. Chem. Soc.* **35** (1913) 1732.
[50]G. S. Forbes and C. E. Norton, *J. Am. Chem. Soc.* **48** (1926) 2278.
[51]J. E. Dubois, A. Monvernay, and P. L. Lacaze, *Electrochim. Acta* **15** (1970) 315.
[52]B. C. Southworth, R. Ostergoung, K. D. Fleischer, and F. C. Nachod, *Analyt. Chem.* **208** (1961).
[53]J. S. Mayell and A. J. Bard, *J. Am. Chem. Soc.* **85** (1963) 421.
[54]M. Finkelstein, R. C. Petersen, and S. D. Ross, *J. Am. Chem. Soc.* **81** (1959) 2361; **82** (1960) 1582.
[55]M. Finkelstein, R. C. Petersen, and S. D. Ross, *Electrochim. Acta* **10** (1965) 465.
[56]H. N. McCoy and W. C. Moore, *J. Am. Chem. Soc.* **33** (1911) 273.
[57]H. N. McCoy and F. L. West, *J. Phys. Chem.* **16** (1912) 261.
[58]H. W. Sternberg, R. E. Markby, I. Wender, and D. M. Mohilner, *J. Am. Chem. Soc.* **89** (1967) 187.
[59]G. Fraenkel, S. H. Ellis, and D. T. Dix, *J. Am. Chem. Soc.* **87** (1965) 1406.
[60]H. W. Sternberg, R. E. Markby, I. Wender, and D. M. Mohilner, *J. Am. Chem. Soc.* **91** (1969) 4191.

[61]U. Schindewolf, private communication; in course of publication.
[62]H. W. Sternberg, R. E. Markby, I. Wender, and D. M. Mohilner, *J. Electrochem. Soc.* **13** (1966) 1060.
[63]A. P. Krapcho and A. A. Bothner-By, *J. Am. Chem. Soc.* **81** (1959) 3658.
[64]J. E. Dubois and A. Bienvenue, *Tetrahedron Letters* **1966**, 1809.
[65]J. E. Dubois, P. L. Lacaze, and A. M. de Fieguelmont, *Compt. Rend.* **262C** (1966) 181.
[66]J. E. Dubois and G. Dodin, *Tetrahedron Letters* **1969**, No. 28, 2325.
[67]H. W. Sternberg, R. Markby, and I. Wender, *J. Electrochem. Soc.* **110** (1963) 425.
[68]R. A. Benkeser, E. M. Kaiser, and R. F. Lambert, *J. Am. Chem. Soc.* **86** (1964) 5272; R. A. Benkeser and S. J. Mels, *J. Org. Chem.* **34** (1969) 3970.
[69]A. Misono, T. Osa, T. Yamagishi, and T. Kadama, *J. Electrochem. Soc.* **115** (1968) 266.
[70]A. Misono, T. Osa, and T. Yamagishi, *Bull. Chem. Soc. Japan* **41** (1968) 2921; see also T. Osa, *Symp. on Synthetic and Mechanistic Electroorganic Chemistry*, Preprints, p. 157, U.S. Army Res. Office, Durham, N.C., October 1968.
[71]T. Asahara, M. Seno, and K. Kaneko, *Bull. Chem. Soc. Japan* **41** (1968) 2985.
[72]W. M. Latimer, K. S. Pitzer, and C. M. Slansky, *J. Chem. Phys.* **7** (1939) 108; see also A. N. Frumkin, *J. Chem. Phys.* **7** (1939) 552.
[73]J. O'M. Bockris and S. D. Argarde, *J. Chem. Phys.* **49** (1968) 5133.
[74]A. N. Frumkin, *Electrochim. Acta* **2** (1960) 351; *J. Chem. Phys.* **7** (1939) 552.
[75]H. F. Halliwell and S. C. Nyberg, *Trans. Faraday Soc.* **59** (1963) 1126.
[76]D. C. Walker, *Can. J. Chem.* **44** (1966) 2226; E. A. Shaede and D. C. Walker, *Chem. Soc. Special Publ. No. 22*, p. 277 (1967); *Anal. Chem.* **39**, 896 (1967); see also G. Hughes and R. J. Roach, *Chem. Comm.* **1965**, 600.
[77]V. A. Yurkov, in *Soviet Electrochemistry*, *Symposium 1959*, Vol. II, p. 85, Consultants Bureau, New York, 1961.
[78a]Various papers in *Solvated Electrons*, Adv. in Chem. series, Vol. 50, Am. Chem. Soc., Washington, D.C., 1965.
[78b]E. C. Hart, *Science* **146** (1964) 19; M. Anbar, *Quart. Rev. Chem. Soc. London* **22** (1968) 579; U. Schindewolf, *Angew. Chem. Int. Ed.* **7** (1968) 190.
[79]R. C. Krutenat and H. H. Uhlig, *Electrochim. Acta* **11** (1966) 469; cf. N. Kobosow and V. Nekrassov, *Z. Electrochem.* **36** (1930) 529.
[80a]B. E. Conway, in *Chemical Physics of Ionic Solutions*, p. 577, Eds., B. E. Conway and R. G. Barradas, John Wiley and Sons, New York, 1966.
[80b]L. I. Kristalik, *J. Electrochem. Soc.* **113** (1966) 1117; R. Parsons, *J. Electrochem. Soc.* **113** (1966) 1118; B. E. Conway, *J. Electrochems. Soc.* **113** (1966) 1120.
[81]N. F. Barr and A. O. Allen, *J. Phys. Chem.* **64** (1959) 928.
[81a]R. R. Hentz and D. W. Brazier, *J. Chem. Phys.* **54** (1971) 2777.
[82]H. A. Laitinen and C. J. Nyman, *J. Am. Chem. Soc.* **70** (1948) 2241.
[83]J. O'M. Bockris and E. C. Potter, *J. Chem. Phys.* **20**, (1952) 614.
[84]J. P. Hoare and S. Schuldiner, *J. Electrochem. Soc.* **99** (1952) 488; **102** (1955) 178; **104** (1957) 564; see also S. Schuldiner and J. P. Hoare, *J. Phys. Chem.* **61** (1957) 705;**62** (1958) 229; **62** (1958) 504.
[85]K. J. Laidler, *Trans. Faraday Soc.* **62** (1966) 2754.
[86]J. O'M. Bockris, in *Modern Aspects of Electrochemistry*, Vol. I, Chapter IV, Butterworths, London, 1954.
[87]J. A. V. Butler, *Electrocapillarity*, p. 42, Methuen, London, 1940.
[88]A. N. Frumkin and A. Slygin, *Acta Physicochim. URSS* **3** (1935) 719; **4** (1936) 991; **5** (1936) 819.

[89]F. Will and C. A. Knorr, *Z. Elektrochem.* **64** (1960) 258; D. Gilroy and B. E. Conway, *Can. J. Chem.* **46** (1968) 875.
[90]M. W. Breiter, *Ann. N.Y. Acad. Sci.* **101** (1963) 709.
[91]J. Rabani, *Solvated Electrons*, Adv. in Chem. series, Vol. 55, p. 242. Ed., R. F. Gould, Am. Chem. Soc., Washington, D.C., 1965.
[92]P. Delahay and G. Perkins, *J. Phys. Coll. Chem.* **55** (1951) 586, 1146.
[93]J. O'M. Bockris and A. M. Azzam, *Trans. Faraday Soc.* **48** (1952) 145.
[94]A. N. Frumkin, *Z. Phys. Chem.* **160** (1932) 116. Cf. B. E. Conway, *Disc. Faraday Soc.* **39** 47 (1964) and in *Progress in Reaction Kinetics*, Chapter 4, Ed., G. Porter, Pergamon Press, London, 1967.
[95]A. Eucken and B. Weblus, *Z. Elektrochem.* **55** (1951) 114; B. E. Conway and E. Gileadi, *Trans. Faraday Soc.* **58** (1962) 2493.
[96]B. E. Conway and J. O'M. Bockris, *J. Chem. Phys.* **26** (1957) 532.
[97]J. O'M. Bockris and B. E. Conway, *Trans. Faraday Soc.* **48** (1952) 724.
[98]U. Schindewolf (Institüt fur Kernverfahrenstechnik, Karlsruhe), private communication to R. Haynes and B. E. Conway.
[99]W. N. Hansen, *Surface Science* **16** (1969) 205; E. Feinleib, *Chem. Phys. Letters* **16** (1966) 1200.
[100]R. Annino, R. Erickson, J. Michaelovich, and B. MacKay, *J. Am. Chem. Soc.* **88** (1966) 4424.
[101]I. Rosenthal, J. R. Hayes, A. J. Martin, and P. J. Elving, *J. Am. Chem. Soc.* **80** (1958) 3050.
[102]Z. R. Grabowski, B. Czochralska, A. Vincenz-Chodkowska, and M. Balasiewicz, *Disc. Faraday Soc.* **45** (1968) 145.
[103]E. Muller, *Z. Elektrochem.* **33** (1927) 253; T. Sekine, A. Yamura, and K. Sugino, *J. Electrochem. Soc.* **112** (1965) 439; see also B. E. Conway and E. J. Rudd, *Trans. Faraday Soc.* **67** (1971) 440.
[104]E. Hayon and A. O. Allen, *J. Phys. Chem.* **65** (1961) 2181.
[105]B. E. Conway and J. O'M. Bockris, *Electrochim. Acta.* **3** (1961) 340.
[106]P. Zuman, *Elucidation of Organic Electrode Reactions*, Academic Press, New York, 1969.
[107]K. E. Heusler and L. Gaiser, *Ber. Bunsenges. phys. Chem.* **72** (1969) 1059.
[108]Y. Y. Gurevich, A. M. Brodskii, and V. G. Levich, *Elektrokhimia* **3** (1967) 1302.
[109]F. S. Dainton, private communication.
[110]R. A. Marcus, *Ann. Rev. Phys. Chem.* **15** (1964) 155.
[111]J. O'M. Bockris, M. A. V. Devanathan, and W. Mehl. *J. Electroanal. Chem.* **1** (1959) 143.
[112]J. O'M. Bockris and A. M. Azzam, *Trans. Faraday Soc.* **48** (1952) 145.
[113]H. W. Nürnberg, *Fortsch. Chem. Forsch.* **8** (1967) 241.
[114]L. Landau, *Phys. Zeit. Sowjetunion* **3** (1933) 664.
[115]J. Jortner, *Rad. Research Suppl.* **4** (1964) 24.
[116]J. Jortner, *J. Chem. Phys.* **30** (1959) 839; see also J. Jortner and S. A. Rice, in *Solvated Electrons*, Chapter 2, Adv. in Chem. series, Vol. 55, Ed., R. F. Gould, Am. Chem. Soc., Washington, D.C., 1965.
[117]A. Bewick and B. E. Conway, in course of publication.
[118]J. D. McIntyre (private communication).

3

Critical Observations on Measurement of Adsorption at Electrodes

H. H. Bauer

University of Kentucky
Lexington, Kentucky

and

P. J. Herman and P. J. Elving

University of Michigan
Ann Arbor, Michigan

I. INTRODUCTION

1. Nature and Scope of the Discussion

A great deal has been written that is relevant to the title of the present chapter; however, the subject has not by any means been exhaustively treated. The present chapter focuses on the operational aspects—How can adsorption at electrodes be studied and what information can be obtained?—and attempts to discuss these questions critically rather than to review comprehensively all of the detailed relevant work that has been done. Where topics have been well and recently reviewed, duplication has been avoided; as a result, much of the detailed discussion deals with selected matters that have not been satisfactorily treated, e.g., application of the Gibbs adsorption equation and use of capacity measurements at solid electrodes; in the remaining discussion, it is the aim to present information which clarifies the central issues concerning choice of methods that are available, how reliable they are, and how they are alike or differ. Adsorption at semiconductor electrodes and from fused salt media will not be explicitly considered.

Adsorption at electrodes differs from adsorption in a more general sense in that the electrical potential of the adsorbent is of explicit interest and can be independently varied; it is therefore implicit in the present discussion that adsorption can be measured as a function of the electrode potential, which is a direct experimental variable. This is not the case, however, with respect to the charge on the electrode.

On the basis of the theoretical significance of zero-charge potentials,[1–3] some authors (see Ref. 4 for complete references) have specifically suggested use of the so-called rational potential scale. Others have more recently stressed[5–10] the use of "charge" as the independent variable, especially for the analysis of data.[11] The fact that isotherms expressed with respect to charge instead of potential (or vice versa in the case of other authors) are claimed to be more symmetric, is in itself not very convincing because of the generally approximate character of those isotherms (see Refs. 12 and 13 for a complete critical discussion of this problem). It should be emphasized that, whatever is the scale by means of which an author analyzes his data, it is always desirable to present the basic results obtained in terms of the "operational potential scale," quoting the specific reference electrode used, so that the reliability of the data and comparison with other results can be examined.

Electrodes also differ from other adsorbents because electron-transfer reactions between phases may occur. It is convenient to distinguish the adsorption of species that are not involved in electron-transfer reactions *at the potential under consideration*, and adsorption of those that simultaneously take part in both electrochemical and adsorption processes; as just implied, it is important to realize that a system of one particular chemical composition may be relevant to one or other of these two classes, depending on the potential of the electrode (see Refs. 14–20 and 292 for recent reviews of the latter class or group).

The first group of phenomena, with which this survey is primarily concerned, includes the adsorption of species that are not electroactive under any known circumstances (and that are possibly adsorbed under conditions where another component of the system is undergoing electrochemical transformation) as well as the adsorption of "electroactive" species at those potentials where the rate of their electrochemical transformation is negligibly small. In practice, it must be recognized that the adsorption of electroactive species can be investigated by the methods to be discussed, mainly at potentials where the electron transfer rate is negligible; otherwise, the phenomena are more akin to other redox processes than to adsorption in the absence of electron transfer, and methods of study are those used to investigate Faradaic processes in general (some points of interest in this connection are discussed in Section I.5 on chemisorbed films).

It is important to recall that the typical observations on adsorption at electrodes have usually been made in systems containing an electrolyte and an uncharged surfactant. The latter is generally adsorbed at potentials where the electrode is positively charged as well as at potentials of negative charge; at greater (positive or negative) polarizations than the range bounded by the adsorption potentials, an ionic double layer subsists and the surfactant is not adsorbed. In the potential region where adsorption prevails, resulting phenomena (decreased surface tension, decreased electrical capacity, and, consequently, decreased cell admittance) can often be validly discussed in thermodynamic terms on the basis that the adsorption process is at equilibrium (see below).

It is difficult, however, to predict theoretically the exact moment when this equilibrium is reached, on starting with a fresh electrode surface in the presence of an organic surfactant,[21–23] because the kinetics for adsorption involve both a diffusion step and an energy barrier (to pass "through") near the interface. Although diffusion is usually rate-controlling and although most particles can be considered to adsorb when reaching the surface (in the case of an adsorbable species), the difficulty is in predicting the time necessary to "rehomogenize" the subsurface just after a monolayer is formed (see Figure 1).

In connection with the adsorption process depicted in Figure 1, it has been experimentally shown (see Ref. 22 for complete references) that the general kinetic patterns for adsorption at air–solution and electrode–solution interfaces are in many respects very similar because of the similar free energies of the two interfaces.[24] Step III in Figure 1 corresponds to the so-called rehomogenization step. The indicated thickness of the subsurface is probably not very realistic in the case of an electrode, because of the expected thickness of a variable double layer.

Steps other than diffusion can be rate-determining.[11,15,26–28] Another particularly intricate situation concerning equilibrium is encountered when a relaxation technique is used in the neighborhood

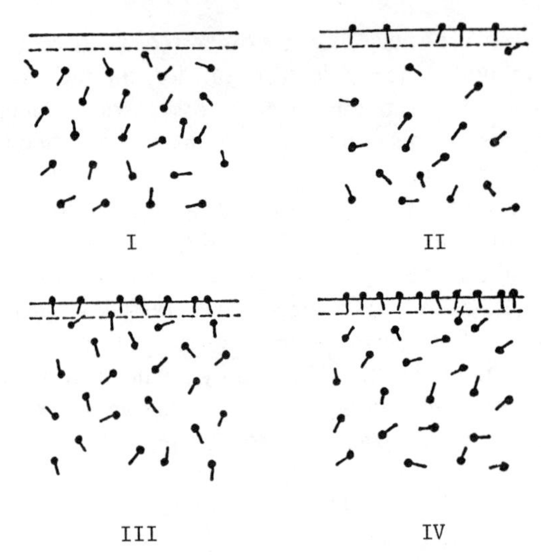

Figure 1. Adsorption from solution. (I) The interface is just created, no molecule is adsorbed, and the concentration is uniform through the whole solution. (II) The adsorption process has just begun; some molecules are adsorbed. There is an important concentration gradient; the role of diffusion is very important. (III) Now the equilibrium is established between the bulk of the solution and the subsurface but the flow of molecules which cross the adsorption energy barrier is still not zero. (IV) The equilibrium is reached. (The dashed line represents the frontier between interface and subsurface.) [Taken with permission from J. F. Baret, *J. Phys. Chem.* **72** (1968) 2755.]

of adsorption–desorption potentials (see Ref. 28 for dispersion effects encountered when using a relaxation technique to study adsorption). In this region, the extent of adsorption changes so rapidly with potential that the superposition of a small-amplitude (millivolts) alternating voltage, as in measurements of admittance or capacity, produces a large differential capacity or admittance, the magnitude of which is partly determined by the relative magnitudes of the signal frequency and of the rate and potential dependence of the adsorption process. It is convenient to use a term other than "adsorption process" to describe this phenomenon and the expression "tensammetric process" is proposed: this corresponds to the usage[29] of "tensammetric waves" for the admittance peaks produced by this type of process, and can thus serve to distinguish unambiguously the quite distinct situations encountered at the adsorption–desorption potentials (the *tensammetric* [peak] *potentials*) and at the potentials of prevailing adsorption, respectively.

So far, certain characteristics of adsorption processes have been discussed without critical examination. Closer examination reveals that it is not always precisely clear what steps are included within such processes; this question is treated in the following section.

2. Definition of "Adsorption"

The general concepts intuitively associated with "adsorption" are straightforward: a heterogeneous process with the relevant species being in a more favored state on the surface than away from it is envisaged. For example, in the case of a surfactant, it is clear that, as stated by Lange,[30] "In a surfactant solution at practical concentrations of use, the ratio of water molecules to surfactant molecules is about 20,000 to 1. At such extreme dilutions the surfactant can hardly exhibit a marked effect unless it accumulates—in other words, is adsorbed—at sites where it is to be effective, i.e., the interfaces." In this connection, we are implicitly inclined to associate the probability of adsorption with the concept of solubility, which has been semiempirically expressed by Traube's laws.[31] In this sense, valuable information has been obtained, for example, by comparing the behavior of homologous series of compounds at mercury electrodes (see Figures 2 and 3).[32,33,153]

Incidentally, although solubility must not be confused with adsorption when one compares the adsorbability (affinity for the electrode) of two organic compounds, it is generally forgotten that the comparison is only meaningful when the two compounds are in the same state (in terms of activity) relative to their own solubility in the solutions involved (see Figure 2). The same basic phenomena explain why adsorption at electrodes, which is so often primarily controlled by solubility rather than by adsorbability, is less selective,[34] in the sense that one cannot hope to develop, on a wide scale, adsorption chromatography with electrodes; in any event, pure adsorption chromatography in liquids does not exist[35] and partition chromatography (based on solubility) is well known to be more selective. Consequently, any factor affecting solubility, e.g., salting out, affects adsorption.[36–38] In any event, the idea of adsorption is still associated with the concept, at equilibrium, of an excess of the adsorbate "on" the surface; before equilibrium is reached, a situation is imagined where the rate of adsorption exceeds that of desorption (kinetic picture of adsorption).

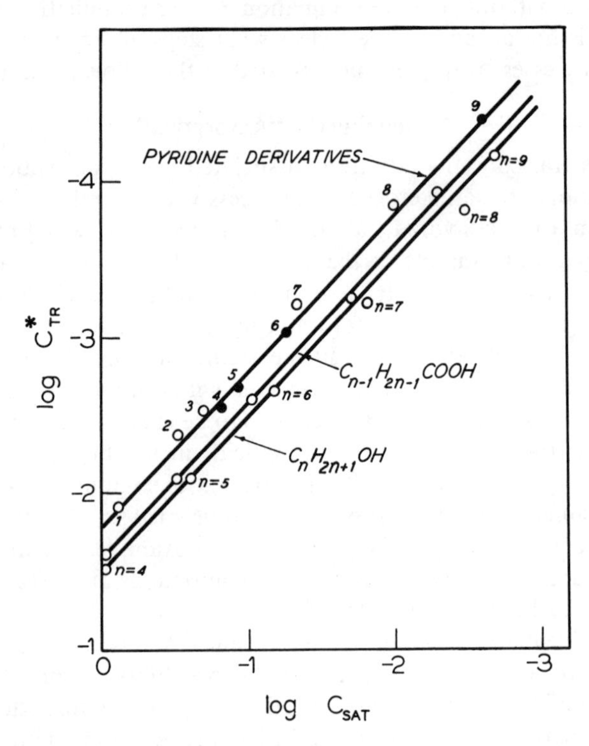

Figure 2. Relation between solubility (log C_{sat}) and the adsorbability (expressed as the value of log C^*_{TR}, where C^*_{TR} is the transition concentration, i.e., that at the first turning point in the Gibbs isotherm at -0.6 V versus s.c.e.) for pyridine homologs in 1 M aqueous NaOH solution at the dropping mercury electrode (the relation can be simply expressed as a particular case of the Gibbs isotherm). 1, pyridine; 2, α-methylpyridine; 3, β-methylpyridine; 4, γ-methylpyridine; 5, γ-chloropyridine; 6, γ-ethylpyridine; 7, α-ethylpyridine; 8, quinoline; 9, γ-benzylpyridine. Comparative data are presented for the water–air interface for normal aliphatic alcohols and acids. [Taken with permission from P. Corbusier, Thesis, Free University of Brussels (1957).]

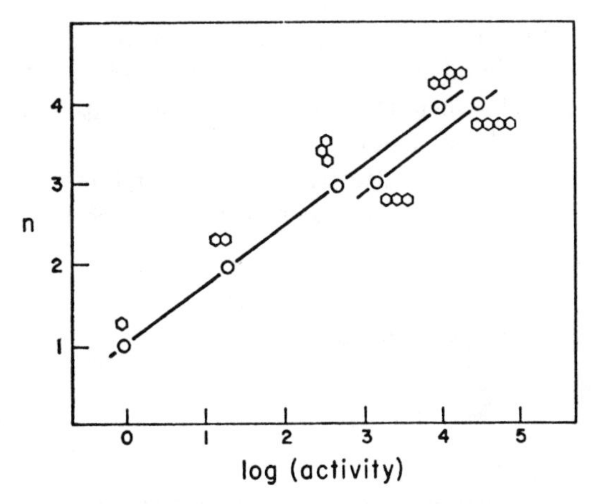

Figure 3. Ring size of aromatic hydrocarbons and their surface activity at mercury. *n* is the number of rings; log (activity) is an arbitrary measure of the surface activity corresponding to the concentration required to give the same lowering of surface tension by each compound. [Taken with permission from B. E. Conway, *Chem. Can.* **12** (1960) 40, based on quoted Russian work.]

Actual situations present complications that do not always allow straightforward application of this idea, as exemplified by the following:

(i) Usually, more than one adsorbable species is present, with consequent competition for the surface between adsorbate and solvent, between adsorbate and ions of supporting electrolytes, between two adsorbates (one of which, for example, may be a contaminant[39,40]), etc. Competition with the solvent and, frequently, with electrolytes is usually ignored; this can lead to confusion and errors.[41]

The importance of electrochemical adsorption as a process involving replacement of solvent by the adsorbed molecule was first recognized by Butler,[25] who took into account the electrostatic energy difference for the adsorbate and an equal volume of solvent. The concept of adsorption as a replacement reaction, in distinction to the situation for the adsorption of gases on solids, has been re-emphasized in more recent work.[26,26a,b,42,43] Although such a model gives us a better idea of one mechanism of adsorption [see the simplified scheme of Figure 4 regarding some of the kinetic and energetic aspects involved (mean adsorption energy[44,45])], it is of relatively little help when, for example, the competitive adsorption of ions and neutral molecules is being considered.

In the case when two adsorbates are present, interesting results have been reported[46,47] on the basis of the observed tensammetric

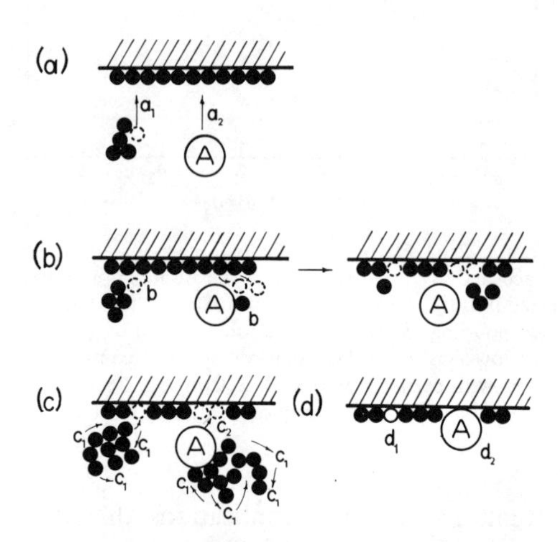

Figure 4. Simplified mechanism of adsorption at electrodes adapted from the views of Bockris and Müller (K. Müller, in *Electrosorption*, p. 135, Ed., E. Gileadi, Plenum Press, New York, 1967). (a) Formation of initial system of the reaction; (a_1) diffusion of a hole from bulk solution to a position adjacent to the adsorbed layer of solvent; (a_2) diffusion of a molecule of A to a position adjacent to the adsorbed layer; (a_1) and (a_2) occur simultaneously in general. (b) Jump of a solvent molecule from an adsorbed position into an empty position of the second layer (this can happen for several molecules). (c) Reorganization of the second layer; (c_1) reorganization of solvent molecules in the second layer; (c_2) approach of the A molecule to the position of the free site(s) in the first layer; (c_1) and (c_2) may occur simultaneously. (d) Reoccupation of the position in the first layer; (d_1) with a reoriented solvent molecule (relaxation); (d_2) with a molecule of A (adsorption of A). Rate-determining step: (a_2) or (b).

peaks of mixtures of two surfactants; the adsorption competition is often complicated by the formation of complexes between the substances, and by the adsorption of such complexes. The co-adsorption of aromatic and aliphatic species has been recently investigated.[48] Various treatments have been given for separation of the competitive adsorption of the charged and uncharged forms of a species simultaneously present in a solution at relative concentrations dependent on the *p*H.[49]

(ii) The terms "on (at, in) the surface (interface, interphase*)" are often ill-defined. Attempts to produce rigorous definitions lead ultimately to a tautology: "close enough to the interface to produce the effects by which adsorption is defined or measured." The essential point is that by "adsorption," in the present context, is meant only one class of phenomena encountered at surfaces; in other contexts, a larger or different class of phenomena, also called "adsorption," is of interest. It is, of course, self-evident in all connections that one must be certain as to precisely what one seeks to measure before attempting to do so, but the present case is a sufficiently confused one to make the point worthy of emphasis.

In the derivation of the Gibbs adsorption equation, "surface excess" is defined in the Gibbs model[50] as the amount of substance present in addition to that which would be there if each phase were completely uniform in composition up to a geometrical, two-dimensional surface of zero thickness. Any "modification of the chemical composition at the boundary between two bulk phases"[51] may be associated with adsorption without any preconceived notion of the exact distribution of matter in the region of heterogeneity. Thus, a "surface excess" of ions may exist at an electrode even though many of them in this "excess" may be tens, even hundreds, of angstrom units from the surface; for example, all parts of the diffuse, as well as of the compact, double layer contribute to the surface excess,[52] a concept which is at the origin of our current picture of the double layer.[53,54] On the other hand, it is not common to regard a species as being adsorbed unless it is within bonding distance of the surface, owing to the facts that intermolecular forces vary inversely with a high power of the distance[55]; in these terms, adsorption from the

*The term interphase may be used for the real three-dimensional region comprising the thickness of the adsorbed layer. This term, however, is less suitable when the diffuse layer is included.

gas phase is generally described as resulting from a direct interaction with the surface.[56–59] This leads to situations where, for example,[60] Gibbs surface excesses are discussed while it is simultaneously postulated that "adsorbed" ions lie closer to the surface than do "unadsorbed" ions; the making of such a connection represents an assumption that is not generally identified as such.

(iii) The electrical characteristics of the electrode are invoked to make a distinction between ions held at the interface by virtue of purely electrostatic forces ("nonspecific adsorption") and species held by other forces[61] as well ("specific adsorption").[62] Though useful, this distinction is too often arbitrary because of the nature of the criteria[62] developed to assign "specificity" in adsorption. Borderline cases may be expected to arise since the differentiation made is somewhat reminiscent of that between ionic and covalent chemical bonding, with the added feature that ion–dipole interactions are frequently present. A recently developed method for measuring specific adsorption in solutions of mixed electrolytes[63,64] avoids some of the numerous hypotheses included in the Grahame-type approach and seems very promising, especially for evaluating the models of discreteness-of-charge effects in the "compact layer."[52,61]

It is thus evident that in order to construct a structural model, consideration of adsorption at electrodes must inevitably result in questions considerably beyond just whether there is a surface excess of the species investigated, e.g., in questions about nature of the bonds involved, distances of the order of molecular dimensions, and orientation of dipoles. The problem of adsorption covers, in fact, the problem of knowledge of the whole interfacial region—the so-called "interphase." However, by examining the more or less "pathological case," we can hope to progress faster. It is evident that different methods of study may be called for, depending upon what feature of the heterogeneous behavior of the species of specific interest.

3. Inferences about Adsorption from Observable Phenomena

The discussion in the previous section demonstrates that measurement of adsorption is not necessarily a straightforward problem. Quite generally, to measure the extent of a phenomenon, we must know, quantitatively, how it is related to the actual experimentally observed effects. At present, our understanding of surface chemistry and heterogeneous processes is such that quantitative relations of the type required, of known rigorous validity, are few and are applicable only under restricted circumstances.

One can devise ways of showing with certainty that a particular species has, at an electrode surface, an effect which is not shown by substances defined as not being adsorbed. However, it is not usually easy to devise ways of doing this quantitatively and of measuring the nature and extent of the adsorptive behavior by an *independent* method, which is often the specific aim; for example, in studying the polarography of an adsorbed species, it is usually desirable to measure the extent of adsorption by a nonpolarographic method. Perhaps the greatest difficulty is the fact that the kinetics of the adsorption processes commonly influence the observations made in electrochemical kinetic studies, whereas direct methods of measuring adsorption are usually most applicable in situations where equilibrium subsists.

It has been insufficiently emphasized that, under typical conditions (one electrolyte plus one surfactant), the study of the degree of inhibition of a Faradaic reaction, e.g., of an "indicator test ion" such as chromate,[65] by adsorption of the surfactant (a typical manifestation of adsorption) is in itself a particularly promising means for testing or even improving our actual knowledge of adsorption phenomena on electrodes, especially regarding the compactness of the film (see, e.g., Refs. 22 and 66–69). However, this approach is rather indirect (*cf.* the subsequent discussion of effects of adsorption on Faradaic processes).

In the present state of knowledge, adsorption at electrodes and its relation to experimental observations represent an area where understanding is lacking to a considerable degree, and investigations are likely to result in conclusions doubtful to a greater extent than is comfortable. The truth of this generalization may be indicated by two examples: Measurements of adsorption by removing an electrode from the solution prior to the adsorption measurement (see below) have been widely made even though it is quite unlikely[75] that the results can be quantitatively valid, since equilibrium is no longer maintained. Much work has been done on the capacity of solid electrodes, including use of such measurements as an indication of the true surface area, when it is evident that what has been measured is not purely a double-layer capacity (see below). In these and other cases, work is proceeding *faute de mieux* by the best means available and is therefore not to be lightly criticized; however, the inherent uncertainties need to be occasionally emphasized.

4. Adsorption of Inorganic Species

The present survey is concerned with available methods for measuring the adsorption of substances other than those commonly employed as supporting or background electrolytes in electrochemistry. This division broadly corresponds to one between organic and inorganic substances and provides a convenient, if loose, terminology. The restriction to organic species, in the scope of this review, embodies a distinction that is commonly made and that should be borne in mind in view of the implications involved. For example, although the methods discussed are also used in the study of double layers composed exclusively of inorganic ions, there are at times important differences in detail and in the theory[60] on which the measurements are based, e.g., in the use of the Gibbs adsorption equation for charged components and in nonthermodynamic assumptions introduced in the calculation of the amounts of adsorbed ions from measurements of double-layer capacities. Furthermore, a considerable and well-reviewed[62,66] body of work is devoted specifically to the nature of the electrical double layer and ways of determining its properties.

However, a distinction between charged and uncharged species is not intended. Organic ions, e.g., anions of carboxylic acids and protonated amines, generally behave more like "organic" species than like ions of the halogens or the alkali metals.

Although in a solution of an organic surfactant, interest is focused on the adsorption on the electrode of the tensioactive compound, it should be noted that the effect of the adjacent ionic layer can also be important, e.g., the abnormal ionic excess which is found there.[10]

5. Chemisorbed Films

Another field that is difficult to define precisely and will not be discussed in the present review involves electrodes with chemisorbed films.* For example, solid–metal electrodes often are covered by a

*This topic has been treated in another volume in this series (see Gileadi and Conway[71]).

layer that may be a metallic oxide, a layer of chemisorbed atoms of oxygen or hydrogen, or something similar[70]; in studies of the evolution of hydrogen or of oxygen, layers of adsorbed atoms are frequently postulated to be present over certain ranges of potential and can be detected experimentally at certain metals. The study of these processes currently uses methods, concepts, and theoretical formulations rather different from those to be discussed here. A full and recent discussion of the fundamental ideas involved with references to the significant literature has been provided by Conway[17] and Delahay[66] and by Gileadi and Conway.[71]

From a practical point of view, it is essential to keep in mind that the presence of such surface films on solid electrodes may often arise from a chemical rather than an electrochemical effect. The "history" of the electrode—its exposure to oxygen or other reactive substances, including those used to form bonds with the electrode support, and the duration of exposure of the electrode to the solution and under what conditions of polarization—may often appreciably determine the electrode performance. The variables are so numerous that one should attempt to characterize the electrode in all important chemical and physical terms before making an electrochemical characterization or measurement.

As may be expected, the adsorptive properties of a solid electrode are very different when the surface is oxidized and when it is reduced; concomitant changes in the polar nature of the surface may even reverse the selectivity of adsorption of polar and nonpolar molecules from solutions containing both types of species.[72,73]

6. Electrochemical Transformations Involving Adsorbed Reactants

Anomalous* results in studies of electrochemical reactions have often been attributed to adsorption of the reacting species. It is in such work that independent and *in situ* methods of measuring adsorption are most urgently needed, especially since, in most cases, adsorption has been inferred on the basis of theories derived for model sequences of reactions, which constitutes a rather fallible procedure. Many electrochemical reactions involve "chemical" steps as well as electron transfer and physical mass transport; often,

*In many reactions at electrodes, particularly of organic substances, adsorption of the reacting species is the *normal* situation.

the predicted behavior of a system with adsorbed reactants is not, in qualitative ways, strikingly different from that of a system where the reactants undergo chemical side-reactions.

The methods to be reviewed can be used to examine the adsorptive properties of electroactive species at potentials not too close to that where reduction or oxidation occurs. Most species that are adsorbed significantly are adsorbed over considerable ranges of potential and it is usual and reasonable to suppose that the parameters characteristic of the adsorptive behavior change in a continuous way with potential except, perhaps, at the tensammetric potentials. Therefore, it seems attractive to measure adsorption away from any redox potentials, attempt extrapolation of the behavior to these potentials, and compare the results with inferences drawn from the study of anomalies in the electrochemical properties at the redox potentials. Apparently, little work has been reported in which such a procedure was applied as an independent and quantitative check of the parameters of the adsorption processes (see the review by Mairanovskii,[15] in which he refers to these phenomena as being a form of "autoinhibition").

On the other hand, it must be kept in mind that in the presence of simultaneous Faradaic and adsorption processes, it is currently still more convenient to study ionic adsorption than that of organic species because of the inherent lack of control with respect to organic species in solution. To quote Barker and Bolzan[19]: "The adsorbed species is then often an ion pair or an undissociated molecule and it is possible to vary the adsorption coefficient over several powers of ten by altering the concentration of a complexing anion in the solution; such systems have also the important advantage that the charge transfer step is usually a rapid process and that only one form of the depolarizer (often the oxidant) is adsorbed to an appreciable extent."

Electrochemisorption of *intermediates* in various complex reaction sequences has received detailed study[71] with respect to isotherms and the kinetics of formation and desorption of adsorbed radicals. Two types of situation arise: one in which dissociatively chemisorbed radicals are produced which suffer electrochemical desorption in Faradaic surface reactions, e.g., in the case of H, or organic radicals at Pt derived from organic substances such as HCOOH or CH_3OH; the second, where the adsorbed species are electrochemically adsorbed or desorbed in a Faradaic surface reaction e.g., $e + H_2O + Pt \rightarrow PtH + OH^-$; $H_2O + Pt \rightarrow$ "PtOH" $+ H^+ + e$; and the corresponding reverse reactions. Reviews have appeared elsewhere.[71]

II. DIRECT METHODS OF MEASURING ADSORPTION

1. Depletion of the Solution

If the introduction of a surface produces a decrease in the concentration of a species in the bulk of the solution, adsorption at that surface has clearly occurred and can be readily measured quantitatively (apart from possible difficulties in defining the "true" surface area of the adsorbing surface). This method is commonly used where the adsorbent is a powder, with the change in concentration of the solution being measured by conventional analytical methods.

Where electrodes are concerned, the method is much less satisfactory because of the relatively small area of most electrodes* in comparison with a powder. Therefore, dilute solutions, large electrodes, and sensitive analytical techniques must be used, involving, e.g., radioactively tagged adsorbate molecules[74] or direct spectrophotometric determination of solution concentration.[75] This approach may, perhaps, be useful in specialized cases, but does not have the general applicability that would permit measurements to be made under widely varying experimental conditions, particularly as far as adsorbate concentration is concerned. For example, it is well known that for powders in the presence of an organic surfactant at high concentration, we have access only to differential isotherms; more quantitative data require the setting up of separate Gibbs isotherms for two gaseous films on the solid, a procedure which involves several hypotheses.[76] In the same way, for the case of electrodes, we have often access only to "relative isotherms."[73]

2. Accumulation on the Electrode: Measurements Made after Removal from Solution

If the electrode, together with the adsorbed substance, can be removed from the solution without change in the amount adsorbed, then the latter can be measured in a number of ways. The most convenient

*Porous, high-area fuel-cell electrodes or sintered powder electrodes used in batteries are important exceptions.

approach involves a radioactively tagged adsorbate. This technique, which was used as early as 1915 by von Hevesy[77] for studying the exchange between a metal and its ions, continues to be used (e.g., Hackerman and Stephens[78] and Balashova and Kazarinov[73]). However, it suffers from obvious disadvantages:[75] adsorption equilibrium can hardly remain unchanged during and after separation of electrode and solution, and adhering solution has to be removed from the electrode (commonly carried out by rinsing and/or blotting); the effect of these undesirable experimental procedures cannot be assessed with certainty unless control experiments on adsorption can be carried out *in situ*. This method, therefore, seems better used in a subsidiary and complementary role than as the main investigating tool; e.g., Schwabe has suggested the possibility of using autoradiography to determine the *distribution* of adsorbate over the surface and has published autoradiographs obtained in this way.[79,80]

Green *et al.*[289] describe an involved procedure, in which the electrode is an endless tape moving through the solution and the radiation from tagged adsorbed molecules is measured after the tape has left the solution; there is provision for determining the amount of solution adhering to the tape by a capacity measurement. The statement is made and supported by theoretical discussion that errors due to disturbance of adsorption equilibrium as the tape leaves the solution are negligible; however, the argument does not carry conviction in the context of the admitted experimental uncertainties and of the loss of control of the electrode potential as the tape leaves the solution.

3. Accumulation on the Electrode: Measurements *in situ*

This approach is obviously the most desirable one and requires only successful implementation. Joliot[81] showed that the electrolytic deposition of polonium could be followed by using electrodes that were, at the same time, windows of ionization counters. This principle has been applied to the study of accumulation resulting from adsorption rather than from deposition.

Cook[82] appears to have first described a method particularly suited to radioactively tagged substances that emit "soft" radiation; this method has been extensively employed by Schwabe[79,80,83] and by Bockris and co-workers.[84–87]

Schwabe[79,80,83] used a cell mounted directly above a scintillation counter; the part of the cell above the window of the counter was a thin (2 mg cm^{-2}) mica foil, on which the metal electrode (50 μg cm^{-2}) had been deposited in vacuum. The intensity of the radiation received by the counter was only slightly decreased by absorption in the mica, e.g., even the weak ^{14}C β-radiation was decreased in intensity by only one-third. With weak (100–400 keV) β-ray emitters present at solution concentrations of 2×10^{-3} M, the measured activity was due almost entirely to species adsorbed at the electrode[79]; correction for the small activity due to the supernatant solution can readily be made.[83] Kinetics, as well as adsorption isotherms, and the mutual replacement of ions were studied.

With substances that emit harder radiation, the measured activity includes an undesirably large contribution from unadsorbed molecules because the solution does not sufficiently absorb such radiation. In such situations, the method of Kafalas and Gatos[88] can be used, in which only a thin film of solution, directly in front of the electrode, can irradiate the counter; the remainder of the solution, which is in reservoirs and channels leading to the electrode-containing compartment, is shielded from the counter. A detailed description and discussion has been given by Schwabe and Schwenke.[89] Recently, a cell has been described[73] in which a large electrode is lowered into a well opposite a counter for measurement.

4. Optical Observation of the Surface Layer *in situ*

Changes occurring at or near the electrode surface during electrochemical reaction have recently been studied by newly introduced optical methods. In ellipsometry,[70,90–92,299] the change in polarization as a result of reflection of a beam of polarized light from the electrode surface provides data about the phase retardation and amplitude decrease of the components of the radiation parallel and perpendicular to the plane of incidence, i.e., about the refractive and absorptive properties of the layer at the surface. Interpretation of such data as a function of electrode potential can provide highly interesting information, e.g., at platinum,[70] the oxide film causes changes in the ellipsometric parameter Δ which closely follow the anodic charge required for film formation, despite initial indications to the contrary.[91,91a] This behavior is also observed in modulated reflectance experiments.[91a] The earlier indications[91] of a critical potential and coverage required for formation of an optically detectable film may have arisen from insufficiently sensitive instrumentation available at the time the initial work was carried out. At 1.1 V E_H, however, a definite change in the optical parameters of the film on Pt occurs[91a] and corresponds to formation of a film from which one electron per site has been passed in the anodic oxidation of the surface. In a study of the passivation of nickel, a film about 60 Å thick formed at lower potentials than those where passivation occurred; however, at the potential marking the onset of passivation, the absorption coefficient of the film abruptly changed.[92]

In attenuated multiple-internal-reflection (or similarly named) spectroscopy, a beam of light reflected by the walls of an optical "waveguide" shows intensity losses at wavelengths corresponding to the absorption spectrum of species outside the waveguide. By using waveguides that are at the same time also electrodes, one may be able to study the nature of the intermediates, the bonding of adsorbed species, and similar phenomena. Preliminary results that establish the feasibility of the technique with infrared[93] and visible[94,95] radiation have been published.

III. MEASUREMENT OF SURFACE TENSION, CHARGE, AND CAPACITY

1. Thermodynamic Theory

The surface tension σ, the charge on the electrode q, and the differential capacity C at an electrode surface, where equilibrium subsists and no charges cross the interface (absence of a Faradaic process), e.g., the well-known ideally polarizable electrode which, *inter alia*, has been reviewed by Mohilner[97] (see also recent comments by Frumkin[96] on this classical concept), are related to one another and to the surface excess Γ_i of a species i by the familiar equations

$$(\partial\sigma/\partial\mu_i)_{E,\ldots} = -\Gamma_i \tag{1}$$

$$(\partial\sigma/\partial E)_{\mu\ldots} = -q \tag{2}$$

$$C = \partial q/\partial E \tag{3}$$

where μ represents chemical potential.

Thus, surface excesses are in principle obtainable for charged species as well as for neutral ones from surface-tension, charge, or capacity measurements (Grahame[60] has pointed out that the greatest accuracy can be achieved by measuring C and integrating back to q and σ), but it is not often sufficiently emphasized that equations (1)–(3) are applicable only under equilibrium conditions and equilibrium does not necessarily subsist under conditions frequently used in electrochemical studies, mainly because of a significant Faradaic process.

Consequently, although the methods to be discussed in this section are able to provide an unequivocal answer to the question as to whether a particular species is adsorbed or not, quantitative characterization is not necessarily possible. For example, it is one thing to infer from an observed change in double-layer capacity that something has happened at the electrode surface; it is quite another matter to determine an adsorption isotherm from double-layer capacity changes. However, when this method is systematically applied to gather a large body of data, valuable information can be obtained (cf. the work of the Russian school, e.g., Refs 32 and 99, and that of Parsons).

In spite of some recent comments[100] to the contrary, it cannot be denied that the Grahame approach for thinking about ionic excess is one of the best established, primarily because of the simplicity of his assumptions, which are supported by experiments of high reliability. Finally, the double layer[53,61,62] is always discussed in terms of the simple but useful concept of inner and outer Helmholtz planes; we use the same concept even in the formulation of sophisticated models. In the case of ions, a new approach has been mentioned (see Section I.2).

Before discussing the different methods of measurement, it is essential to review the derivation and application of the Gibbs adsorption equation, because the latter is quoted in different ways by various workers, and the significance of the surface excess is not always made clear.

Gibbs[50] gave in 1875–78 a detailed derivation along the following lines: The presence of a surface—a plane dividing two phases—is connected with inhomogeneity of the two adjoining phases in the vicinity of the surface; this inhomogeneity bespeaks a greater energy content than if the surface were "absent" and the two phases uniform in composition (the Gibbs model). The total excess in energy ε^s, where s refers to a surface phase, is equated with the presence of excess amounts m_i^s of relevant species and of excess entropy η^s together with the product of surface tension and area (s) of the surface [Gibbs' equation (502) on p. 229 of Ref. 50]:

$$\varepsilon^s = T\eta^s + \sigma s + \mu_1 m_1^s + \mu_2 m_2^s + \cdots \tag{4}$$

After rearrangement (using a Gibbs–Duhem relation), differentiation, and division by s [Ref. 50, equation (508)]:

$$d\sigma = -\eta_s\, dT - \Gamma_1\, d\mu_1 - \Gamma_2\, d\mu_2 \cdots \tag{5}$$

where η_s is defined as the *superficial density of entropy*, η^s/s, and Γ_i is the *superficial density of species i*, m_i^s/s; thus, Gibbs defines "superficial density" as "excess per unit area." As far as the experimental determination of Γ_i is concerned, Gibbs' conclusion is not at first clear: at one point (Ref. 50, p. 230), he states that $T, \mu_1, \mu_2, \ldots$ in equation (5) are independently variable and that, by differentiation, one obtained equations "giving the values of η_s, Γ_1, Γ_2, etc. in terms of the variables just mentioned." In other words, expressions of the type of equation (1) hold, and the superficial density for each component is obtainable by varying its chemical potential and measuring changes in surface tension.

It must be kept in mind that Gibbs derived the equation given here as (5) for the case of two "fluid masses" and not for a fluid and a solid, and certainly not for an electrode [Gouy[101] (1917) ignored this equation and application of the Gibbs equation to an electrode was first made by Frumkin[102] (1923); it is surprising to realize the time chemists required to recognize its importance in electrochemistry (see Refs. 103–105). Actually, the derivation of equation (5) for an electrode is, in principle, not much more difficult]. Gibbs, in considering the application of his equation to two fluids, concluded that such measurements were not practically feasible (Ref. 50, pp. 232 ff) because σ is determined by measuring the curvature of the surface as a function of the difference in pressure across it and it is not possible to make measurements at sufficiently large differences of pressure to permit changes of σ with changes of μ_i to be determined accurately enough to provide the required rate of change ($\partial\sigma/\partial\mu_i$). In other words, this means, in most cases, that the change of curvature with μ_i is negligible (quasiplanar case) and that, since the variance of the system is given by c for a system of $c + 1$ constituents (details in Ref. 106), one degree of freedom is lost which is necessarily compensated for, at constant temperature and pressure, by the loss of the independent variable character of one of the μ_i's.

Finally, Gibbs considered plane surfaces and concluded (Ref. 50, pp. 234ff) that, by placing the surface of reference so that one of the Γ_i's vanishes, a "convenient form of the fundamental equation" is obtained, from which Γ_i's may be obtained relative to this surface of reference, but that in this case, the "surface of tension"* is no longer the same as the "dividing surface." It is also concluded that, when the actual surface of tension is used as the dividing surface, it is only possible to obtain surface excesses Γ_i relative to the surface excess of one of the other species present, unless there is a species whose density in the two homogeneous phases is the same, when the superficial density of a component can be obtained by application of equations such as equation (1) above.

Grahame[60] used equation (5) as his starting point in discussing electrocapillarity, but states that "the actual position of this plane [= dividing surface] is arbitrary except for the dictates of convenience"; he then chooses the plane of reference as being defined by $\Gamma_{\text{solvent}} = 0$, and finally justifies this choice by showing that, with any other surface as reference, "it is only a stoichiometric calculation to find the amount by which any Γ changes for a shift in the position of the surface of reference." This, of course, does not tell us very much about the exact position of this plane. Under those conditions, it is clear that $\Gamma = 0$ does not necessarily mean that adsorption with respect to the actual surface of tension (here, the electrode surface) is absent; the discussion becomes less clear when it is stated that Γ is found to be negative "in very concentrated solutions of salts whose ions are believed to be unadsorbed . . . [due to] . . . a small constant adsorption of solvent which results in the displacement of larger amounts of solute in concentrated than in dilute solutions."

Parsons[108] quotes the Gibbs equation with Γ_i defined as "the number of molecules of . . . *i* in unit area of the interphase," which need not be equivalent to Gibbs's definition, because Parsons[109] uses a surface layer model of the same kind as the one developed by Guggenheim,[110] which avoids the definition of Gibbs surface excess but remains as imprecise because it supposes exact knowledge

*The location of the surface of tension is perfectly well defined theoretically and is in no way arbitrary, although its practical location may be difficult (see Ref. 107 for further discussion).

of the size of the heterogeneous phase; however, for organic species but not for ions, since the heterogeneous part is usually located in a quasimonolayer region, the Gibbs "relativized" excess and the Parsons "concentration" excess are identical (it can also be shown[111] that, whatever the model used, the relative surface excesses are numerically identical). After introduction of electrochemical potentials and various rearrangements, Parsons obtains equations similar to equation (1) except that the Γ's are "surface excesses," which he defines as "linear combinations of two surface concentrations" (the introduction of electrochemical potential represents a step which implies a not so evident hypothesis, as discussed briefly by Hurwitz[112]).

This probably explains why Delahay[66] stated that he followed Parsons,[108] but returned to Gibbs's definition of Γ_i as surface excess, rather than surface concentration, and concluded that only "relative surface excesses" can be determined by thermodynamic arguments. He also pointed out that the use of $\Gamma_{H_2O} = 0$ by Grahame[60] is arbitrary and of no advantage because it does not affect the information that can actually be obtained (however, see subsequent discussion).

In point of fact, the significance of accumulation at the surface and the relative (as opposed to absolute) nature of experimentally obtainable data about accumulation can be discussed in a very straightforward manner on the basis of Gibbs' definition of excess and superficial density* (surface excess per unit area) and of the use of Gibbs–Duhem relations. Restricting equation (5) to constant-temperature as well as constant-pressure conditions, and taking for simplicity a binary system,

$$(d\sigma)_{P,T} = -\Gamma_1\, d\mu_1 - \Gamma_2\, d\mu_2 \tag{6}$$

The Gibbs–Duhem relation is (e.g., Ref. 116)

$$n_1\, d\mu_1 + n\;\; d\mu_2 = 0 \tag{7}$$

*This concept finally leads, with a minimum of effort, to the establishment by a rigorous argument of the main result of the thermodynamics of surfaces (see Everett[113]), i.e., the Gibbs adsorption isotherm, which directly relates superficial properties and the *quasi*independent variables (see subsequent discussion) which control them, where the latter variables are those initially defined for the bulk solution. All of the tentative treatments for elaborate derivations of the Gibbs equation, e.g., Ref. 114, have been until now relatively uninteresting for the experimentalist. The same is, unfortunately, also true of the molecular theories of surface tension.[115]

where n_1 and n_2 are the numbers of moles of species 1 and 2 in the bulk phase. Combining equations (6) and (7), we obtain

$$(\partial\sigma)_{P,T} = -\Gamma_1\,\partial\mu_1 + (n_1/n_2)\Gamma_2\,\partial\mu_1 \tag{8}$$

i.e.,

$$-(\partial\sigma/\partial\mu_1)_{P,T} = \Gamma_1 - (n_1/n_2)\Gamma_2 = \Gamma_{1(2)} \tag{9}$$

where $\Gamma_{1(2)}$ represents the *surface excess (per unit area) of species 1 relative to that of species 2*, which is an invariant of the position of the dividing surface but not of the reference substance 2 (see Ref. 111); it is, for this reason, called a relative surface excess. This definition is readily substantiated by noting that, when $\Gamma_2/\Gamma_1 = n_2/n_1$, $\Gamma_{1(2)}$ becomes zero. In that case, it should be emphasized that the surface tension does not change as μ_1 (or μ_2) is varied; in other words, even though both species may have accumulated at the surface, the surface tension is not altered unless the species have accumulated at the surface in a different proportion than their mole fractions in the bulk phase. This may appear to be intuitively wrong; however, the converse argument is somewhat more intuitively convincing.

Since the forces between molecules of species 1 and 2 for the bulk composition n_1/n_2 are "balanced" in the same way as for the surface composition Γ_1/Γ_2 ($= n_1/n_2$), no additional forces result from such accumulation and the tendency of the surface to expand or contract will not be changed.

Equation (9), then, shows that an infinitesimal rate of change of surface tension with an infinitesimal change in the chemical potential of species 1 is equal to the surface excess of that species relative to that of species 2.

The use of an arbitrarily pre-chosen reference surface is unnecessary and, therefore, confusing. However, one can try (for ions) to compare absolute surface excesses calculated according to a model of the diffuse layer and relative Gibbs surface excesses (measured); it is usual in this case to define the absolute excesses as those at the outer Helmholtz plane[117] and then to relate them, at best, to the measured excesses. This can only be done by first making assumptions on the separability of the double layer, typically into outer and inner layers (with or without specific absorption), and, second, by making use of a model for the inner layer. The approach with a model remains, of course, the only one in which the tentative location of the Gibbs dividing surfaces can be eventually discussed with profit, e.g., when comparing either two specifically adsorbed ions differing in size or dilute and concentrated solutions of the same specifically adsorbed ion, in order to evaluate the importance of solvation in the inner layer in the determination of superficial excesses. In other words, the Gibbs equation is too general in character to account directly for the details and complexity of the double layer, which in itself is based first on a model, supported by different types of experiments and constructed from an understanding of the contradictions between the experimental results, but not from the Gibbs equation as a first step. Consequently, the Gibbs equation will be of immediate applicability only in some special test cases, e.g., with organic surfactants in not too concentrated solutions. When a full monolayer is formed, the surface of division corresponding to $\Gamma = 0$ can in general be identified with the surface of tension. However, even in this case, our knowledge of the interface is limited because ionic adsorption on the organic film is not necessarily negligible[10] and can affect many properties of the interface. In this case of an "organic double layer" (surfactant plus aqueous electrolyte), moreover, we cannot *a priori* use the same models as for an aqueous double layer (aqueous electrolyte alone), because there is at present still no evidence in terms of experimental data that this extrapolation is valid.

Grahame's discussion[60] of zero or negative experimental values of Γ (see above) is, consequently, also less clear than it might be; in fact, a negative experimental value of Γ for a salt means that, in principle, the proportion of salt to water at the surface is less than in the bulk of the solution, because "The dissolved ions, by virtue of ion–dipole attractions, tend to pull the water molecules into the interior of the solution. Additional work must be done against the electrostatic forces in order to create new surface. It follows that in such solutions, the surface layers are poorer in solute than is the bulk solution. The solute is then said to be *negatively adsorbed*...."[118] However, in the case of salts, it is not sufficiently emphasized that, typically, two situations occur. For example, at the water–air interface, it is not possible by use of the Gibbs equation to determine independently the amount of adsorption of one ionic species (for this reason, other techniques are often used[294]). The Gibbs excess must also be corrected by a coefficient which takes into account the electrolytic dissociation[51]; it is generally not always obvious which value to use for this coefficient. This is currently still the subject of a polemical discussion concerning univalent, polyvalent, and associated organic species[119–122]; the problem is that a model of the double layer must always be introduced. In the case of electrodes, however, individual ionic Gibbs superficial excesses are experimentally available when using an electrode reversible to one species (the recent development of ion-specific electrodes is very promising in this regard).

Grahame's discussion is confusing partly because he uses the term "adsorption" in different senses (see Section I.2 on the definition of adsorption), e.g., at some points, he uses Γ as a criterion, which carries no implication about distance from the surface, but he also says that $\Gamma = 0$ does not mean lack of ionic adsorption (Ref. 60, pp. 469–70) and that adsorbed ions lie closer to the surface than do unadsorbed ones (p. 483). However, some interesting ideas[123] have been proposed on the significance of comparing capacity and surface-tension data in terms of surface excesses which would imply different reference planes, i.e., an outer Helmholtz plane and a plane referred to the Gibbs special convention ($\Gamma_{solvent} = 0$), respectively. This has been recently discussed for concentrated solutions.[124]

The concept of an arbitrarily fixed surface of reference may become confusing when it is used in the context of experimental data. For instance, Conway and Barradas[125] describe their experimentally obtained values of Γ_A thus: "The surface reference plane is chosen so that $\Gamma_{H_2O} = 0$ and Γ_A values hence refer to (relative) amounts of organic solute adsorbed in the double layer as conventionally regarded."

Gibbs' own discussion of the experimental determination of surface excesses is initially confusing (see above; it must be kept in mind that Gibbs was not an experimentalist[126]) in the present context because he considered a situation in which the pressure was not maintained constant and pressure differences across the surface could manifest themselves in varying curvature of the surface (see Refs. 51 and 110 for a detailed discussion of the effect of curvature). All the later workers quoted, however, deal explicitly or implicitly with constant-pressure conditions and, therefore, the chemical potentials of all the species present are no longer independently variable; for n species, $(n - 1)$ chemical potentials are independently variable. If one is interested in the adsorption of species N, μ_N will be varied experimentally and, consequently, the chemical potential of at least one of the other components cannot be maintained constant.[17,51] Therefore, to determine surface excesses through application of equation (1), *one chooses a reference species by allowing the chemical potential of that species to vary*; the chemical potentials of species other than that being studied (species N) and that chosen for reference (species $N - 1$) are maintained constant; consequently, the results give $\Gamma_{N(N-1)}$. Commonly, one varies the activity or concentration of one *solute* and keeps constant those of the other *solutes*; no attempt is made to keep constant the activity of the solvent and, consequently, surface excesses are obtained relative to that of the solvent. (This idea has been only relatively recently re-emphasized, e.g., Refs. 127, 133, and 295.) One can, of course, express this symbolically by setting $\Gamma_{\text{solvent}} = 0$ (as taken by Conway and Barradas[125]), and say that one is choosing a surface of reference such that this condition holds; unfortunately, such a manner of expression is confusing rather than explanatory because the surface of tension is, in general, then not identical with the surface of division.

To summarize, there have been variations in the terminology used by various authors to define Γ_i (surface concentration, surface excess, superficial density) and $\Gamma_{i(j)}$ (surface excess, relative surface excess). It would best agree with normal usage of these words and with the physical situation if Delahay's choice[66] were followed, namely, "surface excess [per unit area]" for Γ_i and "relative surface excess" for $\Gamma_{i(j)}$ (terminology and conventions have also been recently discussed by Kipling[128]).

The Gibbs adsorption equation does not specifically consider charged species or charged electrodes. However, the treatment can be applied to such conditions by an analogous set of arguments, as has been very clearly set forth by Parsons[108] and Delahay.[66] An important resulting relationship, the Lippmann equation [equation (2)], shows that the charge on the electrode can be derived from the slope of the electrocapillary curve. Conversely, the surface tension can be obtained from knowledge of the charge as a function of electrode potential.

Equation (3) defines the differential capacity, which is the third measurable quantity from which adsorption can be derived in thermodynamically based treatments.

Details of the type of information obtainable at ideally polarized electrodes and methods of treating the data have been comprehensively discussed by Grahame[60] with respect to ionic species and by Conway and Barradas,[125,129] Frumkin and Damaskin,[99] Gierst *et al.*,[22] Meibuhr,[130] and Blomgren and Bockris,[131,132] *inter alia*, with respect to organic compounds. The surface area per molecule (and hence orientation), energy of adsorption, energy of interaction of molecules within the film, information about effective dipoles, and similar information have been deduced with greater or lesser certainty in the investigations reviewed and reported by the workers quoted.

The reliability of these inferences ultimately depends on the validity of the measurements made of the surface concentration (amount per unit area). The usual procedure is to take the value of Γ as

$$\Gamma = -(1/RT)(\partial\sigma/\partial[\ln C])_{E,\ldots} \tag{10}$$

whereby it is assumed that concentration C equals activity, and to use this Γ as the surface concentration, converted from moles to molecules per unit area. From reference to equation (9), such use of the data is seen to involve two more assumptions (commonly left implicit): first, that $n_1\Gamma_2/n_2$ is negligible compared with Γ_1 (the subscript 2 refers to solvent), and, second, that all the molecules comprised in the calculated Γ lie in the same plane parallel to the interface. The assumption equating concentration and activity is likely to be satisfactory in dilute solutions, where, in addition, Γ_2 is small (for pure solvent, $\Gamma_2 = 0$, unless the molecules in the interface differ in density of packing from those in the bulk phase[293]), as is also n_1/n_2. The last assumption, however, does not inspire the same general confidence, e.g., it rules out adsorption beyond a monolayer. In spite of the many attempts to develop reasonable nonthermodynamic hypotheses[133–135] or pseudothermodynamic ones[136] to interpret Gibbs excesses, our knowledge of the interphase remains limited when the heterogeneity extends beyond a monolayer. However, compared to other areas of research in surface chemistry, progress, independent of thermodynamics, has been most rapid in electrochemistry.

The experimental verification of Gibb's equation either at the water–air interphase[137] or even at solid electrodes[73] is generally encouraging, and agreement of Gibbs isotherms with other measurements is often satisfactory. It is relatively surprising that theory is generally too weak to predict the existence of multilayer adsorption,[138] although some interesting results in this connection have been reported[51] on the base of expanded thermodynamic (statistical mechanical) treatments of solutions.

On the basis of the assumption that the molecules are in a plane parallel to the interface, it is possible to calculate the surface requirement per molecule; then, by comparison with requirements calculated from the structure of the molecule, the orientation of the molecules in the adsorbed state can be inferred, i.e., "flat" or "perpendicular."[125,129] This can lead to an indication of the bonding involved, e.g., π-electrons in the case of aromatic compounds lying "flat," although in this case, the electrocapillary depressions are sometimes abnormally small,[37] as is also the capacity. The care necessary in interpreting such data is evident from the case of piperidine in aqueous solution at mercury electrodes, where adsorption having the characteristics of π-like interactions was indicated; this was due to the transformation of piperidine to pyridine at the positive potentials involved.[296]

An interesting situation is encountered when the calculated Γ changes abruptly at some value of the bulk concentration[139]; this phenomenon may mark a change in orientation[125] of the adsorbed molecules ("flat" to "perpendicular"), but it could also indicate the formation of a second layer of adsorbed molecules. In this respect, the often observed continuous[22] or discontinuous[140] variations of Γ_{maximum}, i.e., maximum slope of Gibbs isotherms, with potential is certainly characteristic of a change in the compact nature of the film or of a change of orientation of components in the film. However, in most descriptions in terms of isotherms, this is generally not taken into account. On the other hand, the "first turning point"[141] is too frequently considered as corresponding to the achievement of a monolayer, *faute de mieux*. In spite of the many attempts to apply thermodynamic theory and measurement to mixed monolayers in various areas of surface chemistry,[297,298] mixed adsorption of organic surfactants has not been generally considered in electrochemistry (but see Refs. 46 and 47); this is an important lacuna.

2. Measurement of Surface Tension

With a liquid electrode (usually mercury; occasionally, gallium[99] and, at higher temperatures, some alloys), the classical Lippmann electrometer can be used. The principal advantage of this method is that rigorous theory is available to support the relation between what is observed and what is needed (the surface tension). However, since the measurements do take time, contamination may occur (a kinetic problem). Typical reproducibility for good measurements is below 0.1 dyne cm^{-1}.

Although results with the Lippmann electrometer obtained by a single investigator are generally claimed to be very precise, e.g., Ref. 142, the electrocapillary curves reported by different investigators—even for neutral salts—frequently do not coincide for a wide potential range, e.g., from the potential of zero charge to −1.5 V, which would seem to indicate that the data are less satisfactory than is generally recognized. In addition, it is generally not stated that, in using Lippmann electrometers at relatively negative potential, a rather high pressure of mercury and a small internal capillary diameter have to be used to maintain the accuracy or even to render the measurement possible; this may partially explain the lack of data at negative potentials. At anodic as well as at strongly cathodic potentials, the reliability of electrocapillary data is sometimes doubtful as a result of a "sticking meniscus." This phenomenon, observed particularly in dilute solutions of nonspecifically adsorbed electrolytes, has been discussed in some detail by Lawrence *et al.*[143]

Somewhat less accurate is the commonly used polarographic capillary, from which the drop time for mercury is closely proportional to surface tension.[144] The comparative accuracy of the two methods has been recently discussed.[133,145] The accuracy of the drop-time technique has been greatly improved by the theoretical consideration[145] of the exact relation between drop time and surface tension by taking into account the counter-pressure; as a result, the claimed precision is 0.1 dyne cm^{-1}.[145] These two methods are so convenient that few attempts to use other methods have been reported; worth noting in this connection, however, are the thorough studies of the geometry of sessile drops and of contact angles by Smolders[146,287,288] and Melik-Gaikazyan.[147] Extensions of this work could make possible a more reliable use of contact-angle measurements with gas bubbles to obtain surface-tension data at solid electrodes.

The general trend now is to use modern computer techniques[145,148,149] to minimize errors and to draw the best electrocapillary curves.*

Detailed, helpful discussions of the use of the Lippmann electrometer have been recently given by Conway and Barradas,[125] and Hansen *et al.*[155]; some real improvements in capillary electrometer technique has also been recently reported by Gordon, Halpern, and Conway.[156]

Reference should be made to the many experimental arrangements now known for drop-time measurements,[22,144,157–162] some of which are automated and even commercially available for the recording of electrocapillary curves (e.g., Refs. 163–165) or for the direct recording of the Gibbs' isotherm with the bulk concentration varying exponentially with time.[166,167]

With solid electrodes, the situation is quite different. Measurements have been made of the contact angle of gas bubbles,[168–173] and of surface hardness measured by scratching and other frictional techniques[174–176] or by stretching.[177,178] However, these methods do not directly measure the surface tension of the electrode–electrolyte interface and there is some doubt about the closeness of the relationship between the measured quantity and the surface tension. In addition, the methods are not convenient; for solid electrodes, measurements of charge and/or capacity may be more promising than attempted measurements of surface tension.

As previously emphasized, equations (1)–(3) and (9) are valid only for equilibrium conditions (see Ref. 33, for example, for a discussion on verification of equilibrium conditions in electrocapillary

*A note of caution must be sounded, e.g., in the case of a surfactant plus an electrolyte, the junction between the normal (electrolyte alone) and depressed (adsorption) electrocapillary curves is sometimes very sharp[36,150–152]; thus, for pyridine in basic medium, the measured cathodic desorption potentials occur in an interval of less than 0.5 mV or even less than 0.1 mV.[36,150,152] (See Figure 5.) In such a situation, if measurements of surface tension are not made close enough to the desorption potential, the smoothing process used to draw the curves may hide the sharp character of a desorption process.

measurements); if equilibrium does not subsist, a change in surface tension, charge, or capacity can nevertheless be taken to show that adsorption of the relevant species occurs, but relative surface excesses cannot be calculated from equation (9); the only type of solution for resolving the problem is too theoretical.[180]

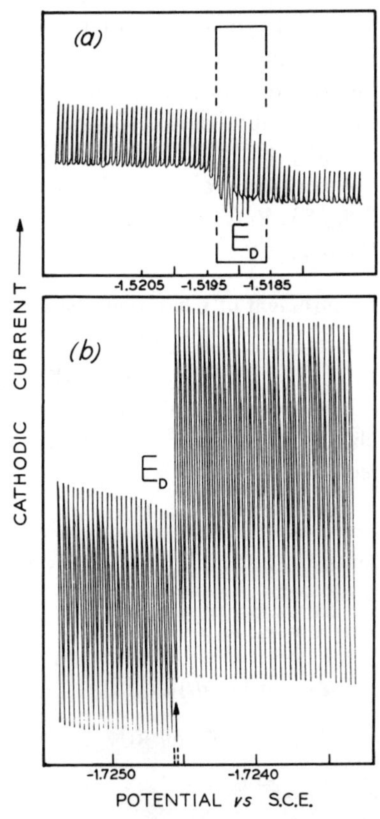

Figure 5. (a) Desorption potential observed on a dc polarogram for a solution 0.175 *M* in pyridine, 0.5 *M* in Na_2SO_4, and 0.01 *M* in NaOH at a slow scan rate ($\sim$1 mV/min^{-1}). (b) Desorption potential for a similar solution, which is 0.95 *M* in pyridine, at a very slow scan rate ($\sim$0.5 mV min^{-1}). (Taken with permission from L. Gierst and P. Herman, unpublished results.)

It can happen that not only does equilibrium not subsist, but results may not even be reproducible. The much lauded reproducibility of the dropping mercury electrode is not always evident, e.g., the drop time may change in a quite irregular manner; such behavior obviously makes quantitative calculations impossible. However, if the irregularity is confined to a particular range of polarizing potential, it is reasonable to assume that a potential-dependent process and not the state of the capillary itself is causing irreproducibility; thus, a qualitative test for adsorption, including the potential range over which it is appreciable, is available. The capillary behavior can frequently be made reproducible by silicone treatment.[181,182] (Various problems associated with the use of a capillary are discussed in Ref. 183; Refs. 154 and 184 discuss the problem of the sphericity of the drop.)

3. Measurement of Charge on the Electrode

The charge on an electrode is not directly measurable; methods that give values of the charge actually involve the measurement and transformation of another electrical variable. For example, in the technique of Loveland and Elving,[185] a potential that varies linearly with time is applied:

$$dE/dt = K \tag{11}$$

where E is the potential, t the time, and K a constant. The differential capacity is then given by

$$C = dq/dE \tag{12a}$$

$$= (dq/dt)(dt/dE) \tag{12b}$$

$$= (1/K)(dq/dt) \tag{12c}$$

$$= i/K \tag{12d}$$

where i is the current. The charge is obtained by integration of the current–time profile:

$$q = \int dq = \int KC\,dt = \int i\,dt \tag{13}$$

Integration of the charging current during the growth of a dropping mercury electrode has also been used for determination of the surface charge density.[186–188]

Alternatively,[189–191] the potential–time function at constant current is also the potential–charge function and the charge can be determined, again indirectly.

As far as measurement is concerned, therefore, it is not worthwhile to discuss charge determination separately from the determination of capacity; the latter is most commonly measured, reported, and discussed. Where charge is used, it has been derived from what is essentially a technique more commonly used for capacity determination.

Before discussing methods for measuring the double-layer capacity, however, it should be stressed that equations (1)–(3) could not, with the same significance, have been written "in the reverse order." A change in surface tension always indicates an adsorption process; a change in differential capacity does not necessarily do so. Thus, passage of current through the interface can change the capacity (Grahame's "pseudocapacity") without affecting the surface tension; integration to obtain the charge on the electrode would then similarly reflect Faradaic and not adsorptive behavior.

Even further, capacities that do arise from adsorption processes may, on integration, give quite erroneous ideas about surface-tension changes. Since equations (1)–(3) are valid only when equilibrium subsists, adsorption processes that are not at equilibrium cannot be quantitatively discussed on the basis of these relations. The typical phenomenon to which this restriction must be applied is that of a tensammetric process. Fortunately, a simple experimental criterion is available for determining the nature of the capacity measured; if the capacity is independent of frequency (or satisfies an equivalent criterion with nonperiodic electrical signals), it is reasonable to assume that it reflects the nature of the double layer in the sense implied by equations (1)–(3) taken together. However, if there is a dependence on frequency, such a conclusion cannot reasonably be drawn; this is also a matter of considerable importance in relation to studies at electrodes other than the d.m.e., because it is not possible to tell whether the only correction to apply for frequency dispersion is one that includes only change (relaxation of adsorption) occurring near the surface of the electrodes (local perturbation of the double layer); in fact, other effects may coexist (see the following section).

4. Measurement of Double-Layer Capacity

It has just been emphasized that quantitative interpretation of electrode capacities in terms of surface excesses of adsorbed substances via equations (1)–(3) presupposes that the capacity corresponds to the double-layer equilibrium and is therefore not a function of the frequency of an applied alternating signal. (If measurements are made with nonperiodic electrical signals, behavior analogous to frequency dispersion is observable but the detection of such dispersion may be much more difficult, e.g., the potential–time relation with current pulses would be nonlinear rather than linear,[192] but the degree of curvature is here a much less sensitive criterion than is the comparison of measured capacities at two frequencies of an alternating signal.)

Techniques for measuring double-layer capacities by means of applied alternating signals are generally grouped in two categories, i.e., bridge methods and others, which share the difficulties associated with small-amplitude signals at audio and subradio frequencies; shielding of circuits and prevention of stray leakages to ground must be given the most careful attention. Bridge methods are frequently claimed to be inherently more accurate than others, e.g., tensammetry, because a null rather than a deflection method is involved. In point of fact, it is most commonly the performance of the individual instrument that is decisive, rather than the type of method employed. Thus, Melik-Gaikazyan[193,194] described a bridge that appeared to give reliable results between 10 and 500,000 Hz; subsequently, it was reported[195] that at low frequencies, leakages to ground produced a spurious frequency dispersion with this apparatus. Hansen *et al.*[155] developed a substitution method based on bridge techniques in order to decrease effects of distributed capacity. Britz and Bauer[196] developed a substitution method based on ac polarographic techniques which achieves a very high order of accuracy and reliability. A number of discussions and descriptions of instrumentation is available in the electrochemical literature: e.g., Breyer and Bauer,[301] Hansen *et al.*,[155] Nancollas and Vincent,[197] Breiter,[198,199] Schmid,[200] and Ramaley and Enke,[192] and further references are given in the latter article; references and standard texts on electrical measurements *per se*, e.g., Buckingham and Price,[201] are also very useful; a very helpful discussion of various bridge circuits and their applicability has been given by Hills and

Payne.[202] Special timing methods are required when bridge measurements are made with a dropping mercury electrode, e.g., as discussed by Nancollas and Vincent.[197] Recently, a self-timing bridge, which detects the phase rather than the amplitude, has been described.[203]

Double-layer capacities have also been frequently measured by techniques that do not employ alternating signals, e.g., oscilloscopic or other recording of current (proportional to capacity) with a fast triangular voltage-sweep,[185,204] potential–time behavior with applied steps or square waves of current,[192,205–207] and current–time behavior with controlled potential (pulses).[208]

Frumkin and Damaskin[99,209] recently reviewed the theoretical basis for using capacity data in studies of adsorption and the information actually gained in such work; no attempt will be made here to summarize their review or to duplicate it. However, it is proposed to discuss work that has been done on the capacity at solid electrodes.

The measurement of double-layer capacities is based on the concept that an electrochemical cell, in which no Faradaic reaction occurs and in which one electrode is much larger than the other, can be represented by a series combination of a purely resistive element (resistance of solution and electrodes) and the purely capacitive double-layer at the small electrode. This model is used to interpret results obtained by the methods mentioned above. The validity of the model can be tested by impedance measurements made at different frequencies, because the calculated series capacity and resistance should be independent of frequency; when they are not, we speak of a frequency dispersion.

Grahame[210] was the first to show that the series-capacity-and-resistance model corresponds to physical reality; earlier workers with all types of electrodes had found frequency dispersion. Grahame could substantiate the model at mercury electrodes in a frequency range between 240 and 5000 Hz, provided that purified solutions (freed of surfactants) were used together with spherical electrodes suspended on a thin support (both dropping mercury electrodes and hanging drops were used). Within the frequency range available with the apparatus used, frequency dispersion was observed if a mercury-pool electrode was used (the effect being greater at large than at small electrodes) or if the mercury drop was "shaded" by its support, e.g., a blunt capillary tip. According to recent observations,[182] it seems that the best conditions are achieved when the drop diameter at the balance of the bridge is at least 4–5 times that of the tip diameter.

Melik–Gaikazyan[194] confirmed Grahame's finding that frequency dispersion could be avoided at spherical mercury electrodes; dispersion appeared if the drop was shaded, if the current line length between experimental and auxiliary electrodes was not uniform, or (at low frequencies) if solution penetrated into the tip of the capillary. The frequency range used was given as 10–200,000 Hz and with decreased accuracy to 500,000 Hz; however, as already mentioned, doubt has been cast on the low-frequency results by the finding[195] that ground leakages were present in the apparatus.

It is widely accepted that measurements at solid electrodes show frequency dispersion (cf. Ref. 211). Yet, very little data has been published to show the magnitude of the effect and no conclusive explanation has so far been given. The work on mercury quoted above indicates that the electrode shape is important and that lack of spherical symmetry may result in frequency dispersion; such an explanation for the frequency dispersion at solid electrodes is supported by the work of Grahame[212] ("... generally ... dispersion is greater on wires and sheets than on spheres ... [because it is] more difficult to satisfy the symmetry requirements ...") and of Borisova and Ershler,[213] who reported that solid droplike electrodes behaved like mercury drops and differed from platelike electrodes of the same solid materials. On the other hand, Grahame *et al.*,[214] despite considerable experimental effort and ingenuity, were not able to prepare solid droplike electrodes that showed no frequency dispersion. Frequency dispersion at solid spherical electrodes has also been reported by Conway and Bockris,[215] Clavilier,[216–218] and Pineaux[219]; Robertson[220] found the same frequency dispersion at plane and hemispherical electrodes.

Perhaps the most popular explanation for frequency dispersion at solid electrodes is the presence of an inhomogeneity or of surface roughness* of the electrode[212–214,221–224]; a model for such an effect has been treated by de Levie.[226] The conception generally held of the effect of roughness on "the probability of adsorption" has been very simply schematized by de Levie[227] as a distance

*Roughness here is not the term correlated with specific adsorption at electrodes and called more precisely "chemical softness."[225]

effect. On the other hand, Laitinen and Gaur[300] found the same frequency effect at macroscopically rough and at polished electrodes. The work at spherical, solid electrodes[214–219] concerns fused droplets, whose surface could hardly be very rough, and yet dispersion was observed (the dispersion was ascribed to the presence of surface films on the electrode); in fact, Clavilier's[217] results show a somewhat large effect at smooth than at rough, etched, spherical electrodes. Moreover, the report[223] that single crystals show no frequency dispersion while polycrystalline materials do, contains figures that merely show a *smaller* dispersion at the single crystals; Clavilier and Pineaux[218] found the same dispersion at monocrystalline and polycrystalline platinum electrodes.

The attribution[228] of observed dispersion to comparatively slow reorientation of solvent molecules held at solid surfaces is not in keeping with the report[229] that the measured capacity is the same at liquid and solid gallium electrodes; the concept has been also contradicted on more *a priori* grounds.[230]

Clearly, the observed frequency dependence at solid electrodes is a problem that still awaits solution. One reason for this may be the fact that most workers have approached the matter as if a single factor is always the major cause for dispersion at solid electrodes. It is obvious from work such as that discussed above that a number of factors can produce dispersion, which must all be controlled if real progress is to be made. It has been rather convincingly shown that dispersion can arise as a result of any one of the following:

(*a*) *Unequal current paths (different resistances) from the auxiliary electrode to various parts of the working electrode.* These may be due to a rough surface,[214,226] to a surface that is inhomogeneous for reasons other than macroscopic roughness,[221] e.g., polycrystallinity,[223] to "shading" of spherical electrodes by their support,[194,210,215] to lack of macroscopic symmetry with nonspherical electrodes,[210,212,213]* and to thin films of solution penetrating into

*Recent theoretical studies[231,232] have shown that the nonuniform current distribution associated with a disk electrode gives rise to an appreciable frequency dispersion. The correct double-layer capacity is obtained by extrapolating to the low-frequency limit, whereas extrapolation to infinite frequency gives an arbitrary value that depends on the various parameters of the system (solution conductivity and diameter of the disk, as well as correct double-layer capacity) and that may in fact be zero. On the other hand, extrapolation to infinite frequency gives the correct value for the series resistance, while extrapolation to low frequencies gives a limiting value (for the resistance) that depends on the various parameters of the system.

the tip of the capillary supporting a drop or between a solid electrode and its insulating covering.[91,211,233–235] An effect such as the latter may subsist at electrodes that are merely dipped into the solution[215,221] without control of the meniscus.

Grahame[210] has pointed out that, if nonuniform current paths are responsible for dispersion, the magnitude of the dispersion will vary markedly with the conductivity (concentration) of the electrolyte; this simple criterion has not yet been systematically applied in any of the attempts to explain dispersion at solid electrodes.

(*b*) *Presence of surface films.* It is generally accepted that films can produce dispersion and that solid metals, when prepared without special precautions, frequently have oxide films. Yet only a few workers[214,215,221] have adopted procedures designed specifically to control the latter factor.

The presence of films and, *a fortiori*, of small amounts of electroactive species, leads to frequency dispersion that will vary with the polarizing potential; this is an excellent criterion for distinguishing such effects from the nonuniform resistances discussed. If such an effect is indeed present, preelectrolysis of solutions and electrolytic pretreatment of electrodes may well result in great improvement.[291]

(*c*) *Presence of trace amounts of surfactants.* Frequency dispersion due to progressive contamination of the surface will be accompanied by a drift of the observed capacity with time. Consequently, purification of solutions with adsorbents might well be used as a standard treatment.[39,40]

Until a frequency-independent capacity is achieved, the measured quantity cannot have ascribed to it the significance associated with a double-layer capacity. This point is obvious and has been clearly made in connection with impedance measurements[210] as well as with impulse methods.[236] Nevertheless, many workers have reported "double-layer capacities" obtained in work where data on frequency dispersion was not reported;[98,229,237–246] such "double-layer capacities" have been used to determine the true surface area of electrodes by comparison with more or less arbitrarily chosen "correct" values for the capacity per unit area.[98,244,245] It has been assumed that the true double-layer capacity can be obtained by extrapolation to infinite frequency,[98,222,223,239,242,243,247] although Grahame[210] has shown that the low-frequency values may in some cases be the more appropriate ones; extrapolated values have been used to subtract the capacitive component from the total admittance at lower frequencies in studies of Faradaic processes.[222,248–250] Impulse methods have been adopted on the assumption that with pulses having sufficiently rapid rise-times, true double-layer capacities (corresponding to values derived from extrapolations to infinite frequency) would be obtained.[98,239,242,243]

In connection with studies of this nature, it is well to recall Grahame's warning that,[210] "...no reliable conclusions can be drawn...so long as a fairly large unexplained frequency effect persists..."

5. Measurement of Tensammetric Processes

Since the admittance or capacity peaks produced by tensammetric processes characterize the regions of potential where prevailing adsorption gives way to prevailing desorption, even the simplest type of ac polarographic apparatus makes possible a quantitative and convenient measurement of the potential range over which a substance is adsorbed under various experimental conditions. Tensammetric peaks may indicate a change in nature of an adsorbed film[251,252] rather than its replacement by the supporting-electrolyte ions, or the replacement of a film of uncharged molecules by one of a charged species.[196,251]

Tensammetric peaks can be observed at solid electrodes as well as mercury, and in nonaqueous as well as in aqueous solutions. However, very little work has been done under either of these experimental conditions (the work has been summarized by Frumkin and Damaskin[99]).

The heights, shapes, and peak potentials of tensammetric waves depend on the nature and concentration of both the surfactant and the supporting electrolyte, and on the frequency of the measuring signal. In principle, the kinetics of the adsorption process and the characteristic parameters of the isotherms should be available from measurements on tensammetric waves. This expectation has been incorporated in several attempted theoretical treatments of tensammetric processes; the little experimental work that has been done has generally been in encouraging agreement with theory.

Two theoretical approaches have been used. In one, far-reaching assumptions are made at the outset about the nature of the adsorption process (Langmuir isotherm,[253,254] Temkin isotherm,[255–257] and others,[99,258] as well as various ways[99] in which adsorption depends on electrode potential or charge) and experimental results are used to determine, at least initially, how well the assumed model holds in practice. In this empirical approach, it should be recalled that the isotherms themselves have first been evaluated by considering the experimental results obtained in the potential region between the tensammetric peaks. At the same time, a considerable effort has been independently made to find a general theoretical pattern to justify the parameters introduced in isotherms and their dependence on potential and charge; extrapolation is then made to tensammetric processes. In this way, a general derivation which avoids the choice of a definite form of the isotherm has been proposed,[259] and shows certain so-called "regularities," e.g., that the free energy of adsorption must necessarily be a quadratic function of potential and how this can be used in the present case to predict the dependence of peak potentials and of peak height on concentration.

Unfortunately, it has not been possible to associate systematically this "generalized quadratic isotherm" with other known isotherms or with coverage and expected shift of zero-charge potential. If, formally, the potential is substituted by the charge in this generalized isotherm, using a particular dependence on potential or coverage (associated with a model of two condensers in parallel), the model obtained (two condensers in series) is physically meaningless. For further details, see Schuhmann,[260] Damaskin,[209] and Frumkin.[13]

Finally, it is evident that the isotherm approach, although empirical, is useful, if only to evaluate or to allow rejection of certain criteria proposed for comparing isotherms or choosing one definite isotherm for a specific situation (also see Ref. 261 concerning choice of isotherm). The typical problem is exemplified by the situation where different isotherms have been proposed, involving conflicting relations between the standard free energy of adsorption $\overline{\Delta G^\circ}$ and coverage θ; according to the Blomgren–Bockris[131] concept, $\overline{\Delta G^\circ}$ is proportional to $\theta^{3/2}$ for neutral molecules[125,129] and to $\theta^{1/2}$ for ions, whereas the Frumkin concept[262] results in $\overline{\Delta G^\circ}$ proportional to θ for small coverage. By a simple mathematical analysis, Damaskin *et al.*[290] showed that in the case of the adsorption of pyridine on mercury, at the present level of accuracy of measurement, the linear relations obtained between $\overline{\Delta G^\circ}$ and $\theta^{3/2}$ for neutral molecules or between $\overline{\Delta G^\circ}$ and $\theta^{1/2}$ for ions do not exclude a linear $\overline{\Delta G^\circ}$–$\theta$ relation and that the calculated Γ_i (surface excess)–concentration curves agree perfectly for the two types of representation.

In the second theoretical approach to tensammetric processes, fewer assumptions about the nature of the adsorption process are made initially and information about adsorption kinetics and isotherms is obtained experimentally[194,258,263,264]; this way of attacking the problem has been very clearly set forth by Parsons.[265]

The only initial assumption made in this second approach is that the charge on the electrode depends only on the amount adsorbed and on the electrode potential. Then, the differential capacity is given by

$$C = (\partial q/\partial E)_\Gamma + (\partial q/\partial \Gamma)_E(\partial \Gamma/\partial E)_\mu \tag{14}$$

At very high frequencies, the amount adsorbed can no longer change periodically with the imposed variations of potential, i.e., $(\partial \Gamma/\partial E)_\mu \to 0$ and $C \to (\partial q/\partial E)_\Gamma$; the latter term is called the "true," "high-frequency," "infinite-frequency," or "constant-coverage" capacity. It has been suggested[66,260] that the last name is preferable because the first is ambiguous and the second and third refer to a particular technique, which suggests implicitly the occurrence of a dispersion which is not necessarily present. At intermediate frequencies, the variation of tensammetric admittance with frequency reflects the nature of $(\partial \Gamma/\partial t)_\mu$. If the assumption is made that the process consists of the adsorption step itself coupled to mass transport of adsorbate and only small changes from equilibrium are considered, it can be inferred from the frequency dependence of the admittance which of the two steps is rate-limiting in the overall process, e.g., application to the behavior of the lower alkyl alcohols has shown that adsorption is considerably faster than diffusion.[194,258,263,266,267]

This approach is *a priori* more attractive than the one in which a particular isotherm and potential dependence of coverage are initially assumed. However, there are limitations and uncertainties. Consideration[263] of a situation in which the adsorbed molecules associate at a finite rate leads to a predicted frequency dependence that differs from the case where only diffusion and adsorption play a part; application to the analysis of experimental results indicates a rate-limiting association of adsorbed molecules with *n*-hexyl alcohol and higher fatty acids.[263,266,267] Thus, interpretation of data is at least to some extent determined by initial assumptions about the nature of the adsorption process. Another drawback is that the extraction of the desired values from the experimental measurements either involves considerable calculation (using an equivalent-resistance-and-capacitance model) or rests solely on the frequency dependence of the phase angle, which does not provide a very sensitive criterion; graphs constructed on theoretical grounds may show quite distinct differences, e.g., a maximum for the rate-limiting association of adsorbed molecules as opposed to a monotonic increase in the absence of association, but experimental results may not be interpretable with certainty because of limitations on the usable frequency range, e.g., a system that "actually" has a maximum may appear not to have one simply because the decrease occurs at frequencies just higher than those available. Consequently, instrumentation is designed to permit work at ever higher frequencies,[267] although the method thereby becomes of less general interest because of the attendant experimental complications.

In principle, tensammetric peak shapes and tensammetric peak potentials provide information about the potential dependence of the adsorption process and interactions between adsorbed molecules; a full discussion is given by Frumkin and Damaskin.[99] Some interesting work has been reported by Lorenz and Müller.[268] However, the assignment of isotherms—and their potential dependence—is still under discussion, so that such investigations remain exploratory in the sense that one cannot currently invoke this approach as a certain way of getting desired information about the adsorption of a particular species.

A source of uncertainty in all studies of adsorption is the time required for equilibrium to be established between the bulk of the solution and the surface of the electrode. The assumption made in the previous discussion, that only small changes from equilibrium occur, implies that the layer of solution next to the electrode is in equilibrium with the bulk of the solution; however, this holds in practice under somewhat surprisingly limited conditions unless forced convection is employed.[265] Under typical circumstances, periods of the order of minutes are required when diffusion alone is producing mass transport,[269,270] with the result that studies at the d.m.e., for example, often involve nonequilibrium conditions. This is a very serious limitation on the convenience and certainty with which quantitatively correct information can be obtained about adsorption processes.

Tensammetric studies, like Faradaic ones, can be carried out by observing rectification effects rather than by measuring the admittance. A theoretical treatment for the rectified component has been given[264] and a measurement reported.[271] It has been pointed out that second-harmonic measurements with controlled alternating current should be advantageous in such work.[272]

IV. ADSORPTION INDICATED BY EFFECTS EXERTED ON FARADAIC PROCESSES

Changes in the electrochemical behavior of a system as a result of the adsorption of a so-called "inhibitor" have been frequently described; recent reviews dealing with this work include those of Reinmuth,[273] Parsons,[265] Reilley and Stumm,[274] Kastening and Holleck,[275] and Nürnberg and von Stackelberg.[276] There is general agreement that the fundamental bases of the effects observed are not as yet fully understood and that various factors are involved in different cases. The course of electrochemical reactions in the absence of inhibitors is fully understood only where a straightforward coupled mass-transport/electron-transfer mechanism is involved, and in isolated cases of more complicated systems that have yielded only to intensive study; it is not surprising, therefore, that the action of inhibitors is not well classified, let alone comprehended.

The range of effects observed is considerable: the appearance, increase, or decrease of an overvoltage; suppression of polarographic maxima; decrease of current in some potential regions with production of a minimum at potentials not far from the electrocapillary maximum; splitting of a polarographic wave into two separate waves; occasionally, a surprising lack of effect (polarographic reduction of thallous ion); and, commonly, a decrease in ac polarographic wave height,[277] but occasionally an increase.[196]

The causes underlying these symptoms may be several: changed potential distribution in the double layer, which is likely to produce changes in rate constant and transfer coefficient; interference with chemical reactions coupled to the electron-transfer step; physically blocked surface, resulting in a changed rate of the electron transfer (in the limit, completely stopped); changes in the rate of one step which are likely to change the mechanism so that a different step becomes rate-determining; and numerous possible combinations of different effects. Interpretation of the action of inhibitors must hence be made individually for each type of inhibitor and/or conditions involved.

It does not appear to be promising, therefore, to attempt to obtain information about the details of the adsorptive behavior of a substance by studying its effect on redox systems. Nevertheless, some attempts in this direction have been made, e.g., analysis[278] of the effect of butyl alcohol on a polarographic maximum on the basis of theory developed by Frumkin and Levich[279] led to the conclusion that the adsorption process was limited by the rate of diffusion of the alcohol to the electrode.

The adsorption of surfactants has been followed by measurement of overvoltage,[280,281] but the lack of complete understanding of overvoltage phenomena makes this a risky procedure.

Perhaps the most successful studies yet reported are those of Schmid and Reilley.[282] Observation of current–time curves under polarographic conditions in the presence of surfactants enabled division of the effects of inhibitors into three groups, based on the morphology of the curves: (a) where the action of the inhibitor is limited by its diffusion; (b) by the position of the adsorption equilibrium; or (c) by the rate of the adsorption step itself. In the first case, estimates could be made of the surface occupied per molecule adsorbed; the results were reasonable.* This work demonstrates forcibly the value of polarographic current–time observations, a notably useful but inexplicably neglected technique.

The many uncertainties involved in studies of the interaction of redox processes and surfactant adsorption may in special cases be counteracted by the unusual nature of the information that may

*Board *et al.*[283] have recently discussed pitfalls in the procedure used and have described a more generally accurate way of obtaining the area occupied per molecule.

be obtainable. For example, there has been a long controversy over the mode of action of inhibitors on the polarographic reduction of inorganic cations. Heyrovsky[284] postulated that the reduction of divalent cations proceeded via a one-electron step followed by disproportionation,

$$M^{2+} + e \rightarrow M^{+} \tag{15}$$

$$2M^{+} \rightarrow M^{2+} + M \tag{16}$$

with only the chemical step being impeded by surfactants; this view was contested by Frumkin,[285] who considered that electron transfer is impeded. There is no compelling reason why both proposed mechanisms should not be valid under different conditions; in fact, it is rather likely that different types of films would have different effects. If a mechanism can be substantiated for a particular case, one might then be able to infer whether the surface film is "gaseous" (and readily penetrable, but decreasing the chance of collisions producing disproportionation) or "compact" (and likely to decrease the rate of uptake of the first electron). In this connection, it has been recently shown[196] that the disproportionation of a nitroaryl free radical is markedly hindered at low surfactant concentrations, while the rate of reduction of the depolarizer was detectably decreased only at higher surfactant concentrations. Lovrecek and Marincic[286] have invoked Heyrovsky's mechanism to explain the influence of phenol or gelatin on the Cd(II)/Cd(Hg) couple as resulting from a decrease in the rate of the disproportionation reaction.[284]

V. SUMMARY

At electrodes that are obtainable as very thin foils, adsorption can be studied in a very direct manner; the use of radioactively tagged adsorbates gives kinetic data as well as isotherm parameters, and spectroscopic studies hold the promise of elucidating the nature of bonding forces and similar aspects of the state of adsorbed species. At electrodes that can reflect light, inferences about the thickness and nature of surface films can be obtained from ellipsometric and relative reflectance studies. These desirable techniques cannot be employed with an electrode the surface of which is being continually renewed, and therefore it will always be necessary to take the greatest care with purification of reagents and preparation (as well as preservation or reproducible resurfacing) of electrodes.

All other methods measure adsorption in a less direct way and there is a correspondingly greater uncertainty involved. Qualitative information is obtainable with certainty; e.g., at solid electrodes, one may observe the change in apparent impedance with time after addition of the compound of interest[181]; but quantitative investigations are fraught with pitfalls; in particular, the slow rate at which true equilibrium is reached between electrode and solution, especially when the adsorbate is present in small amount and is strongly adsorbed, has only been realized comparatively recently[265,270] and casts doubts on the quantitative validity of much of the work that has been done at the d.m.e., which, for other reasons, is so attractive from the experimental point of view.

Accounts in the electrochemical literature too often leave implicit the assumptions made and the corresponding uncertainties, e.g., in ignoring the surface activity of some components of the system, in taking as a "double-layer capacity" a value obtained at a fixed frequency of the measuring signal, in translating relative surface-excesses obtained on the basis of Gibbs' equations into absolute surface concentrations within a monolayer. In this connection, it is useful to refer to modern general accounts of adsorption, e.g., the critical and informative monograph by Kipling.[31]

Thus, in setting out to investigate the adsorptive behavior of a particular species, one should not expect to obtain quantitatively reliable data too readily. With only reasonable assumptions, one may often obtain surface requirements per molecule, and consequently orientations, that are likely to be very good approximations to the truth, but one cannot guarantee that this will be possible with any particular system. After the direct methods, electrocapillary measurements are the most desirable, but one is then limited either to liquid electrodes and some uncertainty about the attainment of equilibrium, or by the question of whether one is actually measuring the surface tension of a solid electrode by the particular technique employed.

Measurements of capacity introduce a degree of uncertainty somewhat greater again than that associated with electrocapillary studies. Finally, inferences drawn from the effect exerted by a surfactant on the course of an electrochemical reaction must at this stage remain rather tentative: yet, it may be in such work that great promise for the future lies.

ACKNOWLEDGMENT

The authors thank the National Science Foundation for helping to support the present study.

NOTE ADDED IN PROOF

Important developments in the area under review since the original manuscript was completed include the following. The problem of discrepancies between electrocapillary and capacity data has been resolved through recognition of the unreliability of the Lipmann electrometer when the contact angle between mercury and glass changes; the method of choice for obtaining electrocapillary data now appears to be the technique of maximum bubble pressure.[302,303] Randin and Yeager[304] have verified experimentally that nonuniform current distribution to a disk electrode results in frequency dispersion of the capacity, and have been able to obtain frequency invariant capacity values by fitting the electrode with a hood that produces a uniform current distribution. Application of capacity measurements to solid electrodes promises to become much more reliable with recognition of this factor, and the ready availability of lock-in amplifiers and of commercial a.c. polarographs incorporating these indicates an increasing use of quadrature-component measurements in a.c. polarography for gathering fundamental data regarding adsorption. An important recent publication is by B. B. Damaskin, O. A. Petrii, and V. V. Batrakov.[305]

REFERENCES

[1]L. I. Antropov, *Kinetics of Electrode Processes and Null Points of Metals*, Council of Scientific and Industrial Research, India, 1960.
[2]A. N. Frumkin, *Svensk. Chem. Tidskr.* **77** (1965) 300.
[3]L. I. Antropov, in *Soviet Electrochemistry* (*Proc. 4th Conf. Electrochem.*), Vol. 1, p. 11, Consultants Bureau, New York, 1961.
[4]S. Argade and E. Gileadi, in *Electrosorption*, Chapter 5, Ed., E. Gileadi, Plenum Press, New York, 1967.
[5]J. O'M. Bockris, M. A. V. Devanathan, and K. Mueller, *Proc. Roy. Soc.* (*London*) **A274** (1963) 55.
[6]R. Parsons, *J. Electroanal. Chem.* **7** (1964) 136.
[7]A. N. Frumkin, *J. Electroanal. Chem.* **7** (1964) 152.
[8]B. B. Damaskin, *J. Electroanal. Chem.* **7** (1964) 155.
[9]R. Parsons, *J. Electroanal. Chem.* **8** (1964) 93.
[10]B. E. Conway, R. G. Barradas, P. G. Hamilton, and J. M. Parry, *J. Electroanal. Chem.* **10** (1965) 485.

[11]A. K. N. Reddy, in *Electrosorption*, Chapter 3, Ed., E. Gileadi, Plenum Press, New York, 1967.
[12]D. Schuhmann, *J. Chim. Phys.* **64** (1967) 1404.
[13]A. N. Frumkin, B. B. Damaskin, and A. A. Survila, *J. Electroanal. Chem.* **16** (1968) 493.
[14]B. J. Piersma and E. Gileadi, in *Modern Aspects of Electrochemistry, No. 4*, p. 57, Butterworths, London, 1966.
[15]S. G. Mairanovskii, *Catalytic and Kinetic Waves in Polarography*, pp. 85ff, 187ff, Plenum Press, New York, 1968.
[16]E. Gileadi and B. E. Conway, in *Modern Aspects of Electrochemistry, No. 3*, p. 347, Butterworths, London, 1964.
[17]B. E. Conway, *Theory and Principles of Electrode Processes*, Chapter 7, Ronald Press, New York, 1965.
[18]G. A. Tedoradze and R. A. Arakelyan, *Soviet Electrochemistry* **4** (1968) 52.
[19]G. C. Barker and J. A. Bolzan, *Z. Anal. Chem.* **216** (1964) 215.
[20]C. Bianchi, L. Formaro, and S. Trasatti, *Chim. Ind.* (*Milan*) **50** (1968) 26.
[21]P. Delahay and I. Trachtenberg, *Office of Naval Research Contract NR-051-259, Tech. Rept. No. 25* (October 1956).
[22]L. Gierst, D. Bermane, and P. Corbusier, *Ricerca Sci. Contrib. Polarog.* **4** (1959) 75.
[23]J. F. Baret, *J. Phys. Chem.* **72** (1968) 2755.
[24]V. S. Krylov and V. G. Levich, *Dokl. Akad. Nauk SSSR* **159** (1964) 409 [English transl.: *Soviet Phys.—Doklady Phys. Chem.* **159**, 1037].
[25]J. A. V. Butler, *J. Phys. Chem.* **33** (1929) 105; see also *Proc. Roy. Soc.* **A122** (1929) 399; see also R. G. Barradas and B. E. Conway, *J. Electroanal. Chem.* **6** (1963) 314.
[26]K. Mueller, in *Electrosorption*, Chapter 6, Ed., E. Gileadi, Plenum Press, New York, 1967.
[26a]R. G. Barradas and B. E. Conway, *J. Electroanal. Chem.* **6** (1963) 314.
[26b]L. G. M. Gordon and B. E. Conway, *J. Phys. Chem.* **73** (1969) 3609.
[27]J. Baret, L. Arman, M. Bernard, and G. Danoy, *Trans. Faraday Soc.* **64** (1968) 2539.
[28]R. D. Armstrong, W. P. Race, and H. R. Thirsk, *J. Electroanal. Chem.* **16** (1968) 517.
[29]B. Breyer and H. H. Bauer, *Alternating Current Polarography and Tensammetry*, Interscience, New York, 1963.
[30]H. Lange, in *Non-Ionic Surfactants*, Vol. 1, p. 443, Ed., J. Schick, Marcel Dekker, New York, 1967.
[31]J. J. Kipling, *Adsorption from Solutions of Non-Electrolytes*, Academic Press, New York, 1965.
[32]A. N. Frumkin and B. B. Damaskin, *Pure Appl. Chem.* **15** (1967) 265.
[33]P. Corbusier, Thesis, Free Univ. of Brussels (1957), pp. 31, 36–39.
[34]H. A. Laitinen, in *Trace Characterization* (*Materials Research Symposium 1966*), p. 105, U.S. Natl. Bur. Stds. Monograph No. 100, 1967.
[35]J. A. Babbit, H. E. Schwarting, and R. J. Gritter, *Introduction to Chromatography*, p. 21, Reinhold, New York, 1968.
[36]L. Gierst and C. Pecasse, in *Polarography 1964*, p. 305, Ed., G. J. Hills, Macmillan, New York, 1966.
[37]B. B. Damaskin, A. A. Survila, and L. E. Rybalka, *Soviet Electrochemistry* **3** (1967) 818.
[38]A. N. Frumkin, *Elektrokapilljarnye javlenija i elektrodnye potentialy*, Univ. of Odessa, 1919.

[39]N. P. Berezina and N. V. Nikolaeva-Fedorovitch, *Soviet Electrochemistry* **3** (1967) 1.
[40]N. P. Berezina and N. V. Nikolaeva-Fedorovitch, *Soviet Electrochemistry* **3** (1967) 222.
[41]Ref. 31, Chapter 3.
[42]E. Gileadi, in *Electrosorption*, Chapter 1, Ed., E. Gileadi, Plenum Press, New York, 1967.
[43]B. J. Piersma, in *Electrosorption*, Chapter 2, Ed., E. Gileadi, Plenum Press, New York, 1967.
[44]I. Zwierzykowska, *Roczniki Chem.* **38** (1964) 1169.
[45]E. B. Weronski, *Electrochim. Acta* **14** (1969) 231.
[46]S. L. Gupta, S. K. Sharma, and R. N. Soni, *Electrochim. Acta* **10** (1965) 549.
[47]S. L. Gupta and S. K. Sharma, *Electrochim. Acta* **10** (1965) 151.
[48]R. A. Arakelyan and G. A. Tedoradze, *Soviet Electrochemistry* **4** (1968) 122.
[49]S. L. Dyatkina and B. B. Damaskin, *Soviet Electrochemistry* **4** (1968) 903; E. Blomgren and J. O'M. Bockris, *J. Phys. Chem.* **63** (1959) 1475; B. E. Conway and R. G. Barradas, *Electrochim. Acta* **5** (1961) 319, 349; B. E. Conway, R. G. Barradas, P. G. Hamilton, and J. M. Parry, *J. Electroanal. Chem.* **10** (1965) 485.
[50]J. W. Gibbs, *The Collected Works*, Vol. I, pp. 219ff, Longmans Green, 1931.
[51]I. Prigogine, R. Defay, and A. Bellemans, *Surface Tension and Adsorption*, p. 1, John Wiley and Sons, New York, 1966.
[52]H. D. Hurwitz, in *Electrosorption*, Chapter 7, Ed., E. Gileadi, Plenum Press, New York, 1967.
[53]J. O'M. Bockris, *Adv. Catalysis* **17** (1967) 351.
[54]Ref. 17, Chapter 3.
[55]A. Scheludko, *Colloid Chemistry*, p. 78, Elsevier, 1966.
[56]H. Saltzburg, J. N. Smith, and M. Rogers, *Fundamentals of Gas–Surface Interactions*, Academic Press, New York, 1967.
[57]E. A. Flood, *The Solid–Gas Interface*, Marcel Dekker, New York, 1967.
[58]*Faraday Soc. Discussions* **40** (1965) 177–231.
[59]D. J. C. Yates, *Adv. Catalysis* **12** (1960) 265.
[60]D. C. Grahame, *Chem. Revs.* **41** (1947) 441.
[61]C. A. Barlow, Jr., and J. R. McDonald, *Adv. Electrochem. Electrochem. Eng.* **6** (1967) 1.
[62]D. M. Mohilner, *Electroanal. Chem.* **1** (1966) 331
[63]H. D. Hurwitz, *J. Electroanal. Chem.* **10** (1965) 35.
[64]R. Parsons and E. Dutkiewicz, *J. Electroanal. Chem.* **11** (1966) 100.
[65]J. M. Chauvier, Memoire de Licence, Free Univ. of Brussels (1967).
[66]P. Delahay, *Double Layer and Electrode Kinetics*, Chapter 2, Interscience, New York, 1965.
[67]Ref. 17, Chapters 4 and 5.
[68]J. P. Lambert, Thesis, Free Univ. of Brussels (in preparation).
[69]Ref. 15, p. 93.
[70]J. P. Hoare, *The Electrochemistry of Oxygen*, John Wiley and Sons, New York, 1968, pp. 31–32.
[71]E. Gileadi and B. E. Conway, in *Modern Aspects of Electrochemistry*, Chapter 3, p. 347, Eds., J. O'M. Bockris and B. E. Conway, Butterworths, London, 1964; B. E. Conway, E. Gileadi, and M. Dzieciuch, *Electrochim. Acta* **8** (1963) 143; B. E. Conway, in *Progress in Reaction Kinetics*, Vol. 4, p. 399, Ed., G. Porter, Pergamon, London, 1967.
[72]Ref. 31, pp. 166ff.
[73]N. A. Balashova and V. E. Karazinov, *Electroanal. Chem.* **3** (1969) 135.

[74]N. A. Balashova, V. A. Ivanov, and V. E. Karazinov, *Dokl. Akad. Nauk SSSR* **115** (1957) 336; *C.A.* **52** (1958) 8719*i*.
[75]B. E. Conway, R. G. Barradas, and T. Zawidzki, *J. Phys. Chem.* **62** (1958) 676.
[76]S. J. Gregg and S. W. Sing, *Adsorption, Surface Area, and Porosity*, Academic Press, New York, 1967.
[77]G. von Hevesy, *Physik. Z.* **16** (1915) 52; *C.A.* **9** (1915) 2485.
[78]N. Hackerman and S. J. Stephens, *J. Phys. Chem.* **58** (1954) 904.
[79]K. Schwabe, *Isotopentechnik* **1** (1961) 175.
[80]K. Schwabe, *Electrochim. Acta* **6** (1962) 223.
[81]F. Joliot, *J. Chim. Phys.* **27** (1930) 119.
[82]H. D. Cook, *Rev. Sci. Instr.* **27** (1956) 1081.
[83]K. Schwabe, *Chem. Technik* **10** (1958) 469.
[84]E. A. Blomgren and J. O'M. Bockris, *Nature* **186** (1960) 305.
[85]H. Dahms, M. Green, and J. Weber, *Nature* **196** (1962) 1310.
[86]H. Wroblowa and M. Green, *Electrochim. Acta* **8** (1963) 679.
[87]H. Dahms and M. Green, *J. Electrochem. Soc.* **110** (1963) 1075.
[88]J. A. Kafalas and H. C. Gatos, *Rev. Sci. Instr.* **29** (1958) 47.
[89]K. Schwabe and W. Schwenke, *Electrochim. Acta* **9** (1964) 1003.
[90]A. K. N. Reddy, M. A. V. Devanathan, and J. O'M. Bockris, *J. Electroanal. Chem.* **6** (1963) 61.
[91]A. K. N. Reddy, M. Genshaw, and J. O'M. Bockris, *J. Electroanal. Chem.* **8** (1964) 406.
[91a]R. Greef, *J. Chem. Phys.* **51** (1969) 3148; *cf.* M. A. Genshaw and J. O'M. Bockris, *J. Chem. Phys.* **5** (1969) 3149; B. E. Conway and L. Laliberte, in *Faraday Society Symposium on Optical Studies on Adsorbed Layers, London (1970)*; J. McIntyre and D. M. Kolb, *ibid.*, in press.
[92]A. K. N. Reddy, M. G. B. Rao, and J. O'M. Bockris, *J. Chem. Phys.* **42** (1965) 2246.
[93]H. B. Mark and B. S. Pons, *Anal. Chem.* **38** (1966) 119.
[94]W. N. Hansen, R. A. Osteryoung, and T. Kuwana, *J. Am. Chem. Soc.* **88** (1966) 1062.
[95]W. N. Hansen, T. Kuwana, and R. A. Osteryoung, *Anal. Chem.* **38** (1966) 1810.
[96]A. N. Frumkin, *J. Electroanal. Chem.* **18** (1968) 328.
[97]Ref. 62, p. 249.
[98]R. J. Brodd and N. Hackerman, *J. Electrochem. Soc.* **104** (1957) 704.
[99]A. N. Frumkin and B. B. Damaskin, in *Modern Aspects of Electrochemistry, No. 3*, p. 149, Butterworths, London, 1964.
[100]J. O'M. Bockris, in *Electrosorption*, Foreword, Ed., E. Gileadi, Plenum Press, New York, 1967.
[101]A. Gouy, *Ann. Phys. Paris* **7** (1917) 129.
[102]A. N. Frumkin, *Z. Physik. Chem.* **103** (1923) 55.
[103]D. A. McInnes, *Trans. Am. Electrochem. Soc.* **71** (1937) 67.
[104]J. G. Crowther, *Famous American Men of Science*, Penguin Books, 1944, p. 233.
[105]F. G. Donnan, *Commentary on the Scientific Writing of J. W. Gibbs*, Vol. 1, Yale Univ. Press, 1936.
[106]Ref. 51, p. 85.
[107]Ref. 51, p. 5.
[108]R. Parsons, *Modern Aspects of Electrochemistry, No. 1*, p. 103, Ed., J. O'M. Bockris, Butterworths, London, 1954.
[109]R. Parsons and M. A. V. Devanathan, *Trans. Faraday Soc.* **49** (1953) 404.
[110]E. A. Guggenheim, *Trans. Faraday Soc.* **36** (1940) 397.
[111]Ref. 62, pp. 272, 304.
[112]Ref. 52, p. 160.

[113]Ref. 51, introduction by D. H. Everett.
[114]J. K. Eriksson, *Adv. Chem. Phys.* **6** (1964) 142.
[115]A. Harasima, *Adv. Chem. Phys.* **1** (1958) 123.
[116]W. J. Moore, *Physical Chemistry*, pp. 117, 143, Prentice-Hall, Englewood Cliffs, N.J., 1955.
[117]Ref. 62, p. 301.
[118]Ref. 116, p. 503.
[119]D. K. Chattoraj, *J. Phys. Chem.* **70** (1966) 2687.
[120]D. K. Chattoraj, *J. Colloid Sci.* **26** (1968) 379.
[121]B. H. Bijsterbosch and H. J. van den Hull, *J. Phys. Chem.* **71** (1967) 1169.
[122]D. K. Chattoraj, *J. Phys. Chem.* **72** (1968) 1835.
[123]D. C. Grahame and R. Parsons, *J. Am. Chem. Soc.* **83** (1961) 1291.
[124]A. N. Frumkin, R. V. Ivanova, and B. B. Damaskin, *Dokl. Akad. Nauk SSSR* **157** (1964) 1202 [English transl.: *Soviet Phys.—Doklady Phys. Chem.* **157**, p. 822].
[125]B. E. Conway and R. G. Barradas, *Electrochim. Acta* **5** (1961) 319.
[126]M. Rukeyser, *Willard Gibbs*, p. 233, Doubleday, New York, 1942.
[127]E. A. Guggenheim, *Papers of Joint Mtg. Soc. Chim. Phys. and Faraday Soc.* (*Bordeaux, 1947*), p. 11, Butterworths, London, 1949.
[128]Ref. 31, pp. 192–4.
[129]R. G. Barradas and B. E. Conway, *Electrochim. Acta* **5** (1961) 349.
[130]S. G. Meibuhr, *Electrochim. Acta* **10** (1965) 215.
[131]E. Blomgren and J. O'M. Bockris, *J. Phys. Chem.* **63** (1959) 1475.
[132]E. Blomgren and J. O'M. Bockris, *J. Phys. Chem.* **65** (1961) 2000.
[133]R. G. Barradas, P. G. Hamilton, and B. E. Conway, *Collection Czech. Chem. Commun.* **32** (1967) 1796.
[134]J. A. V. Butler, A. Wightman, and W. H. McLennon, *J. Chem. Soc.* **1934**, 528.
[135]R. Parsons and M. A. V. Devanathan, *Trans. Faraday Soc.* **49** (1953) 673.
[136]E. A. Guggenheim and N. K. Adam, *Proc. Roy. Soc.* (*London*) **A139** (1933) 218.
[137]A. W. Adamson, *Physical Chemistry of Surfaces*, p. 87, Interscience, New York, 1967.
[138]J. H. de Boer, *The Dynamical Character of Adsorption*, p. 60, Oxford Univ. Press, 1968.
[139]Ref. 31, pp. 107–108.
[140]H. W. Nuernberg and G. Wolff, *Collection Czech. Chem. Commun.* **30** (1965) 4006; see also R. G. Barradas, P. G. Hamilton, and B. E. Conway, *Collection Czech. Chem. Commun.* **32** (1967) 1790.
[141]C. H. Giles and I. A. Easton, *Adv. Chromatogr.* **3** (1966) 78.
[142]K. Doblhofer, *J. Electrochem. Soc.* **116** (1969) 77C.
[143]J. Lawrence, R. Parsons, and R. Payne, *J. Electroanal. Chem.* **16** (1968) 193.
[144]P. Corbusier and L. Gierst, *Anal. Chim. Acta* **15** (1956) 254.
[145]R. G. Barradas and F. M. Kimmerle, *Can. J. Chem.* **45** (1967) 109.
[146]C. A. Smolders and E. M. Duyvis, *Rec. Trav. Chim.* **80** (1961) 635.
[147]V. I. Melik-Gaikazyan, V. V. Voronchikhina, and E. A. Zakharova, *Soviet Electrochemistry* **4** (1968) 426.
[148]R. G. Barradas, F. M. Kimmerle, and E. M. L. Valeriote, *J. Polarog. Soc.* **13** (1967) 30.
[149]D. M. Mohilner and P. R. Mohilner, *J. Electrochem. Soc.* **115** (1968) 261.
[150]L. Gierst and P. Herman, *Z. Anal. Chem.* **216** (1966) 238.
[151]J. Heyrovsky, F. Sorm, and J. Forejt, *Collection Czech. Chem. Commun.* **12** (1947) 11.
[152]C. Pecasse, unpublished results (1964).
[153]B. E. Conway, *Chem. Can.* **12** (1960) 40.

[154]G. S. Smith, *Trans. Faraday Soc.* **47** (1951) 63.
[155]R. S. Hansen, D. J. Kelsh, and D. H. Grantham, *J. Phys. Chem.* **67** (1963) 2316.
[156]L. G. M. Gordon, J. Halpern, and B. E. Conway, *J. Electroanal. Chem.* **21** (1969) P3; see also L. G. M. Gordon and B. E. Conway, *J. Electroanal. Chem.* **15** (1967) 7.
[157]B. Nygard, *Acta Chem. Scand.* **15** (1961) 1039.
[158]B. Nygard, E. Johansson, and J. Oloffson, *J. Electroanal. Chem.* **12** (1966) 564.
[159]J. W. Hayes, *Anal. Chem.* **37** (1965) 1444.
[160]G. C. Barker and I. L. Jenkins, *Analyst* **77** (1952) 685.
[161]R. C. Propst and M. H. Goosey, *Anal. Chem.* **36** (1964) 2382.
[162]B. Kastening, *Z. Elektrochem.* **68** (1964) 979.
[163]J. Riha, in *Advances in Polarography*, Vol. I, p. 210, Ed., I. S. Longmuir, Pergamon, New York, 1960.
[164]A. J. Bard and H. B. Herman, *Anal. Chem.* **37** (1965) 317.
[165]C. Cachet, I. Epelboin, S. Grimnes, and J. C. Lestrade, *J. Chim. Phys.* **65** (1968) 306.
[166]L. Gierst and P. Herman, unpublished results.
[167]F. Dewiest, Memoire de Licence, Free Univ. of Brussels (1968).
[168]H. G. Moeller, *Z. Physik. Chem.* **65** (1909) 226; *C.A.* **3** (1909) 1239.
[169]A. N. Frumkin, A. Gorodezkaya, B. Kabanov, and N. Nekrasov, *Zh. Fiz. Khim.* **1** (1932) 255; *C.A.* **26** (1932) 4996.
[170]B. Kabanov and A. N. Frumkin, *Z. Physik. Chem.* **165** (1933) 433.
[171]A. Gorodezkaya and B. Kabanov, *Zh. Fiz. Khim.* **5** (1934) 418; *C.A.* **28** (1936) 4286^6.
[172]B. Kabanov and N. Ivanishenko, *Acta Physicochim. URSS* **6** (1937) 701; *C.A.* **32** (1938) 4855^9.
[173]I. P. Tverdovskii and A. N. Frumkin, *Zh. Fiz. Khim.* **21** (1947) 819; *C.A.* **42** (1948) 2160*b*.
[174]P. Rehbinder and E. Wenström, *Acta Physicochim. URSS* **19** (1944) 36.
[175]R. E. D. Clark, *Trans. Faraday Soc.* **42** (1946) 449.
[176]J. O'M. Bockris and R. Parry-Jones, *Nature* **171** (1953) 930.
[177]A. Pfuetzenreuther and G. Masing, *Z. Metallk.* **42** (1951) 361; *C.A.* **46** (1952) 3428*i*.
[178]T. S. Beck, *CITCE, 19th Mtg., Detroit, 1968*, Extended Abstracts, p. 214.
[179]R. E. Johnson, *J. Phys. Chem.* **63** (1959) 1655.
[180]Ref. 51, p. 394.
[181]G. Dryhurst, private communication.
[182]G. Tessari, P. Delahay, and K. Holub, *J. Electroanal. Chem.* **17** (1968) 69.
[183]R. G. Barradas and J. L. A. French, *Anal. Chem.* **39** (1967) 1038.
[184]M. Knudsen, *Ann. Physik* **47** (1915) 697.
[185]J. W. Loveland and P. J. Elving, *J. Phys. Chem.* **56** (1952) 250, 255.
[186]J. N. Butler and M. L. Meehan, *J. Phys. Chem.* **70** (1966) 3582.
[187]S. Vavricka, L. Nemec, and J. Koryta, *Collection Czech. Chem. Commun.* **31** (1966) 947.
[188]G. Lanco and R. A. Osteryoung, *Anal. Chem.* **39** (1967) 1866.
[189]H. Angerstein-Kozlowska and B. E. Conway, *J. Electroanal. Chem.* **7** (1964) 109.
[190]L. Gierst, *Anal. Chim. Acta* **15** (1956) 262.
[191]L. Gierst and P. Mechelynck, *Anal. Chim. Acta* **12** (1955) 79.
[192]L. Ramaley and C. G. Enke, *J. Electrochem. Soc.* **112** (1965) 943.
[193]V. I. Melik-Gaikazyan and P. I. Dolin, *Dokl. Akad. Nauk SSSR* **66** (1949) 409; *C.A.* **43** (1949) 6522*g*.
[194]V. I. Melik-Gaikazyan, *Zh. Fiz. Khim.* **26** (1952) 560; translation supplied by D. C. Grahame.

[195]B. B. Damaskin, *Zh. Fiz. Khim.* **32** (1958) 2199; *C.A.* **53** (1959) 11064*i*.
[196]D. Britz and H. H. Bauer, *Electrochim. Acta* **13** (1968) 347.
[197]G. H. Nancollas and C. A. Vincent, *J. Sci. Instr.* **40** (1963) 306.
[198]M. W. Breiter, *J. Electrochem. Soc.* **109** (1962) 42.
[199]M. W. Breiter, *J. Electroanal. Chem.* **7** (1964) 38.
[200]G. M. Schmid, *J. Electrochem. Soc.* **115** (1968) 1034.
[201]H. Buckingham and E. M. Price, *Principles of Electrical Measurements*, English Universities Press, 1955.
[202]G. J. Hills and R. Payne, *Trans. Faraday Soc.* **61** (1965) 316.
[203]J. B. Hayter, *J. Electroanal. Chem.* **19** (1968) 181.
[204]G. C. Barker and R. L. Faircloth, in *Advances in Polarography*, Vol. I, p. 313, Ed., I. S. Longmuir, Pergamon, New York, 1960.
[205]B. D. Cahan and P. Ruetschi, *J. Electrochem. Soc.* **106** (1959) 543.
[206]J. J. McMullen and N. Hackerman, *J. Electrochem. Soc.* **106** (1959) 341.
[207]J. S. Riney, G. M. Schmid, and N. Hackerman, *Rev. Sci. Instr.* **32** (1961) 588.
[208]C. C. Krischer and R. A. Osteryoung, *J. Electrochem. Soc.* **112** (1965) 735.
[209]B. B. Damaskin, *Russian Chem. Revs.* **34** (1965) 752.
[210]D. C. Grahame, *J. Am. Chem. Soc.* **68** (1946) 301.
[211]H. H. Bauer, M. S. Spritzer, and P. J. Elving, *J. Electroanal. Chem.* **17** (1968) 299.
[212]D. C. Grahame, *J. Electrochem. Soc.* **99** (1952) 370C.
[213]T. I. Borisova and B. V. Ershler, *Zh. Fiz. Khim.* **24** (1950) 337; *C.A.* **44** (1950) 6747*c*.
[214]D. C. Grahame, R. E. Ireland, and R. C. Petersen, Office of Naval Research, Contract N8-ONR-66903, *Tech. Rept.* No. 22, March 23, 1956.
[215]J. O'M. Bockris and B. E. Conway, *J. Chem. Phys.* **28** (1958) 707.
[216]J. Clavilier, *Compt. Rend.* **257** (1963) 3889.
[217]J. Clavilier, *Compt. Rend.* **261** (1965) 2647.
[218]J. Clavilier and R. Pineaux, *Compt. Rend.* **260** (1965) 891.
[219]R. Pineaux, *Compt. Rend.* **258** (1964) 1790.
[220]W. D. Robertson, *J. Electrochem. Soc.* **100** (1953) 194.
[221]L. Ramaley and C. G. Enke, *J. Electrochem. Soc.* **112** (1965) 947.
[222]G. J. Hills and K. E. Johnson, *J. Electrochem. Soc.* **108** (1961) 1013.
[223]Chuan-sin Tsa and S. Iofa, *Dokl. Akad. Nauk SSSR* **131** (1960) 137 [English transl.: *Soviet Phys.—Doklady Phys. Chem.* **130–135**, 231].
[224]J. N. Sarmousakis and M. J. Prager, *J. Electrochem. Soc.* **104** (1957) 454.
[225]D. J. Barclay, *J. Electroanal. Chem.* **19** (1968) 318.
[226]R. de Levie, *Electrochim. Acta* **10** (1965) 113.
[227]R. de Levie, *Adv. Electrochem. Electrochem. Eng.* **6** (1967) 387.
[228]J. O'M. Bockris, W. Mehl, B. E. Conway, and L. Young, *J. Chem. Phys.* **25** (1956) 776.
[229]D. I. Leikis and E. S. Sevastyanov, *Dokl. Akad. Nauk SSSR* **144** (1962) 320 [English transl.: *Soviet Phys.—Doklady Phys. Chem.* **142–144**, 495].
[230]D. C. Grahame, in *Soviet Electrochemistry* (Proc. 4th Conf. Electrochem.), Vol. 1, p. 31, Consultants Bureau, New York, 1961.
[231]H. H. Bauer and D. Britz, unpublished work.
[232]J. Newman, *J. Electrochem. Soc.* **117** (1970) 198.
[233]E. A. Ukshe, N. G. Bukun, and D. I. Leikis, *Russian. J. Phys. Chem.* **36** (1962) 1260.
[234]D. I. Leikis, E. S. Sevastyanov, and L. L. Knots, *Russian. J. Phys. Chem.* **38** (1964) 997.
[235]K. Rozenthal and B. Ershler, *Zh. Fiz. Khim.* **22** (1948) 1344; translated by R. Parsons.
[236]W. Lorenz, *Z. Elektrochem.* **58** (1954) 912.

[237]A. D. Graves and D. Inman, *Nature* **208** (1965) 481.
[238]H. A. Laitinen and C. G. Enke, *J. Electrochem. Soc.* **107** (1960) 773.
[239]H. A. Laitinen and D. K. Roe, *Collection Czech. Chem. Commun.* **25** (1960) 3065.
[240]M. V. Perfilev, S. F. Palguev, and S. V. Karpachev, *Soviet Electrochemistry* **1** (1965) 74.
[241]A. A. Rakov, T. I. Borisova, and B. V. Ershler, *Zh. Fiz. Khim.* **22** (1948) 1390; *C.A.* **43** (1949) 2522*h*.
[242]G. M. Schmid and N. Hackerman, *J. Electrochem. Soc.* **109** (1962) 243.
[243]G. M. Schmid and N. Hackerman, *J. Electrochem. Soc.* **110** (1963) 440.
[244]S. Schuldiner, *J. Electrochem. Soc.* **99** (1952) 488.
[245]S. Schuldiner, *J. Electrochem. Soc.* **101** (1954) 426.
[246]C. Wagner, *J. Electrochem. Soc.* **97** (1950) 71.
[247]D. L. Hill, G. J. Hills, L. Young, and J. O'M. Bockris, *J. Electroanal. Chem.* **1** (1959) 79.
[248]P. J. Hillson, *Trans. Faraday Soc.* **50** (1954) 385.
[249]J. E. B. Randles and K. W. Somerton, *Trans. Faraday Soc.* **48** (1952) 937.
[250]K. J. Vetter, *Z. Physik. Chem.* **199** (1952) 285.
[251]B. Breyer and H. H. Bauer, *Australian J. Chem.* **8** (1955) 472.
[252]B. Breyer and H. H. Bauer, *Australian J. Chem.* **8** (1955) 480.
[253]T. Berzins and P. Delahay, *J. Phys. Chem.* **59** (1955) 906.
[254]W. Lorenz and F. Möckel, *Z. Elektrochem.* **60** (1956) 507.
[255]P. Delahay and D. M. Mohilner, *J. Phys. Chem.* **66** (1962) 959.
[256]P. Delahay and D. M. Mohilner, *J. Am. Chem. Soc.* **84** (1962) 4247.
[257]P. Delahay, *J. Phys. Chem.* **67** (1963) 135.
[258]A. N. Frumkin and V. I. Melik-Gaikazyan, *Dokl. Akad. Nauk SSSR* **77** (1951) 855; English translation made available by D. C. Grahame.
[259]B. B. Damaskin and G. A. Tedoradze, *Electrochim. Acta* **10** (1965) 529.
[260]Ref. 12, p. 1410.
[261]Ref. 12, p. 1411.
[262]A. N. Frumkin, *Z. Physik. Chem.* **116** (1925) 466.
[263]W. Lorenz, *Z. Elektrochem.* **62** (1958) 192.
[264]M. Senda and P. Delahay, *J. Am. Chem. Soc.* **83** (1961) 2763.
[265]R. Parsons, *Adv. Electrochem. Electrochem. Eng.*, **1** (1961) 1.
[266]W. Lorenz, *Z. Physik. Chem.* **18** (1958) 1.
[267]W. Lorenz, *Z. Physik. Chem.* **26** (1960) 424.
[268]W. Lorenz and W. Müller, *Z. Physik. Chem.* **25** (1960) 161.
[269]P. Delahay and I. Trachtenberg, *J. Am. Chem. Soc.* **79** (1957) 2355.
[270]P. Delahay and C. T. Fike, *J. Am. Chem. Soc.* **80** (1958) 2628.
[271]G. C. Barker, *Trans. Symp. Electrode Processes*, *Philadelphia*, *1959*, p. 325, Ed., E. Yeager, John Wiley and Sons, New York, 1961.
[272]H. H. Bauer and A. K. Shallal, *Nature* **214** (1967) 381.
[273]W. H. Reinmuth, *Adv. Anal. Chem. Instr.* **1** (1960) 241.
[274]C. N. Reilley and W. Stumm, in *Progress in Polarography*, Vol. I, p. 81, Ed., P. Zuman, Interscience, New York, 1962.
[275]B. Kastening and L. Holleck, *Talanta* **12** (1965) 1259.
[276]H. W. Nürnberg and M. von Stackelberg, *J. Electroanal. Chem.* **4** (1962) 1.
[277]Ref. 29, pp. 269–70.
[278]T. A. Kryukova and A. N. Frumkin, *Zh. Fiz. Khim.* **23** (1949) 819; *C.A.* **43** (1949) 8911*e*.
[279]A. N. Frumkin and V. G. Levich, *Zh. Fiz. Khim.* **21** (1947) 1183, 1335; *C.A.* **42** (1948) 3244*b*, 5355*b*.
[280]R. S. Hansen and B. H. Clampitt, *J. Phys. Chem.* **58** (1954) 908.

[281]Z. A. Soloveva, *Russian J. Phys. Chem.* **34** (1960) 254.
[282]R. W. Schmid and C. N. Reilley, *J. Am. Chem. Soc.* **80** (1958) 2087.
[283]P. W. Board, D. Britz, and R. V. Holland, *Electrochim. Acta* **13** (1968) 1633.
[284]J. Heyrovsky, *Faraday Soc. Disc.* **1** (1947) 212.
[285]A. N. Frumkin, *Dokl. Akad. Nauk SSSR* **85** (1952) 373; *C.A.* **46** (1952) 10956*i*.
[286]B. Lovrecek and N. Marincic, *Electrochim. Acta* **11** (1966) 237.
[287]C. A. Smolders, *Rec. Trav. Chim.* **80** (1961) 650.
[288]C. A. Smolders, *Rec. Trav. Chim.* **80** (1961) 699.
[289]M. Green, D. A. J. Swinkels, and J. O'M. Bockris, *Rev. Sci. Instr.* **33** (1962) 18.
[290]B. B. Damaskin, A. A. Survila, S. Ya. Vasina, and A. I. Fedorova, *Soviet Electrochemistry* **3** (1967) 729.
[291]D. Tytgat, Thesis, Free Univ. of Brussels (1969).
[292]E. Gileadi, *J. Electroanal. Chem.* **11** (1966) 146.
[293]A. W. Adamson, *J. Chem. Educ.* **44** (1967) 710.
[294]J. E. B. Randles, *Adv. Electrochem. Electrochem. Eng.* **3** (1963) 1.
[295]J. Llopis, *Trans. Symp. Electrode Processes, Philadelphia 1959*, p. 306, Ed., E. Yeager, John Wiley and Sons, New York, 1961.
[296]R. I. Kaganovitch and B. B. Damaskin, *Soviet Electrochemistry* **4** (1968) 221.
[297]D. O. Shah and R. W. Capps, *J. Colloid Sci.* **27** (1968) 319.
[298]P. Joos, *Bull. Soc. Chim. Belges* **76** (1967) 591.
[299]J. L. Ord and D. J. DeSmet, *J. Electrochem. Soc.* **113** (1966) 1258.
[300]H. A. Laitinen and H. C. Gaur, *J. Electrochem. Soc.* **104** (1957) 730.
[301]Ref. 29, Chapter 3.
[302]D. J. Schiffrin, *J. Electroanal. Chem.* **23** (1969) 168.
[303]J. Lawrence and D. M. Mohilner, Spring Mtg. Electrochemical Soc., Los Angeles, May 1970, *Extended Abstracts*, p. 802, no. 303.
[304]J.-P. Randin and E. Yeager, *J. Electrochem. Soc.* **118** (1971) 711.
[305]B. B. Damaskin, O. A. Petrii, and V. V. Batrakov, *Adsorption of Organic Compounds on Electrodes*, Plenum Press, New York, 1971.

4

Transport-Controlled Deposition and Dissolution of Metals

A. R. Despić and K. I. Popov

University of Belgrade
Belgrade, Yugoslavia

I. PHENOMENA INVOLVED

Metal deposition and dissolution usually involve more than a simple discharge or formation of hydrated metal ions. In most practical cases, reactions of complex dissociation or formation occur simultaneously as well as deposition and incorporation of foreign substances (impurities or additives), etc. They all affect the structure and appearance of the deposit obtained. In particular, the shape of the deposit, the surface roughness, and surface coarseness (see Section 2) are found to depend strongly on the path of the flow of the electric current (traditionally termed "primary current distribution") and the path and rate of flow of the species involved, directly or indirectly, in the electrolytic process. Thus, the last factor has a profound effect on the morphology of the deposit and may be a cause of seemingly unrelated phenomena:

(a) *Amplification of surface irregularities* has been known to appear in electrode processes of high specific rate when deposition is carried out at low concentrations of simple ions in solution.[1,2] A special example of such a case is shown in Figure 1.

Since surface roughness is defined as the ratio between the true and the apparent surface area, in the case when it is made up of elevations with triangular cross section, it is equal to the inverse

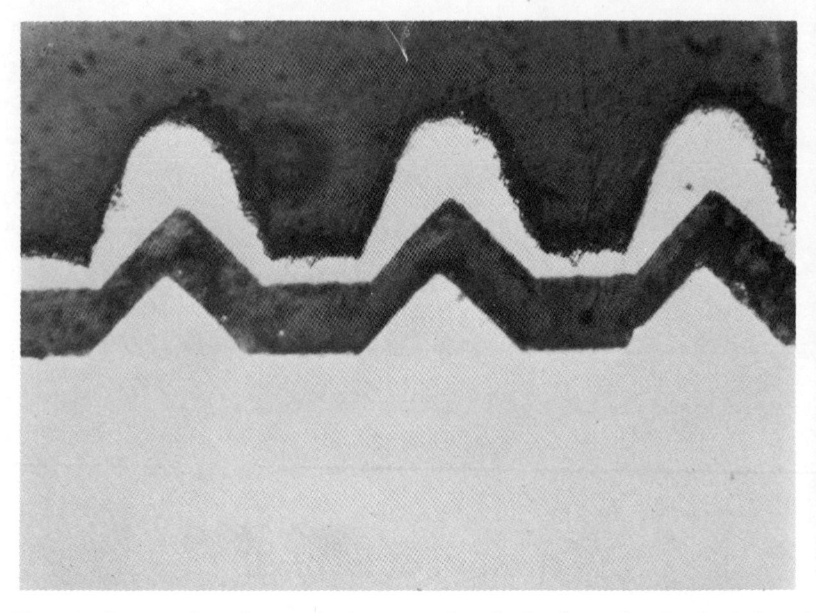

Figure 1. Cross-section micrograph of a copper deposit plated out of a stirred solution of $CuSO_4$ (1.0 N) onto a surface with triangularly shaped elevations (750 ×) (from Despić and Popov[3]).

cosine of the elevation angle. This is seen to undergo a considerable increase during plating.

(b) *Dendritic growth* represents the extreme case of the above phenomenon, when at some points of the surface, protrusions occur, penetrating deeply into the solution. Needlelike, spadelike, or pine-treelike deposits of well-defined crystalline nature are obtained (Figure 2). A large number of phenomenological studies have been reported (see Refs. 4 and 5).

(c) *Growth of whiskers*, i.e., threadlike deposits (Figure 3), is known to occur when deposition is carried out from baths containing substances which exhibit a tendency for preferential adsorption on certain crystal planes.[7,8]

(d) *Powdery deposits* are formed in many metal depositions in a current density range in which sufficient difficulties in the transport of depositing species are encountered.[10]

(e) *Leveling* is the phenomenon opposite to the ones described above, when surface irregularities are ironed out at prolonged

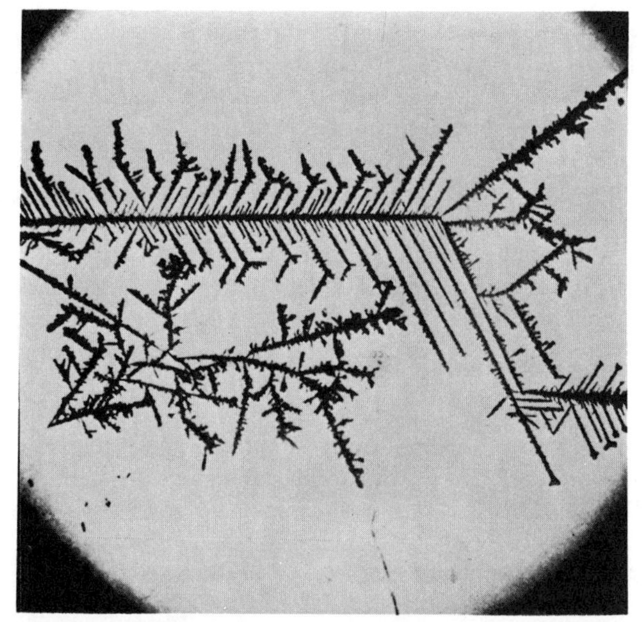

Figure 2. Photomicrograph of a pine-treelike dendritic deposit of silver from silver nitrate solution (from Wranglen[4]).

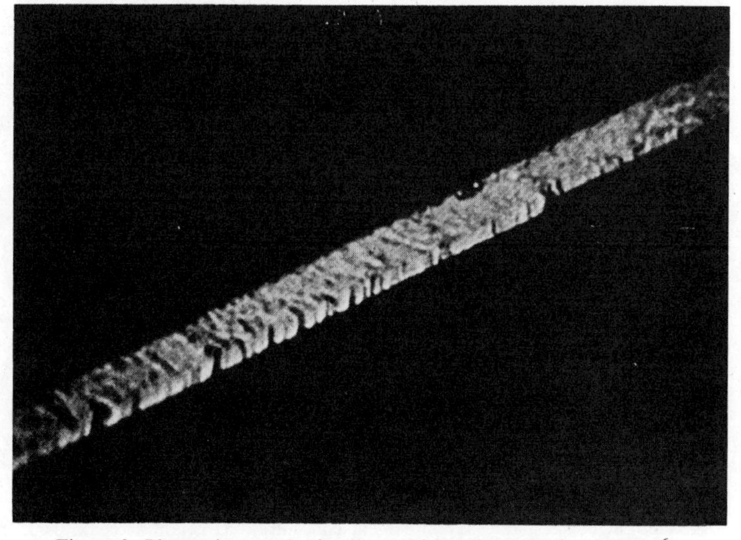

Figure 3. Photomicrograph of a silver whisker (from Graf and Weser[6]).

deposition. An example of this phenomenon is shown in Figure 4. The effect is obtained when plating is carried out from concentrated metal salt baths containing strongly adsorbing organic substances as additives in small concentrations.[11]

(f) *Electropolishing* has been known, since the work of Jacquet,[12] to occur when a metal surface is subjected to anodic dissolution in special baths containing small concentrations of agents with a strong complexing affinity for the metal ions, in an electrolyte forming insoluble products with the same ions. A great number of practical applications of this phenomenon have been described in the electroplating literature.[13] A typical example of this effect is shown in Figure 5.

(g) *Prevention of dendritic growth* by application of pulsating current (i.e., half-rectified ac) for deposition has recently been reported in several cases as being a process with great practical potential.[14–16]

The present status of our understanding of these phenomena shall be reviewed below. A common feature in all of them will be shown to be the difficulty in transporting some species involved in

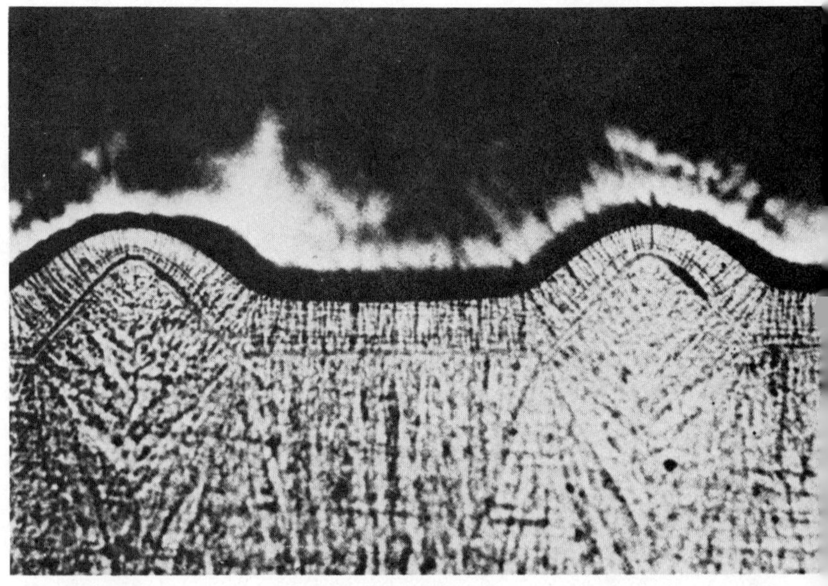

Figure 4. Semibright nickel at low leveling-agent concentration; height at peak, 0.00275 cm (from Ref. 11).

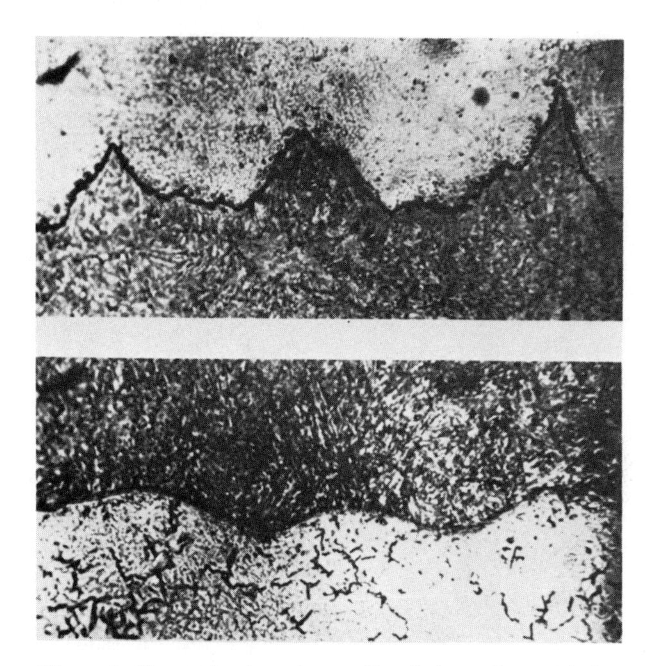

Figure 5. Change in the microprofile of the surface of lathe-worked steel (a), upon electropolishing (b) (from Ref. 9).

the process and the possibility for a unified treatment of all these transport-controlled phenomena will be outlined at the end.

II. AMPLIFICATION OF SURFACE IRREGULARITIES

Any solid metal surface, which represents a substrate for metal deposition, possesses a certain roughness. In addition, it may appear coarse or smooth, and this is not necessarily related to the roughness. Figure 6 shows cases of surfaces with (a) equal roughness and profoundly different coarseness and (b) vice versa. It is the level of coarseness which determines the appearance of metal deposits, while even with considerable roughness, if below the visual level, the surface may appear smooth.

It is convenient to define the surface coarseness as the difference in thickness of the metal at the highest and lowest points above an arbitrary reference plane facing the solution. [In early models used to describe the surface by periodic functions (Figure 7), this is equal to twice the amplitude of the function.]

Historically, it was first realized that under certain conditions of *dissolution*, the surface coarseness tends to decrease[17] (cf. electropolishing). Krichmar[18] was the first to point out that in some cases of *deposition*, under conditions in a way analogous to the previous ones, an inverse effect should take place, i.e., in prolonged cathodic reduction, at conditions close to those of complete diffusive control of the process, the amplification of both the surface roughness and the surface coarseness arises.

Taking a sinusoidal profile for the electrode,

$$H = H_0 \sin(2\pi x/a) \tag{1}$$

(cf. Figure 7), he assumed that the rate of increase in surface coarseness must be given as

$$\frac{dH_0}{dt} = \frac{M}{2\rho nF}(i_\lambda - i_\nu) = \frac{MD}{2\rho}\left[\left(\frac{\partial c}{\partial y}\right)_\lambda - \left(\frac{\partial c}{\partial y}\right)_\nu\right] \tag{2}$$

where M and ρ are the molecular weight and the density of the metal, respectively, and i_λ and i_ν represent the current densities at the highest and lowest points of the profile, respectively, which are

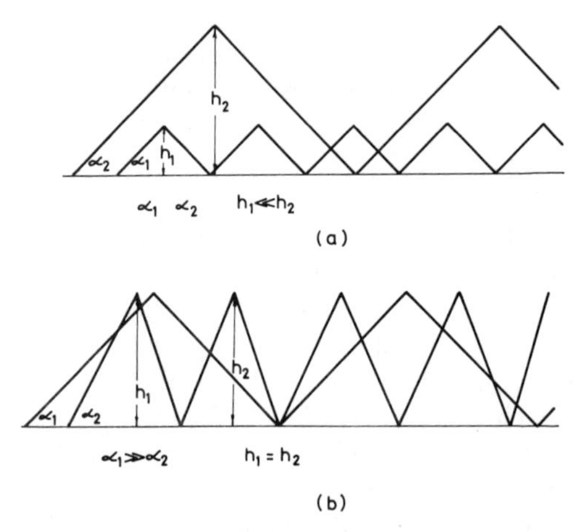

Figure 6. Models of surfaces with (a) equal surface roughness and different coarseness and (b) vice versa.

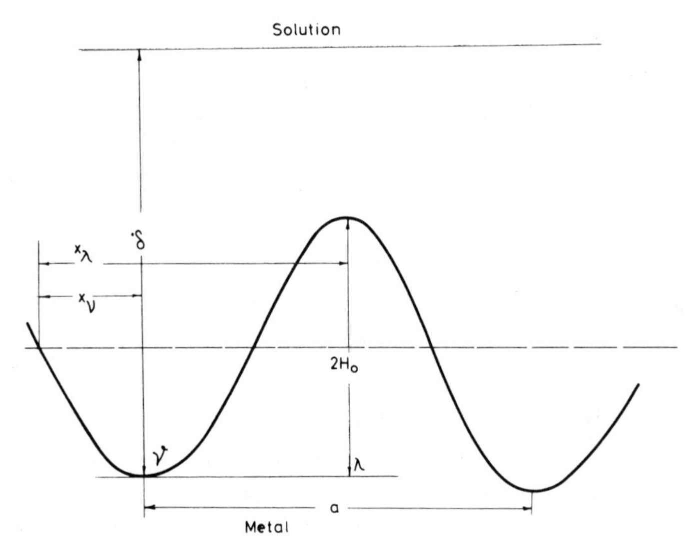

Figure 7. Sinusoidal profile of a model surface for considering transport-controlled phenomena, as used by Wagner[17] and Krichmar.[18]

proportional to the diffusion fluxes of the depositing ions at these points (the two terms inside the square brackets).

To find the concentration gradients $(\partial c/\partial y)_\lambda$ and $(\partial c/\partial y)_\nu$, the solution of the two-dimensional form of Fick's second law is needed, and this is found to be

$$C - C^\circ = (C_0 - C^\circ)\left[\frac{y}{\delta} - H_0\left(\frac{1}{\delta} - \frac{\gamma C_0}{C_0 - C^\circ}\right) \times \frac{\sinh(2\pi y/a)}{\sinh(2\pi\delta/a)}\right] \sin\frac{2\pi x}{a} \tag{3}$$

where δ is the thickness of the Nernst diffusion layer* and γ is a constant, while C° and C_0 are the concentration of the depositing species in the bulk of solution and the average concentration at the surface, respectively.

*According to the Nernst concept of the diffusion layer, δ is given as the distance between the electrode and the point in the solution at which the bulk concentration is attained *when it is assumed that the concentration gradient is constant with distance and equal to that at the electrode surface.*

Differentiating equation (3) with respect to y at points λ and v for the case of $H_0 \ll a$, and replacing in equation (2), one obtains

$$\frac{dH_0}{dt} = \frac{2Mi_L H_0 \pi}{\rho n F a}\left(1 - \frac{C_0 \gamma \delta}{C_0 - C^\circ}\right) \coth \frac{2\pi\delta}{a} \tag{4}$$

where

$$i_L = nFD(C^\circ - C_0)/\delta \tag{5}$$

would be the diffusion-controlled current density for the corresponding situation with a flat surface.

The constant γ was derived to be approximately as

$$\gamma \simeq (\partial C_0/\partial \delta)(1/C_0) \tag{6}$$

i.e., to represent a relative variation of the average surface concentration with the diffusion layer thickness.

If the current density–potential relation is determined by the usual expression,

$$i = C_0 a \exp[(\alpha_c F/RT)\Delta\varphi] \tag{7}$$

i.e., when deposition is activation-controlled and the average current density is constant, then,

$$\gamma = (C_0 - C^\circ)/C_0 \delta \tag{8}$$

To obtain an experimental proof of the derived theory, equation (4) was integrated to give

$$H_0(t) = H_0(0) e^{t/\tau} \tag{9}$$

where

$$\tau = (\rho n F a i_L / 2\pi M i^2) \tanh(2\pi\delta/a) \tag{10}$$

with

$$i_L = nFDC^\circ/\delta = BC^\circ \tag{11}$$

This was checked experimentally by carrying out deposition of silver from an iodide electrolyte onto an electrolytic copper surface profiled on a lathe. Sinusoidal profiles were obtained by electropolishing the samples in 7.5 M H_3PO_4 for 2–3 min. The amplitudes of the profile were measured before and after depositions of given durations, by examining the metallographic samples under a micro-

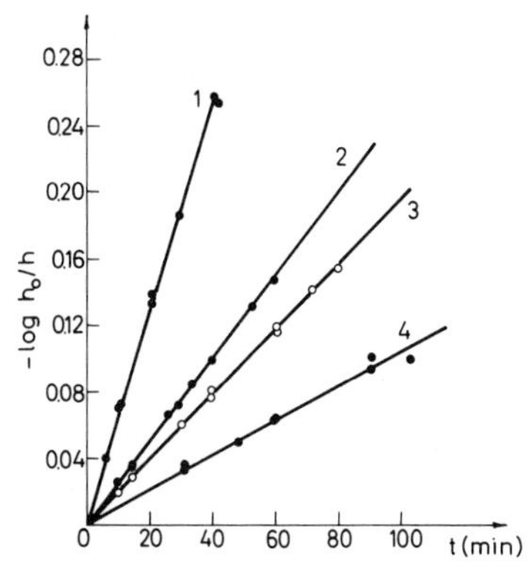

Figure 8. The dependence of the height of a sinusoidal elevation on time of deposition (from Krichmar[18]).

scope. The curves of Figure 8 were thus constructed, and these support the derived equation (9).

Simpler mathematics were used in another, independently derived theory of the same phenomenon by Despić *et al.*[2,20] which, however, also took into account some additional factors.

A somewhat different and more complete presentation is given here. Consider the model of a surface irregularity shown in Figure 9. It is buried deep in the diffusion layer δ, which is determined by a steady linear diffusion to the flat portion of the surface.

Any point at the surface moves into the solution at a rate given by

$$dy/dt = (M/\rho nF)i(x) \tag{12}$$

if $i(x)$ is the local current density.

To derive this at the level of microscopic surface irregularities, one has to take into account the effect of the radius of curvature r of the surface on the free energy of the final state of the reaction, G_r°. This is given by a Kelvin-type effect, whereby

$$(G_i^\circ)_r = (G_j^\circ) + (2\sigma V/r) \tag{13}$$

and σ is the surface tension at the interface concerned (cf. the Barton–Bockris theory[19]).

One should note that in general two cases may arise, leading to different effects of this factor on the rate of reaction. The first is the

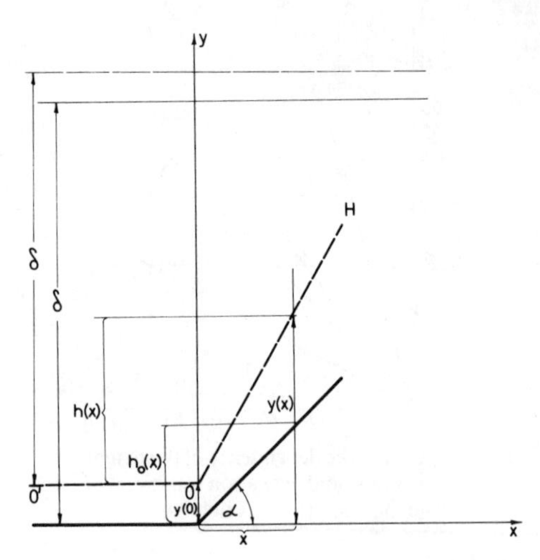

Figure 9. Model of metal deposition under diffusion control onto an elevation rising at an angle α from a surface plane.

case in which the affected state is the final state of the rate-determining step. As exemplified by the reduction of silver ions (Figure 10), it is seen that the shift of the potential energy curves for the two states behind the energy barrier to higher values of the free energy produce changes in the activation energy $\Delta G^{\circ\ddagger}$ in both the anodic and cathodic directions, amounting to $-2\beta\sigma V/r$ and $2(1-\beta)\sigma V/r$, respectively.

One can derive that, in this case, the current density must be given by

$$i(x) = i_0\left\{\frac{C_i^{\mathrm{M}^+}}{C_e^{\mathrm{M}^+}}\exp\left[-\frac{2\beta\sigma V}{rRT}\right]\exp\left[-\frac{\beta F\eta}{RT}\right] - \exp\left[\frac{2(1-\beta)\sigma V}{rRT}\right]\exp\left[\frac{(1-\beta)F\eta}{RT}\right]\right\} \tag{14a}$$

where i_0 is the exchange current density at the flat surface, and the ratio of the concentration of the M^+ species at the current density i, $C_i^{\mathrm{M}^+}$, to that at equilibrium, $C_e^{\mathrm{M}^+}$, allows for the concentration polarization with respect to the ions in solution.

The second case is that at which the affected state is not the rate-determining step. As shown in Figure 10, exemplified by the two one-electron reaction steps of the reduction of Zn^{2+} ions, the shift in the free energies of Zn and Zn-adatom species does not produce any effect on the activation energy barrier of the rate-determining step. Instead, it results in an increase in the concentration of the univalent intermediate in equilibrium with the final state. Hence, the anodic partial current is increased, but the cathodic remains unaffected. Taking this into account in the derivation of the current density, one obtains

$$i(x) = i_0\left\{\frac{C_i^{M^{z+}}}{C_e^{M^{z+}}}\exp\left(-\frac{\beta F\eta}{RT}\right) - \exp\left(\frac{2\sigma V}{RTr}\right)\exp\left[\frac{(2-\beta)F\eta}{RT}\right]\right\} \quad (14b)$$

The ratio of concentrations of the ionic species to be reduced may be written as

$$C_i^{M^+}/C_e^{M^+} = [1 - (i/i_L)] \quad (15)$$

where i_L is the limiting current density. Introducing (15) into (14a) and (14b) and solving for $i(x)$, one obtains general equations applicable to any point at the surface,

$$i(x) = i_L\frac{i_0[K'(r)f_c(\eta) - K''(r)f_a(\eta)]}{i_L + i_0K'(r)f_c(\eta)} \quad (16a)$$

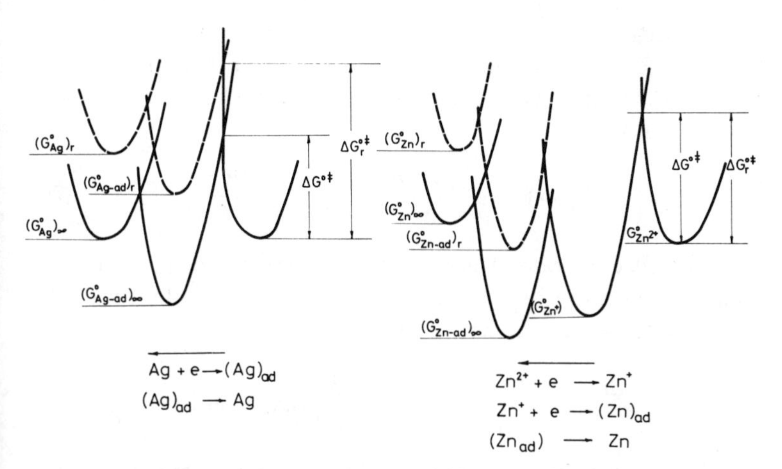

Figure 10. Potential energy diagrams for the reduction of silver and zinc ions. $\Delta G^{\circ\ddagger}$—activation energy at a flat surface; $\Delta G_r^{\circ\ddagger}$—activation energy at a curved surface with radius of curvature r.

or

$$i(x) = i_L \frac{i_0[f_c(\eta) - K(r) f_a(\eta)]}{i_L + i_0 K(r) f_c(\eta)} \tag{16b}$$

where

$$f_c(\eta) = \exp(-\alpha_c F\eta/RT); \qquad f_a(\eta) = \exp(\alpha_a F\eta/RT) \tag{17}$$

where α_c and α_a are the cathodic and anodic transfer coefficients, respectively, and

$$K' = -2\beta\sigma V/r; \qquad K'' = 2(1 - \beta)\sigma V/r; \qquad K = 2\sigma V/r \tag{18}$$

The rate of growth at three parts of the surface should now be considered:

(a) *At the flat part of the surface* (cf. Figure 9), the i_L is that for steady-state linear diffusion, i.e.,

$$i_{L(l)}(0) = nFDC^\circ/\delta \tag{19}$$

and the Kelvin term in equation (16) equals one.

(b) *At the side of an irregularity*, even when neglecting a possible lateral diffusion flux supplying the depositing ions, the rate of motion of any point must be larger than that at which the flat surface moves. This is because the point is closer to the diffusion layer boundary, i.e., the effective diffusion layer is thinner and hence the diffusion flux and the resulting current density are larger. The limiting current density is given as

$$i_{L(l)}(x) = \frac{nFDC^\circ}{\delta + y(0) - y(x)} = \frac{nFDC^\circ}{\delta - h(x)} \tag{20}$$

The effective rate of growth of the elevation at the point x is equal to the rate of motion relative to the rate of motion of the reference plane, i.e., the flat surface,

$$\frac{dh(x)}{dt} = \frac{dy(x)}{dt} - \frac{dy(0)}{dt} = \frac{M}{\rho nF}[i(x) - i(0)] \tag{21}$$

The current densities are given by equation (16) with appropriate values of i_L and with the Kelvin factor being equal to one in both cases, since the side of an irregularity is also assumed flat.

Replacing and rearranging, one obtains

$$\frac{dh(x)}{h(x)\,dt} = \frac{MDC^\circ}{\rho} \frac{1 - [f_a(\eta)/f_c(\eta)]}{\{[nFDC^\circ/i_0 f_c(\eta)] + \delta\}\{[nFDC^\circ/i_0 f_c(\eta)] + \delta - h\}}$$

$$= \frac{A}{B(B - h)} \tag{22}$$

This can be conveniently termed as the amplification factor, for it defines the relative rate of increase of surface coarseness.

At the beginning of deposition, as long as $\delta \gg h$, this is seen to be constant and defined by an interplay of activation and diffusion parameters. Integration of this equation under these conditions yields an exponential dependence of the elevation of the given point on time,

$$h(x) = h_0(x)e^{t/\tau} \tag{23}$$

where

$$\tau_i = \frac{\rho}{MD} \frac{[\delta + nFDC_0/i_0 f_c(\eta)]^2}{[1 - f_a(\eta)/f_c(\eta)]C^\circ} \tag{24}$$

This is in accordance with the findings of Krichmar. However, the time constant τ has a clearer meaning. Also it applies both to reversible and irreversible deposition, i.e., deposition close to the equilibrium electrode potential and far from it.

The increase in surface roughness can also be derived from the above equations if the same model of the surface is used. The roughness factor, i.e., the ratio of the true to the apparent surface area in this two-dimensional presentation, is

$$f_R = \frac{\overline{OA}}{\overline{OB}} = \tan\alpha = \frac{(x^2 + h^2)^{1/2}}{x} = \left(1 + \frac{h^2}{x^2}\right)^{1/2}$$

$$= \left[1 + \left(\frac{h_0}{x}\right)^2 \exp\left(2\frac{t}{\tau}\right)\right]^{1/2} \tag{25}$$

(c) *At a tip of an irregularity*, there is an additional reason for the increased rate of motion. The lateral diffusion flux cannot be neglected and the situation can be approximated by assuming spherical diffusion conditions, with the limiting current density given by

$$i_{L(s)} = nFDC^*/r \tag{26}$$

where r is the tip radius and C^* is the concentration of the diffusing species in the vicinity of the tip, at a distance sufficiently great as to be considered as the boundary of the spherical diffusion field. For small enough radii, this is practically identical in position with the tip height. If the concentration in the linear diffusion layer is considered to be changing linearly with distance, then

$$C^* = C^\circ h/\delta \tag{27}$$

Hence, from (27), (26), and (19), it follows that

$$i_{L(s)} = i_{L(l)}(0)h/r \tag{28}$$

The amplification factor is obtained, again using equation (21), as

$$\frac{dh(x)}{h(x)\,dt} = \frac{MDC^\circ i_0}{\rho\delta r}\left[\frac{f_c(\eta) - K(r)f_a(\eta)}{i_{L(l)}(0)(h/r) + i_0 f_c(\eta)} - \frac{r}{h}\frac{f_c(\eta) - f_a(\eta)}{i_{L(l)}(0) + i_0 f_c(\eta)}\right] \tag{29}$$

Figure 11 demonstrates the dependence of the amplification factor on r at different overpotential values with other parameters corresponding to a rather typical situation ($M = 50$; $D = 10^{-5}$; $C^\circ = 10^{-6}$; $i_0 = 10^{-3}$; $\rho = 8$; $\delta = 10^{-2}$; $\sigma = 1500$; $V = 7$; $T = 298°\text{K}$; $h = 10^{-4}$).

It is seen (Figure 11) that the amplification factor is at maximum at certain r values large enough for the Kelvin effect to be negligible and yet small enough compared with h that the second term in (23) can be neglected. It is for such conditions that the equation was solved explicitly by Diggle *et al.*[20] for the time dependence of the protrusion height.

The solution was shown to be

$$\frac{h(x)}{i_0 n} + r\ln h(x)\frac{f_c(\eta)}{i_{L(l)}(0)u} = \frac{M}{\rho n F}t + \frac{h_0(x)}{i_0 u} + r\ln h_0(x)\frac{f_c(\eta)}{i_{L(l)}(0)u} \tag{30}$$

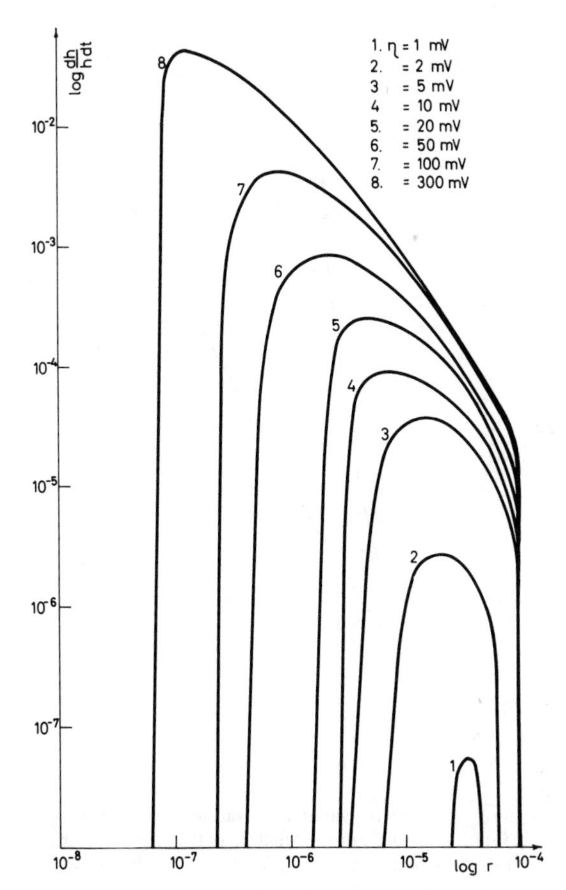

Figure 11. The variation of the amplification factor with the radius of curvature of the surface irregularity.

where

$$u = f_c(\eta) - K(r) f_a(\eta) \tag{31}$$

Figure 12 shows the time dependence of the height of an irregularity with a given tip radius, obtained by solving equation (30) on an analog computer. The dashed curve represents the dependence obtained from equation (23). Two limiting cases arise:

(i) In early stages of growth, at low values of $h(x)$, the terms of equation (24) linear in $h(x)$ vanish and an exponential relationship between height and time, such as that given by equation (23), is

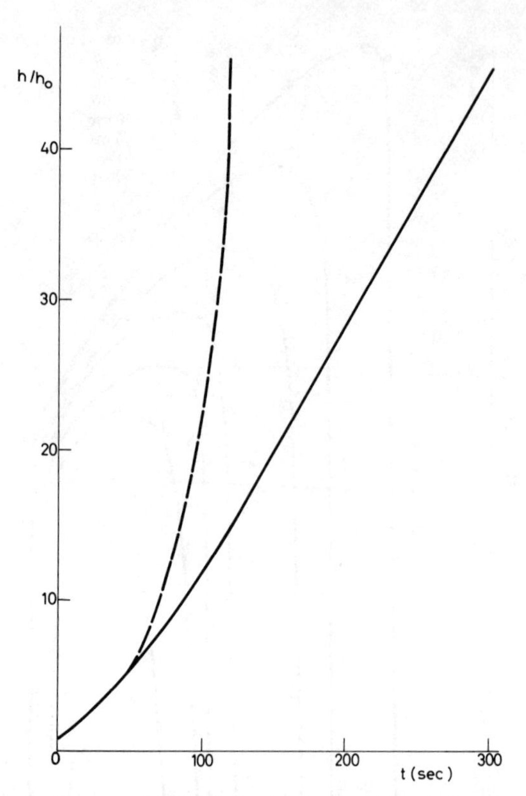

Figure 12. Relative height of a surface irregularity of tip radius $r = 10^{-4}$ cm as a function of the time of placing an overpotential pulse of 12 mV on the electrode.

obtained again. However, the time constant is now

$$\tau_s = \frac{\rho}{MDC^\circ[1 - K(r)f_a(\eta)/f_c(\eta)]}\delta r \tag{32}$$

Comparing this with τ_l of equation (24), we can see that under conditions of negligible Kelvin effect and, as in most cases $\delta \gg nFDC^\circ/i_0 f_c(\eta)$,

$$\tau_s = \tau_l r/\delta \tag{33}$$

This indicates that the tip exhibits a tendency to grow much faster ($r/\delta \ll 1$) than the sides, i.e., it exerts a pulling effect on the sides which should result in a change in shape of the elevation from a triangular to a paraboloidal one. This was indeed foreseen by Barton and Bockris.[19]

(ii) In a later stage of growth, it is the term linear in $h(x)$ in equation (30) which may become significant and even dominant. However, the solution of the equation for such a case shows that when this arises, the growth at the tip should be entirely activation-controlled. This is an unlikely situation for $h < \delta$ and hence it need hardly be taken into consideration.*

It is obvious that in reality the time constant of the exponential period of amplification should have some intermediate value.

To test the validity of the above equations, Despić and Popov[3] have carried out experiments on diffusion-controlled metal deposition on a well-defined, triangularly shaped surface profile, through a diffusion layer of well-defined thickness $\delta \gg h$. A phonograph record negative was used as a substrate upon which a layer of an agar containing copper sulfate–sulfuric acid solution of thickness 0.5–2 mm was placed and left to solidify. As the current was passed and the layer was depleted of copper ions, an increase in the height of the triangular ridges was observed. Metallographic samples were made in wax and cross sections of the deposit were photographed under the microscope. Figure 1 was thus obtained. As the deposition was carried out at constant overpotential values during different time intervals, the height and the angle between the ridge side and the flat surface were measured. As required by equations (20) and (22), the dependences of $\log h$ and $\log[\tan^2 \alpha - 1]$ on time are shown in Figures 13 and 14 to be linear, exhibiting approximately the same time constants. The latter are found to be of the expected order of magnitude, between those indicated by equations (24) and (32).

(d) *Time dependence of the overall current at constant overpotential.* A direct consequence of the increase in roughness and coarseness of a surface should be an increase in the deposition current at a constant electrode potential. This was observed and explained in a qualitative manner by Ibl and Schadegg.[1] The increase in current occurs because (a) the true surface area increases and (b) the limiting current density at any elevated point of the surface

*The observed constant velocity of growth of protrusions (e.g., dendrites) is usually attained outside the hydrodynamic boundary layer ($h > \delta$), and is defined by another equation (cf. next section).

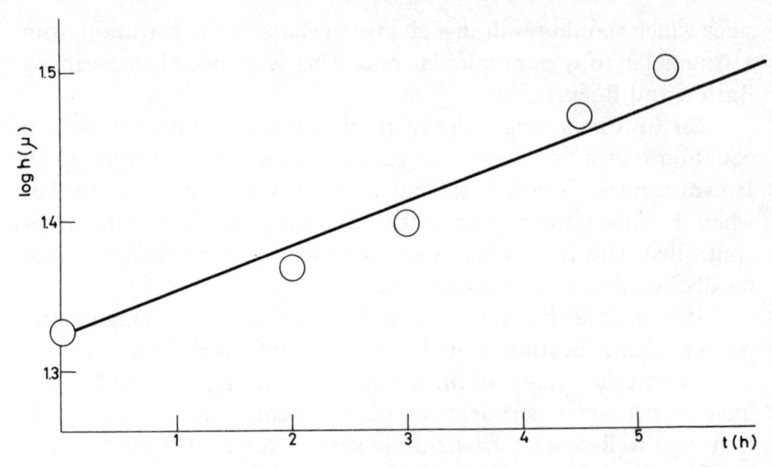

Figure 13. Measured change in the height of an irregularity with time (from Despić and Popov[3]).

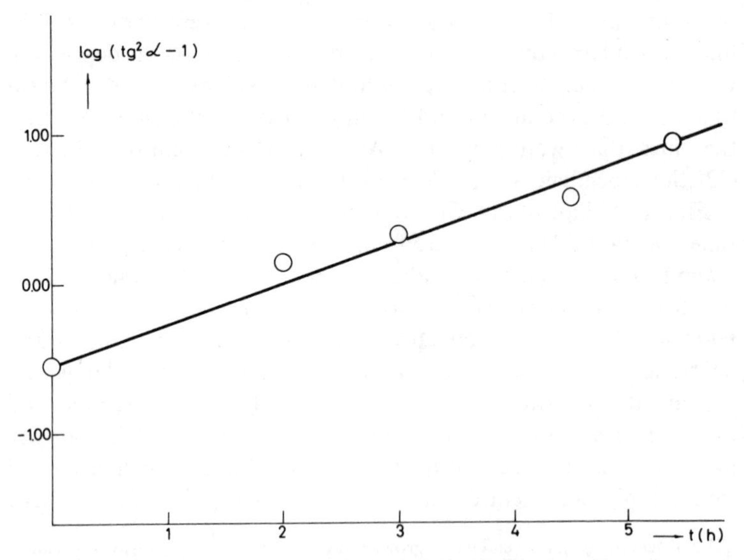

Figure 14. Measured increase in surface roughness during deposition of copper onto a triangularly shaped surface irregularity (from Despić and Popov[3]).

increases with time as the point penetrates into the diffusion layer.

A quantitative treatment of this phenomenon can be made on the basis of the above theory of amplification.

To assess this, one should note that the change in height with time at any point x with respect to a reference plane, $dh(x)/dt$, reflects the difference in current densities of deposition at these two positions [see equation (21)]. $i(0)$ is a time-independent quantity. Hence, the time dependence of $i(x)$ is obtained by differentiating (23) and replacing it in (21). Thus,

$$i(x, z, t) = i(0) + \frac{\rho nF}{M}\frac{dh(x)}{dt} = i(0) + \frac{\rho nF h_0(x)}{M\tau} e^{t/\tau} \tag{34}$$

One should note the three kinds of laws operating at three different portions of the surface. At the flat portion, the current density is independent of time, and at the sides of irregularities and at their tips, it is exponential in time but with different time constants. Let the fractions of the surface pertaining to the three categories be denoted as S_F, S_S, and S_T. Then, the overall current density will be

$$\begin{aligned} i = S_F i(0) &+ S_S\left[i(0) + \frac{\rho nF}{M\tau_l}(h_0)_S \exp\left(\frac{t}{\tau_l}\right)\right] \\ &+ S_T\left[i(0) + \frac{\rho nF}{M\tau_s}(h_0)_T \exp\left(\frac{t}{\tau_s}\right)\right] \end{aligned} \tag{35}$$

where $(h_0)_S$ and $(h_0)_T$ represent the surface integrals or the "average" elevations of the sides and the tips, respectively. Since the sum of the fractions equals one, rearranging (35), one obtains

$$\ln[i - i(0)] = \ln S_S \frac{(h_0)_S}{\tau_l} \exp\left(\frac{t}{\tau_l}\right) + S_T \frac{(h_0)_T}{\tau_s} \exp\left(\frac{t}{\tau_s}\right) + \ln \frac{\rho nF}{M} \tag{36}$$

A linear relationship between $\ln[i - i(0)]$ and time can be expected in the case where one of the first two terms on the right-hand side of equation (36) vanishes. This is to be expected for the second term since the fraction of the surface pertaining to elevation tips is bound to be extremely small. Hence, in such a case,

$$\ln[i - i(0)] = \frac{1}{\tau_l} t + \ln \frac{\rho nF S_S (h_0)_S}{M\tau_l} \tag{37}$$

Considering equation (24), the slope of the function should be proportional to the bulk concentration and to the inverse of the square of diffusion layer thickness.

An experimental investigation has indeed revealed the exponential dependence of the current density with time. A typical graph of the current–time relation obtained in the same system as that described previously (cf. p. 213) is shown in Figure 15. The initial decrease in current is due to the formation of a diffusion layer equal to the hydrodynamic boundary layer determined by the thickness of the agar gel. The $\log[i - i(0)]$ plots are fairly linear with time if $i(0)$ is taken to be negligibly small. This is justified by the result of microscopic investigations, since in most cases the deposit at flat portions of the surface was much thinner than at triangular elevations. The values of τ_l are shown in Figures 16 and 17, and are seen to be linear with $1/C_0$ and with δ^2.

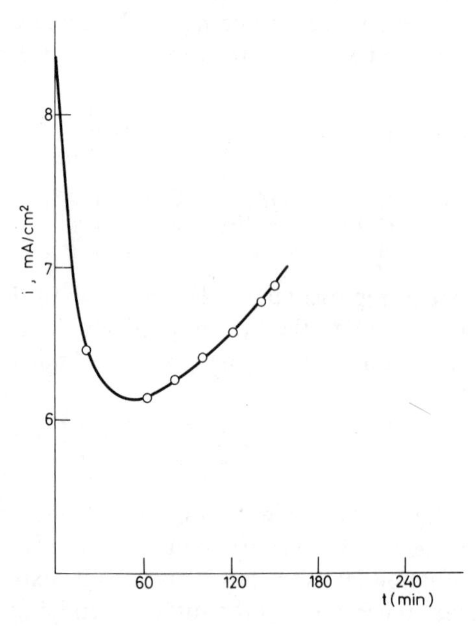

Figure 15. Dependence of the current of copper deposition at constant overpotential ($\eta = -600$ mV) through a fixed hydrodynamic layer on time ($\delta = 0.75$ mm; $C^{\circ} = 0.5 \times 10^{-3}$ mole cm^{-3}). (From Despić and Popov.[21])

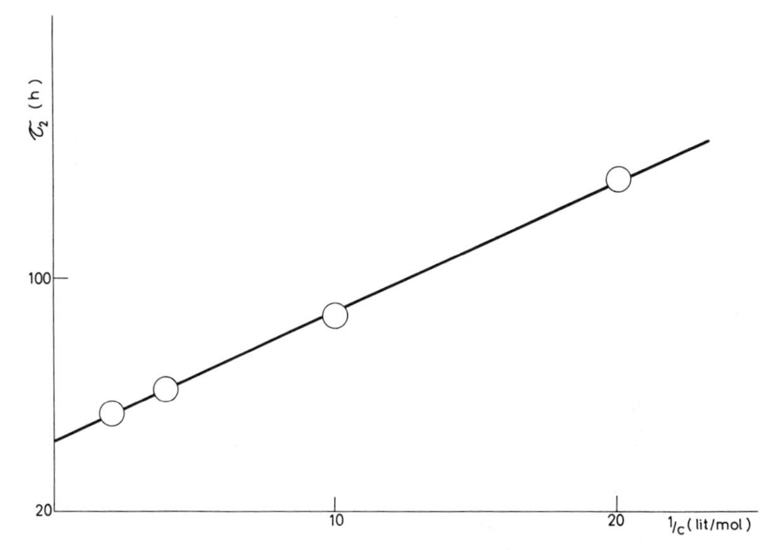

Figure 16. Concentration dependence of the time constant for the amplification of the deposition current density at constant overpotential ($\eta = -300\,\mathrm{mV}$; $\delta = 1\,\mathrm{mm}$); (from Despić and Popov[21]).

Summarizing, the above experiments give fair support to the theory given and hence support the hypothesis that the lateral diffusion fluxes do not play an important role in the initial stages of the deposition process.

However, one could also say that both theories suffer from the fact that they apply when the diffusion layer is much wider than the height of the irregularity, i.e., to a relatively late stage of deposition when the diffusion layer is fully developed and smoothed out so as to follow the macroprofile of the surface. Because of this limitation, it appears from both theories that the amplification factor should be smaller the thicker the diffusion layer, and hence it should decrease with time if the latter is increasing. Yet, the most interesting period, in which the amplification factor could qualitatively be shown to be at a maximum, is the period when it is comparable with H_0 or h_0. It is clear that at the beginning of the process, as the diffusion layer starts developing, it should first follow the microprofile. It is the moment when it starts separating from it and flattening out that the differences in the effective diffusion layer thickness are the largest, and consequently the amplification factor must be at a maximum.

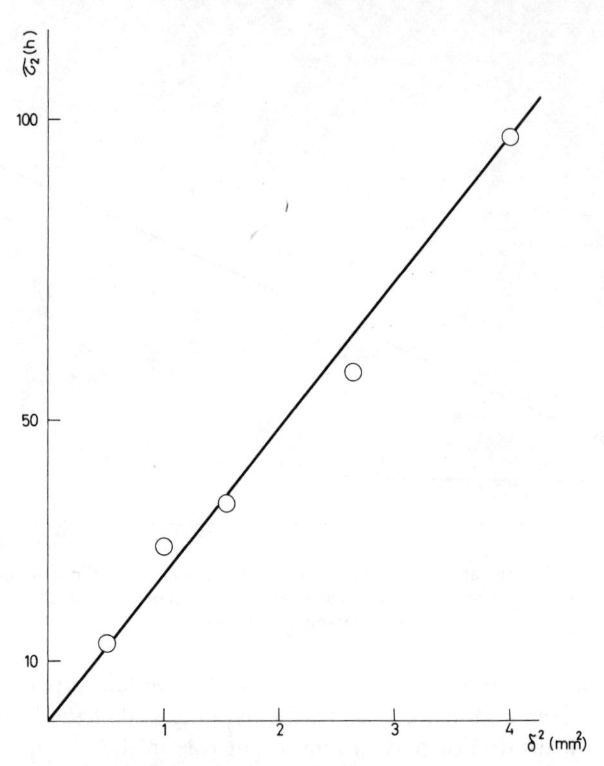

Figure 17. The dependence of the time constant for the amplification of the deposition current density at constant overpotential on the thickness of the hydrodynamic layer ($\eta = -600$ mV; $C^\circ = 0.5$ M $CuSO_4$); (from Despić and Popov[21]).

An attempt to treat this problem in a quantitative manner is outlined at the end of this chapter.

III. APPEARANCE AND GROWTH OF DENDRITIC DEPOSITS

Systematic investigation of the problem of dendritic growth in electrocrystallization is of relatively recent origin. The first extensive accumulation of data on a number of systems (Pb, Ag, Sn, Cd, in a variety of electrolytes) is due to Wranglen.[4]

He systematized the results both according to the conditions of appearance of dendrites in terms of the current density range and according to structural characteristics of the outgrowth. The latter are described in terms of the lattice direction from which the stem and the branches are formed, in terms of their spatial arrangement (flat or two-dimensional, *2D*, and three-dimensional, *3D*), in terms of the number and levels of side branches (primary, *P*, secondary, *S*, tertiary, *T*), and, finally, in terms of the angles the side branches make with respect to the dendrite stem. Thus, e.g., Figure 2 represents *S2D* 110 60° and *S3D* silver dendrites. A series of studies followed[22–31] in an attempt to elucidate basic factors governing this type of deposition. Two groups of questions arise upon considering all of this experimental evidence. The first relate to the conditions of appearance of dendrites upon deposition and their frequency of incidence and rate of propagation. The second are concerned with the shape of the crystals formed and the direction of growth, both of the original dendrite stem and of the side branches of the first and higher orders.

It is their special shape and manner of extension which distinguishes dendrites from the other types of outgrowth caused by the factors described in the preceding section. This indicates a definite effect of the structural properties of the metal superimposing upon other effects to govern the final outcome of the deposition.

1. Conditions of Appearance and Factors Determining Frequency of Incidence and Rate of Growth of Dendrites

Wranglen[4] found that the minimum current density at which dendrites appear is related to the specific rate of the deposition process, as represented by the activation overvoltage needed for a given current density. Thus, changing the anions in the order of increasing overvoltage, the tendency toward formation of dendrites is greatly retarded, so that ever-higher current densities are required. Coarser forms are developed and secondary dendrites are suppressed.

In a thorough investigation of the growth of silver dendrites from a $AgNO_3$, KNO_3, $NaNO_3$ melt, Barton and Bockris[19] gained considerable new knowledge concerning this phenomenon. They found that: (a) a certain critical overvoltage must be exceeded in order to provoke dendritic growth; (b) the growth exhibited a certain induction period before it became visible (at overpotentials not much greater than 3 mV, the induction time could be several hours); (c) a critical current density for the dendritic growth is directly proportional to the concentration of the depositing ions; (d) the incidence of dendrites increases with increasing overall current density; (e) once the dendrites are started, they grow at a constant velocity for the given potential; (f) not all dendrites grow at the same rate under the same potential and concentration, but there appears to be an upper limit of velocity; (g) for one applied potential, the dendrites that grow fastest are usually those that were initiated at higher potentials; (h) for a particular dendrite, successive increments of overvoltage at first cause the velocity to increase, but a point is usually reached where the configuration of the tip is changed either to form a fanlike growth of miniature dendrites or to split into two separate branches; in the former case, the rate of advance is much less than that of the parent dendrite; (i) the cessation of growth of a dendrite upon decreasing overvoltage is accompanied by a change in the shape of the tip from parabolic to prismatic.

Many of these conclusions found further support in the experiments of Reddy[32] in an AgCl–KCl–LiCl electrolyte melt. He also found considerable pseudocapacitance values (700 μF cm^{-2}) which could not be ascribed to concentration polarization with respect to the depositing ions in the melt.

Finally, in experiments similar in type to those of Barton and Bockris on the growth of zinc dendrites from alkaline zincate solutions, Despić *et al.*[2,20] confirmed, using an aqueous electrolyte, most of the above conclusions. The only striking difference was in the number of dendrites appearing at any overpotential above the critical one. Bunches of dendrites were usually obtained and single dendrites could not be produced.

2. General Theory of the Appearance and Growth of Dendrites

The accumulated experimental evidence made possible a quantitative, nonspecific theory of the appearance and propagation of dendritic outgrowth showing that most of the observed phenomena result from difficulties in the transport of the depositing ions. The essential structure of the theory was laid by Barton and Bockris.[19] They assumed that a basic reason for a much faster growth of a dendritic protrusion than that of the rest of the electrode surface lay in the fact that conditions of spherical diffusion applied around its tip while linear diffusion prevailed elsewhere. For this to be true, some minimum concentration changes, reflected in a concentration overpotential, are necessary. On the other hand, the larger the spherical diffusion flux, the smaller the radius of curvature of the tip. This would favor an infinite thinning of the dendrite during the growth. However, as the tip radius becomes very small, a Kelvin-type surface-energy effect enters into play, changing the reversible potential to more negative values and thus reducing the effective concentration and activation overpotential as the driving forces for the tip growth. Hence, an optimum tip radius exists for which the rate of growth is at a maximum.

The equations of Barton and Bockris were derived assuming the tip to be a static hemisphere and the process to be so fast that the activation overpotential is always much smaller than RT/F (with the total overpotential in the reversible region). The first assumption made it difficult to explain the observed steady growths at rates lower than the maximum. Barton and Bockris suggested that dimensional stability of the tip can be attained at any value of velocity if the tip was of a paraboloidal shape, but did not investigate the properties of the diffusion layer in such a case. This was done by Hamilton,[33] who extended the Barton–Bockris theory to the case of a *moving* paraboloid representing the dendrite tip. He showed that in such a case, the necessary conditions of stability are also attained at conditions of radius and velocity other than the optimum ones.

Finally, Despić *et al.*[2,20] extended the theory to cover deposition at lower specific rates, i.e., at higher overpotentials, larger than RT/F, and this for two possible types of mechanisms relating to univalent and divalent ion reduction. They also offered quantitative relations for the initiation of dendritic growth, giving meaning to the induction period. They showed that the penetration of the dendrite into the diffusion layer, even without spherical diffusion conditions of supply, leads to a much faster growth of the tip than of the flat surface. Also, they found reasons for the different frequency of incidence of dendrites at different electrodes, in the existence of a wider or a narrower *range* of optimum tip radii rather than a particular value.

A synthesis of these results is given below.

(*i*) *Initiation of Dendritic Growth*

Dendritic growth should be initiated under the same set of conditions and for the same reasons that lead to the amplification of any surface irregularity (see preceding sections). Hence, it should be governed by the same laws as those derived earlier.

The induction period can be interpreted as the period needed for the exponential, i.e., the avalanche-like, nature of the amplification suddenly to make the protrusion pierce the hydrodynamic layer boundary and become visible. Hence, it is directly related to the time constant τ_s [see equation (32)].

If one defines as the initiation time t_i the time obtained by extrapolation of the linear dendrite length versus time relationship to zero, as done in Figure 26, one can deduce t_i by the following reasoning:

The height of a dendrite at any time is given by

$$h - h_0 = (M/\rho nF)(i_{\text{tip}})_{\text{lin}}(t - t_i) \tag{38}$$

where $(i_{\text{tip}})_{\text{lin}}$ is the constant tip current density, providing for the growth of dendrite outside the hydrodynamic boundary layer, given by equation (54a) or (54b). At any visible h, h_0 can be neglected.

At a certain time t_d, the rate of growth of the exponentially growing protrusion becomes equal to that of the linearly growing dendrite. This happens as the protrusion reaches a certain height h_d. The two rates $(i_{\text{tip}})_{\text{exp}}$ and $(i_{\text{tip}})_{\text{lin}}$, as given by equations (34) and (54), when $i_{\text{tip}} > i(0)$ and $ri_0 f_c(\eta) \gg nFDC^\circ$, are seen to be interrelated by

$$(i_{\text{tip}})_{\text{exp}} = (i_{\text{tip}})_{\text{lin}}(h_0/\delta)\exp(t/\tau_s) \tag{39}$$

Hence, the two are equal when $(h_0/\delta)\exp(t_d/\tau_s) = 1$, so that the time t_d is given as

$$t_d = \tau_s \ln(\delta/h_0) \tag{40}$$

Considering equation (23), we see that $h_d = \delta$. Replacing these values in equation (38) with $(i_{\text{tip}})_{\text{lin}}$ from equation (34) and rearranging, one obtains

$$t_i = \tau_s\left(\ln\frac{\delta}{h_0} - 1\right) = \frac{\delta r[\ln(\delta/h_0) - 1]}{VDC^\circ[1 - K(r)f_a(\eta)/f_c(\eta)]} \tag{41}$$

Note that so far the surface irregularities have been considered to have a tip of a given radius of curvature and the effect of this on their rate of growth was considered. The question arises as to whether there should be a tendency for this radius to change and which direction this change will take place. It appears that there should be some decrease in radius when screw dislocation pyramids are increasing in height. The distance between two successive steps on a spiral is given[34] by $4\pi r$. Hence, as the dendrite precursor height increases, the step distance decreases and this determines the radius of curvature. This tendency is, of course, counteracted by an increasing Kelvin effect and hence the change should stop as the optimum value of the tip radius is attained.

Two phenomena seem to distinguish dendrite initiation from simple amplification of surface coarseness: (a) there seems to exist a certain critical overpotential value below which the dendrites do not grow; and (b) the dendrites exhibit a highly ordered structure and grow and branch in well-defined directions.

Several possible reasons may be considered for the existence of the critical overpotential η_c. One could maintain that this is simply an overpotential needed to make the initiation time reasonably short for the dendrites to appear in the lengthiest experiments; however, this could be questioned, because η_c depends on the nature of the depositing metal (contrast, e.g., zinc and silver), while the time constant of equation (32) appears to be a nonspecific entity.

The second attitude is that η_c is a critical value needed to start rotating a screw dislocation as a precursor of an elevation and a protrusion. Some calculations indicate, however, that this value is much smaller than the experimentally observed values.

The most likely explanation of the existence of η_c seems to be found in an examination of the relation between the rates of growth of dendrites and of a flat surface under diffusion control. This is discussed below, as is the effect of crystal structure on the shape and direction of growth of the dendrites.

(ii) Dendrite Propagation

During a period of growth, dendrites are found to propagate at a constant velocity and tip radius irrespective of the shape. The form of the surface which can satisfy this is found from the condition that each point at the surface must move with the same velocity v in the direction of the y axis in growth (see Figure 18). This implies the rate of deposition perpendicular to the surface v_n is

$$v_n = v \sin \theta \tag{42}$$

In the case of spherical diffusion, this rate, being at each point

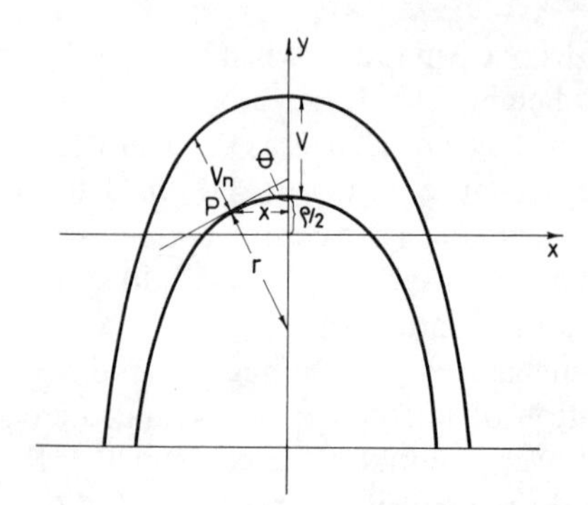

Figure 18. Schematic representation of a paraboloidal dendrite tip (from Barton and Bockris[19]).

inversely proportional to the radius of curvature, is given by

$$v_n = \frac{VD(C_e^{M^{z+}} - C_i^{M^{z+}})}{r} = \frac{VD\,\Delta C}{x}\cos\theta \tag{43}$$

where $C_e^{M^{z+}}$ and $C_i^{M^{z+}}$ are concentrations in the bulk of the solution and at the surface, respectively.

Hence, from (42) and (43), using $(\sin\theta)/(\cos\theta) = \tan\theta = dx/dy$, and rearranging, one obtains

$$dy = (-v/VD\,\Delta C)x\,dx \tag{44}$$

Integrating equation (44), the function describing the surface is obtained as

$$y = \tfrac{1}{2}\rho - \tfrac{1}{2}x^2 \tag{45}$$

where

$$\rho = VD\,\Delta C/v \tag{46}$$

Thus, the surface of a steadily advancing dendrite tip must be paraboloidal in shape, with a well-defined relation between the characteristic tip radius ρ and the constant velocity of propagation. It may of course be different from the maximum one derived below.

The velocity of propagation is related to the tip current density by equation (12). Also, the tip current density under mixed activation and diffusion control is given by equations (14a) and (14b). To find the concentration of the depositing species at the surface, $C_i^{M^{z+}}$, it is necessary to solve the diffusion equation for a moving surface,

$$\left(\nabla^2 + \frac{v}{D}\frac{\partial}{\partial y}\right) C^{M^{z+}}(x, y) = 0 \tag{47}$$

with the appropriate boundary conditions.

One can show that for a constant rate of propagation to be maintained at a constant potential, the concentration of the species at the surface must decrease from the tip down the sides of the stem. However, around the tip itself, this change in concentration is small enough that a position-independent value $C_i^{M^+}$ may be taken as a boundary condition.

Ivantsov[35] has found an exact solution of this equation for the surface given by equation (45). According to this, the concentration field is given by

$$C^{M^+}(x, y) = C_e^{M^+} + (C_i^{M^+} - C_e^{M^+})f \tag{48}$$

where

$$f = \frac{E_i((-v\rho/2D)\{(z/\rho) + [(z^2 + r^2)/\rho^2]^{1/2}\})}{E_i(-v\rho/2D)} \tag{49}$$

and E_i are the exponential integrals

Considering equality (46), it is seen that

$$f = \frac{E_i((-V\,\Delta C/2)\{(z/\rho) + [(z^2 + r^2)/\rho^2]^{1/2}\})}{E_i(-V\,\Delta C/2)} \tag{50}$$

The current density is related to the flux of the substance as

$$i = zFD(C_e^{M^+} - C_i^{M^+})\,\nabla f \tag{51}$$

Differentiating f in equation (49), introducing the result into (51), and solving for $C_i^{M^+}/C_e^{M^+}$, gives

$$C_i^{M^+}/C_e^{M^+} = (i_L - i)/i_L \tag{52}$$

with

$$i_L = \frac{zFDC_e^{M^{z+}}}{\rho/2}\,\frac{\exp(-V\,\Delta C/2)}{E_i(-V\,\Delta C/2)} \tag{53}$$

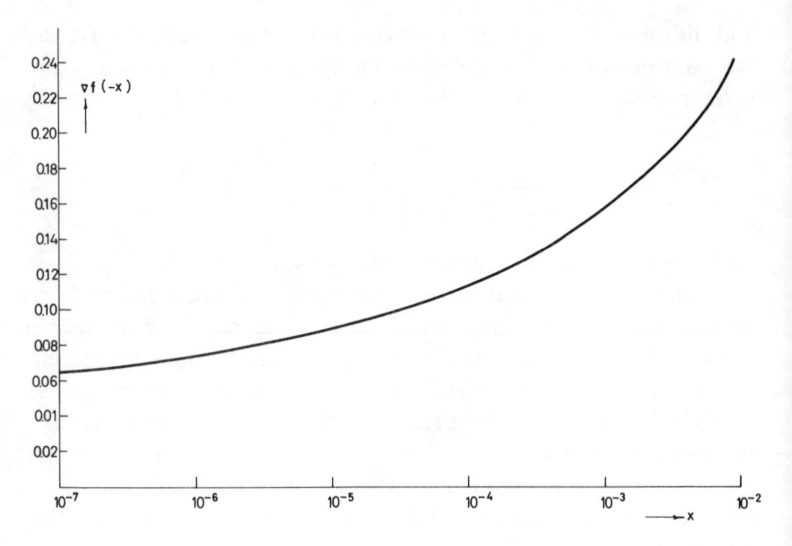

Figure 19. The dependence of the correction factor for paraboloidal diffusion on the variable $x = \mathrm{V}\,\Delta C/2$.

This is seen to differ from the usual expression for a limiting current density of spherical diffusion to a sphere of radius $\rho/2$, by the factor $\nabla f' = [\exp(-V\,\Delta C/2)]/E_i(-V\,\Delta C/2)$.* Figure 19 represents the dependence of this factor on the variable $x = V\,\Delta C/2$, i.e., on the concentration of the ions in solution.

Since equation (52) is identical with (15), introducing the value of the limiting current density given by (53) into (16a) and (16b), and rearranging, one obtains for the tip current densities for the two types of discharge mechanisms described earlier (cf. Figure 10)

$$i_n' = \frac{f_c(\eta)\exp(-1/r_n') - f_a(\eta)\exp(1/r_n')}{(K'/i_0) + f_c(\eta)r_n'\exp(-1/r_n')} \tag{54a}$$

and

$$i_n'' = \frac{f_c(\eta) - f_a(\eta)\exp(1/r_n'')}{(K''/i_0) + f_c(\eta)r_n''} \tag{54b}$$

*Note that i_L is not strictly a limiting current density since ΔC is found in $\nabla f'$. Still, this is close enough to the former for $C_i^{\mathrm{M}^{z+}} < C_e^{\mathrm{M}^{z+}}$.

where

$$i_n' = i/K', \qquad i_n'' = i/K''$$

$$r_n' = \frac{RT\rho/2}{2\beta\sigma V} = \frac{RT\rho/2}{2(1-\beta)\sigma V}, \qquad r_n'' = \frac{RT\rho/2}{2\sigma V}$$

$$K' = RTzFDC_e^{M^{z+}}\,\nabla f'/2\beta\sigma V = RTzFDC_e^{M^{z+}}\,\nabla f'/2(1-\beta)\sigma V$$

$$K'' = RTzFDC_e^{M^{z+}}\,\nabla f'/2\sigma V$$

Equations (54a) and (54b) are identical with those obtained by Despić *et al.*,[2] but include the normalization constants K' and K'' containing the Hamilton factor $\nabla f'$, and the reduced radii containing the parameter of the paraboloidal surface $\rho/2$.

Moreover, for small values of η ($\ll RT/F$) and large values of r_n' ($\gg 1$), equation (54a) reduces to one analogous to that of Barton and Bockris, i.e.,

$$-\eta = \frac{i}{i_0}\frac{RT}{F} + \frac{i(\rho/2)RT}{\nabla f'\,DC_e^{M^{z+}}F^2} + \frac{2\sigma V}{F(\rho/2)} \tag{55}$$

Figure 20 shows the dependence of the tip current density on tip radius calculated using equations (54a) and (54b) for different constant overpotentials. Figure 20(a) is obtained when parameters characteristic of the deposition of silver were employed in (54a), while Figure 20(b) represents the case of zinc, with the appropriate parameters used in (54b).

Two conclusions can be drawn by analyzing the graphs of Figure 20:

1. The curves showing the relation between tip c.d. (i.e., growth rate) and the tip radius for different overpotentials are found to cross a line indicating the value of the limiting current density $(i_{L,l})_n$ for linear diffusion onto a flat surface at overpotentials coinciding remarkably well with the experimentally observed critical overpotentials. Hence, the criterion for a critical current density of the dendrite tip which grows under activation control, past which the conditions for dendritic growth are established, and the criterion for critical overpotential, are defined as

$$i_c = i_0[f_c(\eta) - f_a(\eta)] \geq i_{L,l} \tag{56}$$

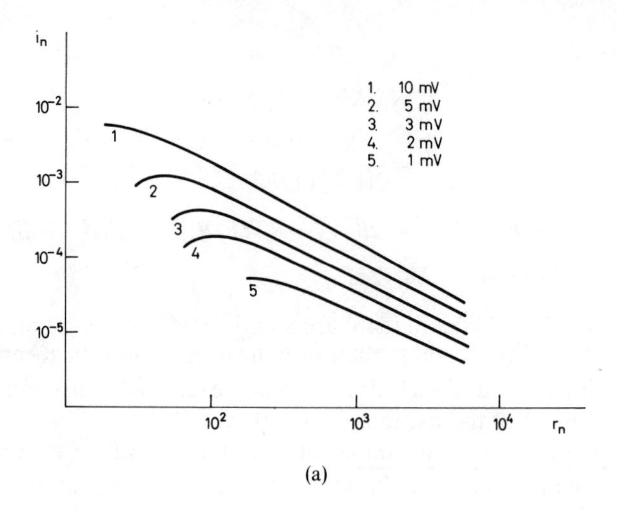

(a)

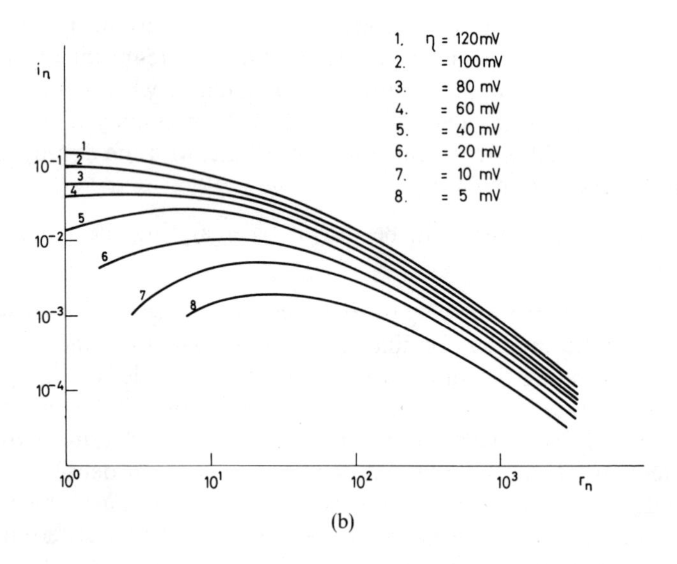

(b)

Figure 20. The dependence of the tip current density on tip radius for the deposition of (a) silver and (b) zinc at different constant overpotentials.

and hence,

$$\eta_c \geq (RT/F)i_{L,l}/i_0 \qquad \text{for} \quad \eta \ll RT/F \tag{57a}$$

$$\eta_c \geq (RT/F)\ln(i_{L,l}/i_0) \qquad \text{for} \quad \eta \gg RT/F \tag{57b}$$

The critical current density depends solely on the conditions of diffusion to the flat surface, while the critical overpotential depends also on i_0 for the deposition process.

2. The difference in mechanism, and even more so the difference in i_0, results in a significant difference in the range of tip radii in which dendrite formation can occur for the two systems: those for silver show sharp peaks of growth rate in a narrow range of tip radii, while those for zinc show broad plateaus over a wide range of tip radii. This explains the observed difference in the character of the deposit for the two metals: the dendrites of silver are propagating rapidly at low overpotentials but are scarce, since the probability of the protrusions falling into the right range of radii is small; zinc dendrites are propagating slowly, but they are more abundant because of the high probability of the appearance of a protrusion with radius in the required range.

3. Comparison with Results of Experiments

A series of experiments gave fair support to the above theory. The measured total current in the predendrite period could be interpreted as being exponentially dependent on time, in terms of the theory of amplification of surface roughness given in the preceding section. The relationships obtained are presented in Figures 21 and 22 as functions of zincate concentration and overpotential, respectively. The concentration and overpotential dependences of the time constants were found to follow the derived equation (24), with $\partial\tau/\partial(1/C) = 50$ sec cm^{-3} and $\partial\tau/\partial\{1/[1-\exp(2F/RT)]\} = 6.5 \times 10^5$ sec. This indicates a δ value approximately equal to 6×10^{-2} cm in both cases. This is not outside the limits that could be expected given the conditions of the experiments.

One should mention that in the small range of change of the total current with time observed by Diggle *et al.*,[20] an interpretation in terms of a quadratic time dependence could be equally well justified. This could be explained as resulting from an increasing

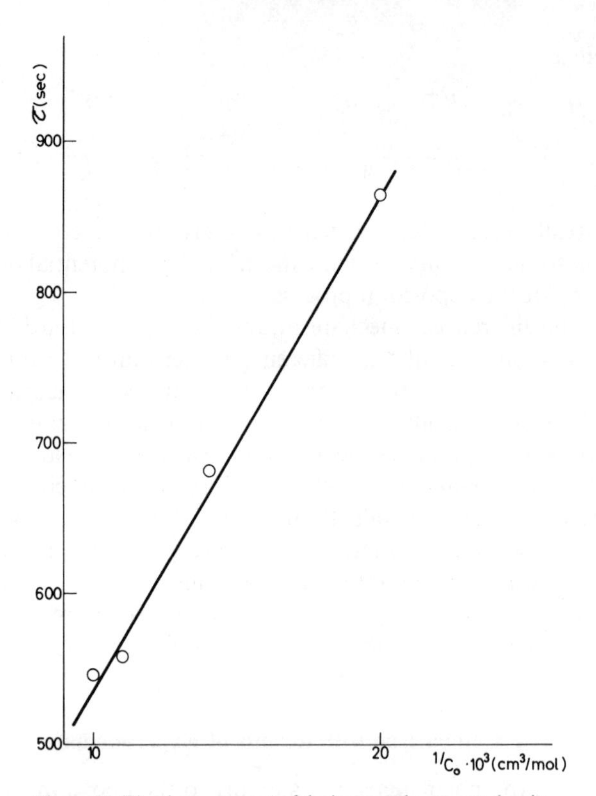

Figure 21. The time constant of the increase in current density of zinc deposition from zincate solutions as a function of reciprocal concentration of zincate ions (plotted from the data of Diggle *et al.*[20]).

surface area of the sides of growing dendrites if it was assumed that they are initiated in numbers linearly dependent on time (continuous initiation). In the present case, the evidence is insufficient for deciding between the two explanations. Should the second one be finally accepted, the meaning of the time constant would, however, be much less clearly defined.

The initiation times extrapolated from the observed growth data are recorded by Diggle *et al.*[20] as functions of concentration of zincate in solution and of overpotential. Equation (41) defines these dependences. From this, it follows that plots of log t_i versus log $C°$ and $1/t_i$ versus $\exp[(2F/RT)\eta]$ should be linear with slopes -1 and

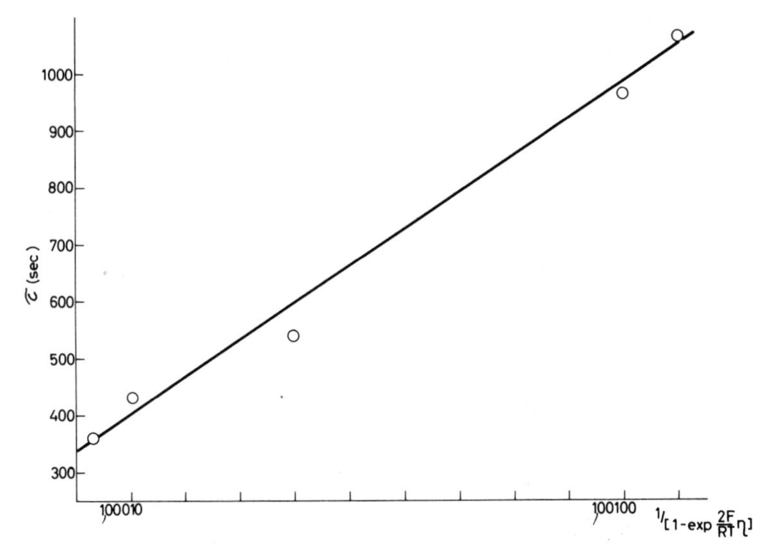

Figure 22. The time constant of the increase in current density of zinc deposition as a function of overpotential (calculated from the data of Diggle *et al.*[20]).

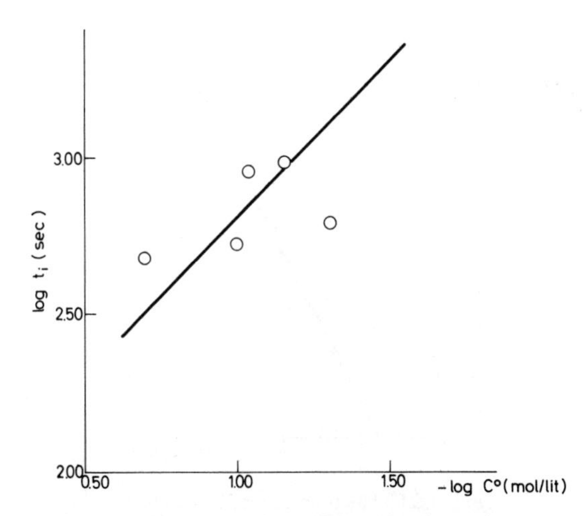

Figure 23. Initiation time of dendrite formation as a function of the concentration of zincate in solution (from the data of Diggle *et al.*[20]).

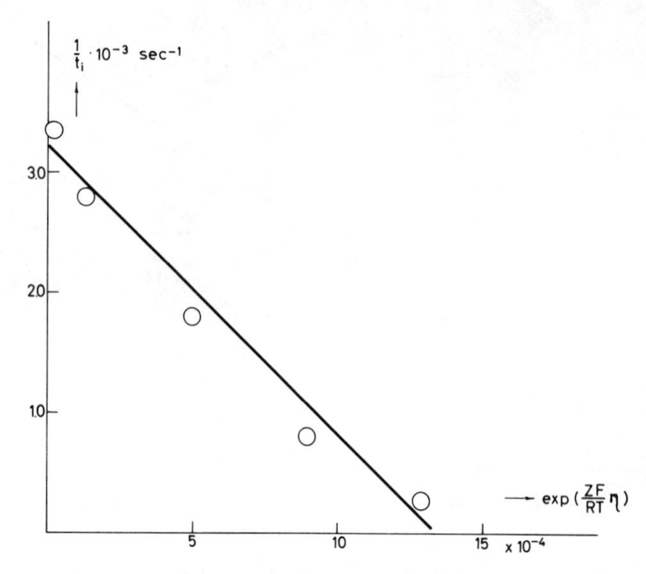

Figure 24. Initiation time of dendrite formation as a function of overpotential (from the data of Diggle *et al.*[20]).

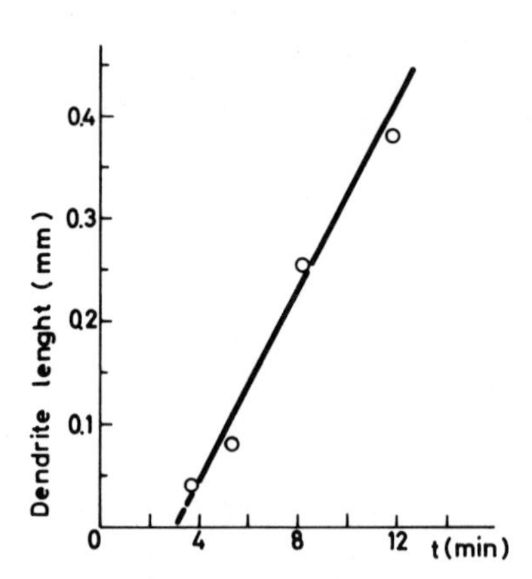

Figure 25. The time dependence of the length of a silver dendrite grown from a molten salt solution of silver ions (from Barton and Bockris[19]).

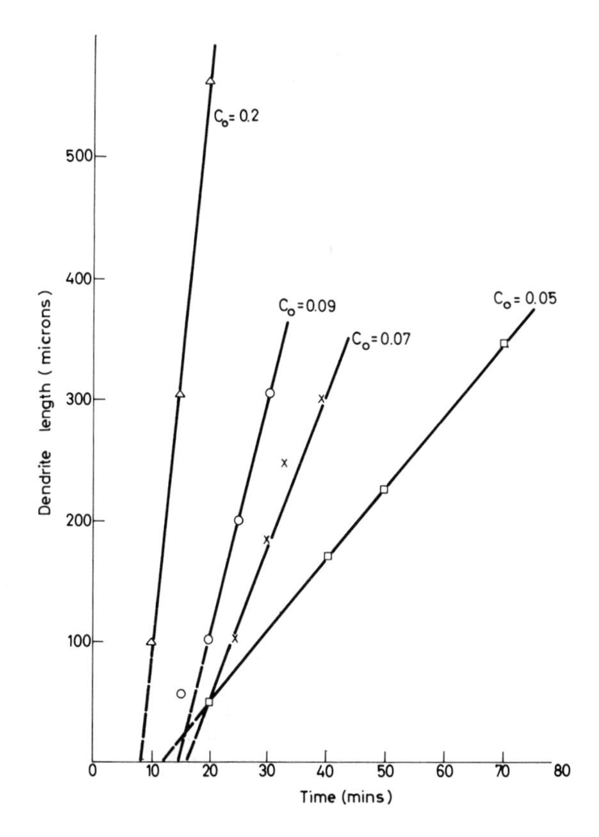

Figure 26. The time dependence of the length of zinc dendrites grown from aqueous zincate solution (from Diggle *et al.*[20]).

$VDC^{\circ}K(r)/\delta r[\ln(\delta/h_0) - 1]$, respectively. These plots are seen in Figures 23 and 24. The large deviations of some points can be explained in terms of the irreproducible nature of δ from one experiment to another, since δ is determined by natural convection.

As seen in Figures 25 and 26, the length of the microscopically observed dendrites varies linearly with time both in the case of silver and in the case of zinc.

For a given constant and large enough tip radius, and considering the concentration dependence of K', K'', and i_0, equations (54a) and (54b) suggests that plots of $\log i$ versus $\log(C_e^{M^{Z+}})$ should

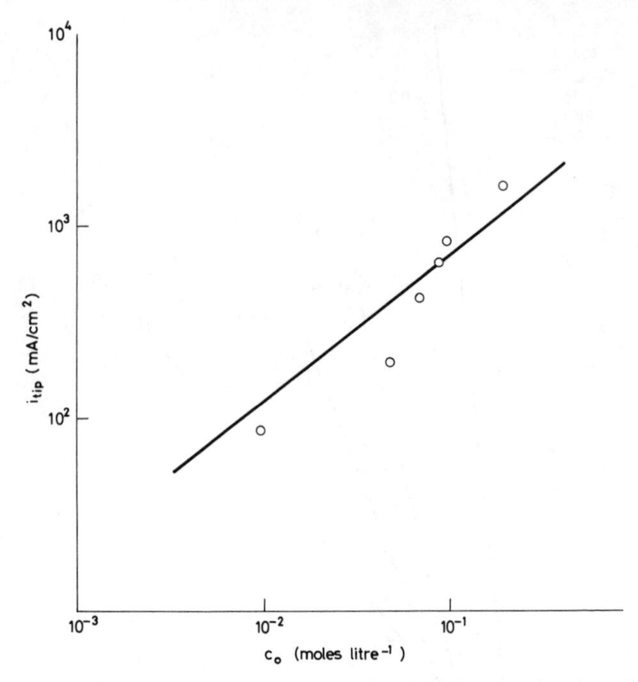

Figure 27. The tip current density at zinc dendrites as a function of zincate concentration in solution. The straight line with a slope of 0.75 is the theoretical relation. (From Diggle *et al.*[20])

be linear, with a slope of $(1 - \beta/2) \sim 0.75$. The available data for zinc dendrites[20] are plotted in this manner in Figure 27.

The dependence of the rate of growth on overpotential can also be compared with theory. For sufficiently large overpotentials ($\gg 40$ mV) and tip radii, equation (54b) can be arranged to give

$$1/i = (1/i_0)\exp[(\alpha_c F/RT)\eta] + (r_n/K'') \tag{58}$$

or approximately

$$\log i = \log i_0 - (\alpha_c F/2.3RT)\eta \tag{59}$$

The plot of $\log i_{\text{tip}}$ versus η is shown in Figure 28. From the slope, an i_0 value of 220 mA cm^{-2} is obtained, which is in the expected range.

Finally, an Arrhenius type of dependence of the tip current density on temperature is observed, as shown in Figure 29. From

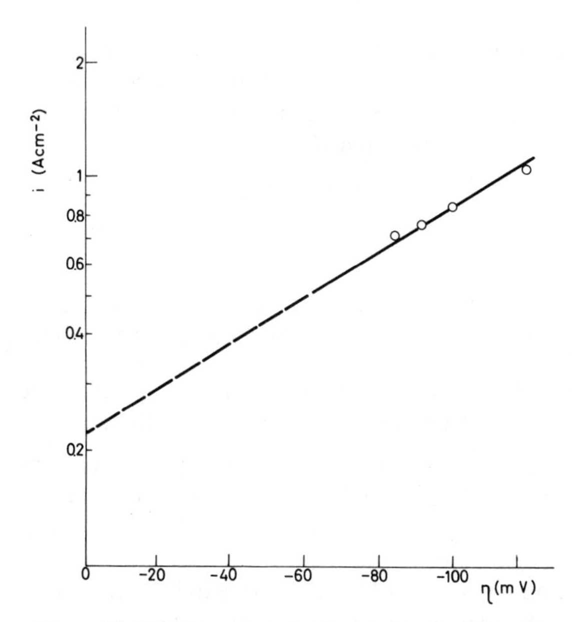

Figure 28. The tip current density at zinc dendrites as a function of overpotential at a zincate concentration of 0.1 mole liter^{-1} and temperature of 35°C (from Diggle *et al.*[20]).

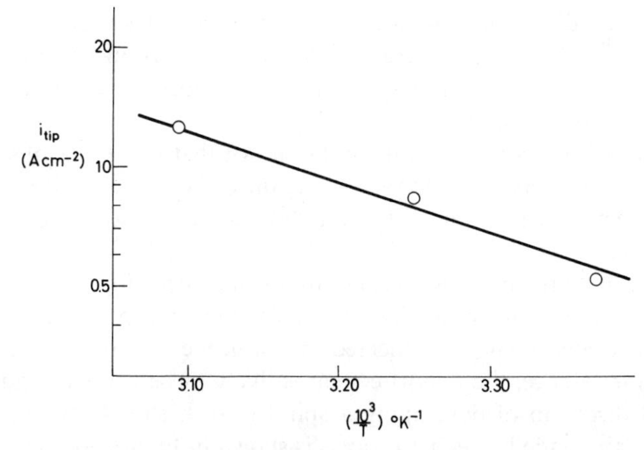

Figure 29. Temperature dependence of the tip current density of zinc dendrites (C° = 0.1 mole liter^{-1}, $\eta = -100$ mV); (from Diggle *et al.*[20]).

this, an activation energy at the overpotential of $\eta \approx -100$ mV can be calculated as 6 kcal mole^{-1}. Analyzing equation (54b), one can see that at such a high overpotential, at which $i_0 f_c(\eta)$ can be considered to be much larger than K''/r_n, it reduces to

$$i = zFDC^\circ \nabla f'/(\rho/2) \tag{60}$$

i.e., the current density is controlled by paraboloidal diffusion to the tip.

Hence, the temperature dependence is that of the diffusion coefficient, and, hence, the activation energy is in the expected range.

4. The Effect of the Crystal Structure of the Depositing Metal on the Direction of Growth and Shape of Dendrites

The direction of growth of dendrites, the regularity of their shape, and the appearance of well-defined and fairly regular branching cannot be explained by the nonspecific factors discussed above. The crystal structure of the depositing metal and some specific crystallization phenomena obviously have to be taken into account.

So far, the overall activation energy of the deposition process was assumed to be independent of position and of the direction in which a surface irregularity extends. Yet, it is clear that there should be a preferential way of building the discharged adatoms into the crystal lattice which should result in favoring a certain direction in a spherical diffusion field for the growth of the structure. The principle of minimum energy of formation introduced for the nucleation processes by Pangarov *et al.*[41–45] should be operative in any crystal-building process.

If this is accepted, then it can be shown that two types of sites are particularly favorable for extending the crystal lattice. These are (a) the screw-dislocation sites, and (b) the separation planes in twinning.

The first are favorable because they can provide for continuous growth without the necessity for nucleation, which requires an additional energy change reflected in the needed degree of supersaturation. Hence, the dendrite stem is likely to be oriented in the natural direction of developing a spiral growth site. A pyramidal dendrite tip made by such a process is shown in Figure 30. Yet, one should have in mind that the rate of initial amplification and growth within the linear diffusion layer at the surface, given by equations

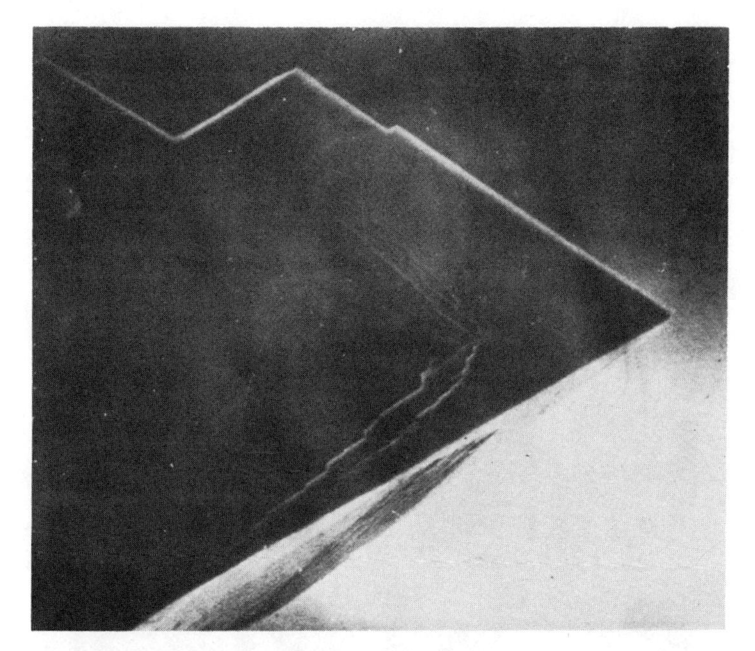

Figure 30. The microphotograph of a zinc dendrite tip (850 ×) (from Diggle *et al.*[20]).

(54a) and (54b) as determined by the supply of depositing ions by diffusion, is that for the growth perpendicular to the surface. If the direction of growth is inclined with respect to the direction of the diffusion field, the rate of growth would be less, the more is the deposition process diffusion-controlled.

In the case of complete diffusion control, the rate of growth should be

$$v_n = v \cos \theta = (VDC^\circ/\delta)(h/r) \cos \theta \tag{61}$$

Hence, it is obvious that those screw dislocations should be favored, the direction of growth of which is close to the direction of the diffusion field. Hence, an angular distribution of dendrites around the direction of the diffusion field should be expected, with a maximum in that direction.

Since $\cos \theta$ is not a particularly steep function, the distribution function need not be too narrow. Indeed this is seen to be true in Figure 31 for deposition of zinc from zincate solutions.

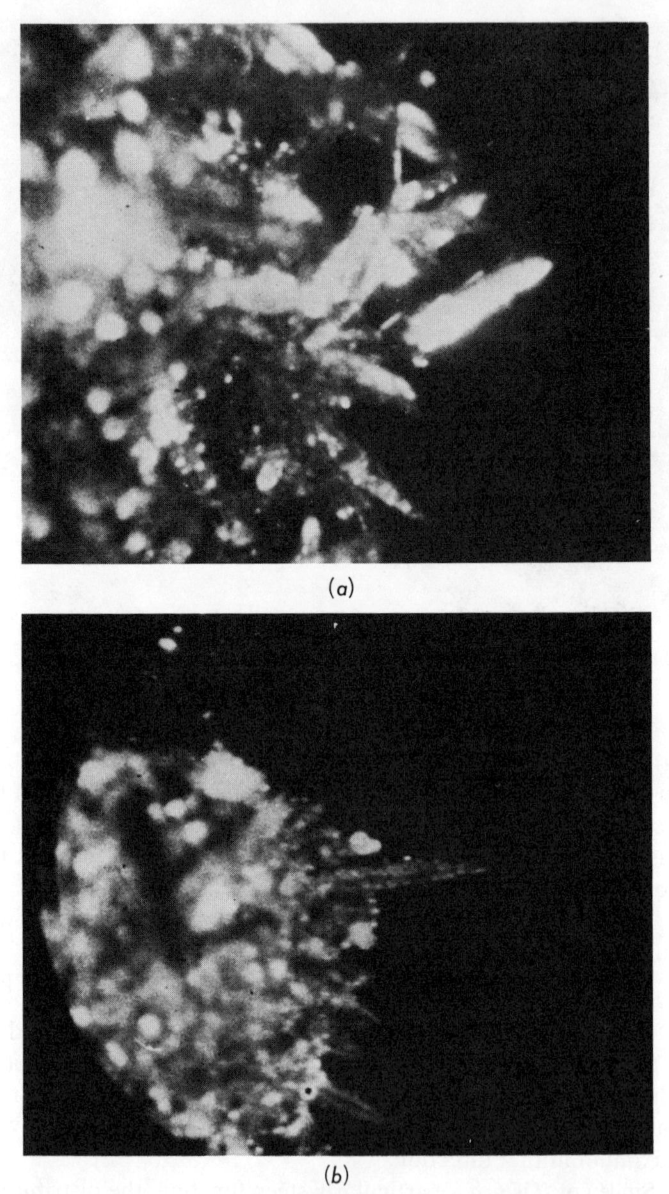

(a)

(b)

Figure 31. Dendritic growth in the deposition of zinc from zincate solution ($C^{\circ} = 0.01$ mole liter^{-1}, $\eta = -100$ mV; $T = 35°C$; $100\times$) (from Diggle *et al.*[20]).

The regularities in the appearance of side branches and, in particular, the well-defined angle between branches of different order and the stem found in most dendrites seem to have their origin in the phenomenon of twinning. Many investigations of crystallographic properties of dendrites reported the existence of twin structures.[31,36,37,40] It is well known that the separation plane provides a particularly suitable location for incorporation of atoms into a crystal lattice, in the form of an "indestructible reentrant groove."[38,39] As shown in Figure 32, as layers of atoms advance in the directions determined by twinning laws, an edge is constantly renewed, in which new layers can be started by one-dimensional nucleation. Little work is required for the latter, so that this mechanism of lattice formation should be competitive with the one based on screw dislocations. The direction of motion of the reentrant edge with respect to the crystal plane is well defined by the basic twin properties and is always the same for a given system. This results in a defined angle always appearing between the direction of growth of the dendrite stem and its branches.

The thorough study of morphological properties of dendrites made by Wranglen has shown that layer growth also plays an

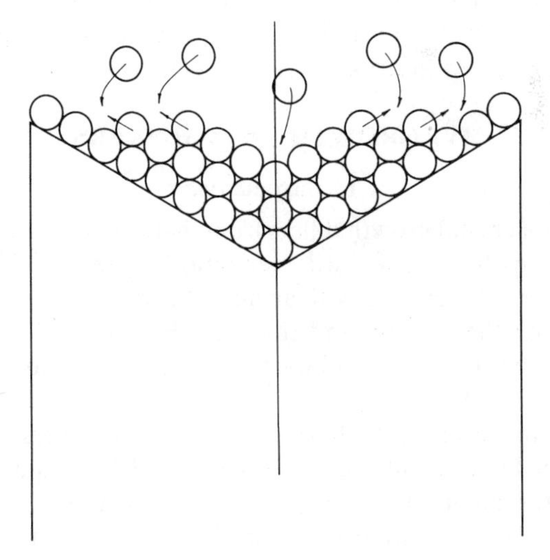

Figure 32. Schematic representation of the formation of an "indestructible reentrant groove."

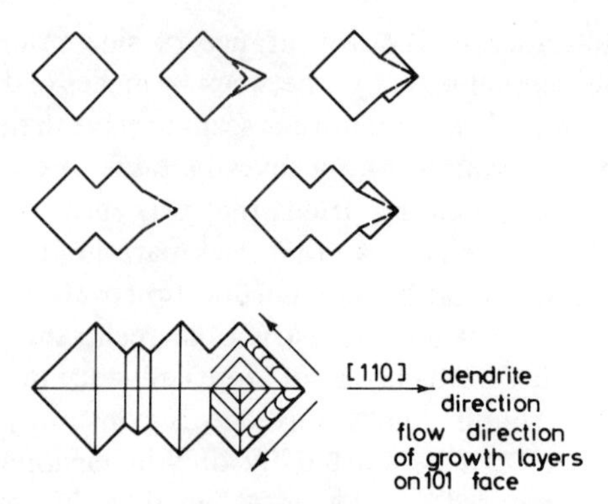

Figure 33. Schematic representation of the growth of dendrites and development of the dendrite habit (from Wranglen[4]).

important role. This, however, takes place in the later stage of formation of the dendrite habit, i.e., in completing the structure as shown in Figure 33, and hence has little effect on the direction of growth and general outlay of the dendrite.

Two problems seem to remain without answer at the present moment: (a) What is the detailed mechanism of formation of the indestructible reentrant groove? and (b) What makes it appear often at regularly spaced locations along the dendrite stem?

IV. GROWTH OF WHISKERS

1. Introduction

This form of crystal growth differs from that of dendrites in as much as (a) it tends to have a still larger ratio between the longitudinal and the lateral dimensions with an almost perfect preservation of the latter during the growth, and (b) it exhibits no tendency to side-branching. Impurities or additives in solution seem to be essential for its appearance.

Gorbunova *et al.*[7,44] have grown silver whiskers from fairly concentrated silver nitrate solutions ($>0.3\,N$) containing oleic acid, gelatin, albumin, and heptyl, octyl, and nonyl alcohols, but they could also be grown from silver nitrate solutions made up from ordinary chemicals which were "aged" by standing for several weeks before use.[45] Copper whiskers could also be grown from solutions containing a variety of organic additives as well as from those containing hydroxylamine ions in the presence of chlorides.[48,49]

A critical current density seems to exist for each whisker, which is related to its lateral dimensions. Figure 34 shows the dependence of the current density at the whisker tip on what may be called the whisker radius.[7,45] At larger tip radii, an almost constant and radius-independent current density is obtained. The critical current density also depends on the concentration of the additive, as shown in Figure 35. Depending on the latter, whiskers have been grown in the current density range from about 0.3 to 25 A cm^{-2}. In the absence of any substance deliberately added to the solution, whiskers could also be grown, but at a much lower rate (current density of the order of 0.05 A cm^{-2}) which could be correlated with the concentration of impurities.[8]

The critical current density is independent of metal ion concentration in the range between 1 and 6 *N* and, at least in the gelatin solutions, of *p*H in the range 2–8. Yet, at metal ion concentrations below about 0.3 *N*, at gelatin concentration of 0.1 g $liter^{-1}$ ($\sim 10^{-8}$ moles cm^{-3}), whiskers become very difficult to grow.[7]

A few more phenomena should be noted: (a) while growing exclusively in one direction only, whiskers dissolve anodically at a

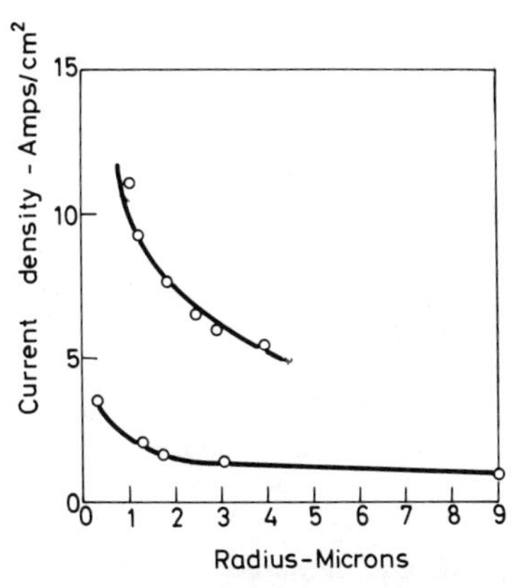

Figure 34. Dependence of critical current density on whisker radius in solutions containing oleic acid at two different concentrations (from Price *et al.*[8]).

practically uniform rate at all sides[8] and at an overpotential much smaller than that needed for growth; (b) a higher overpotential is needed temporarily for the initiation of growth (or continuation after interruption) than for growth at a steady state; (c) if the growth is interrupted for a longer period of time, then it may be continued at the tip, but usually assuming a new direction, or else it may be completely prevented and a new whisker started elsewhere; the minimum time required for complete cessation of further growth is found to depend on the concentration of the additive, as shown in Figure 36; (d) if a constant rate of growth is maintained, by a constant current flow through the cell to the individual whisker tip, fluctuations of overpotential are observed.

Finally, one should note that whiskers differ from other crystals of the same metal in two respects at least: (i) they have an increased electrical resistivity (2–3 times that of crystals deposited in the absence of additives) and an increased tensile strength (3000–8000 lb in.$^{-2}$ compared to a few hundred lb in.$^{-2}$ observed in large, pure silver single crystals).

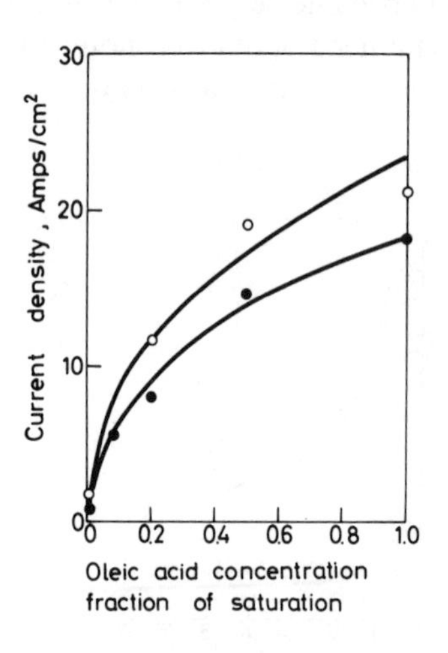

Figure 35. Dependence of critical current density on the concentration of oleic acid at currents of 9×10^{-8} A (upper curve) and 4.8×10^{-7} A (lower curve) (from Price *et al.*[8]).

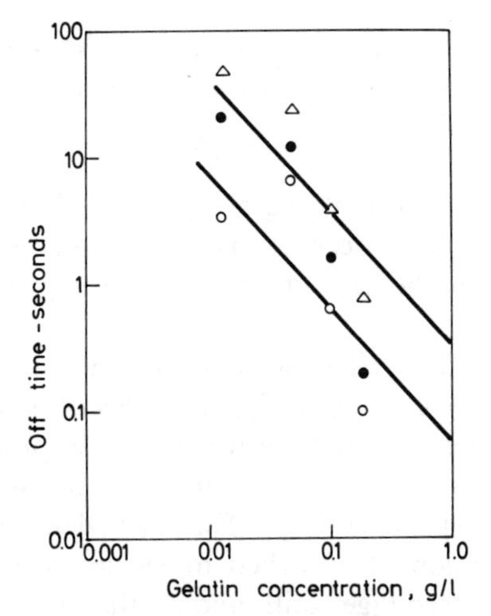

Figure 36. Effect of impurity concentration on the time required for contamination of the tip of a whisker during the period of no current flow (from Price *et al.*[8]).

2. Theory of Whisker Growth

A model of the growth mechanism was developed by Price, *et al.*[8] which gives a good account of most of the phenomena observed. The basic assumption of the model is that molecules of impurities or additives are so strongly adsorbed at all but one crystal plane and at such a concentration as to completely block the deposition and extension of the lattice. On the one plane, however, the process of adsorption is competitive with that of metal deposition burying the adsorbed molecules and, at a steady state, a sufficiently low surface coverage of foreign molecules is maintained for growth to be possible. The latter is assumed to occur by continuous nucleation and movement of steps over close-packed surface. Indeed, the appearance of some whiskers, as e.g., the one shown in Figure 3, suggests a repeated one-dimensional nucleation of the type shown in Figure 32, and the extension of the step in two directions to the edge of the crystal. The growth must stop at this edge since agglomeration of metal atoms beyond it would be connected with an increased free energy.

The energy requirements also define the smallest radius of curvature an advancing step may have. This is

$$\rho_c = \gamma/\Delta G_v = \gamma V/10^7 zF\eta \tag{62}$$

where γ is the work of building a crystal edge of unit length and ΔG_v is the free-energy decrease per unit volume accompanying deposition. In cathodic deposition, the latter is proportional to the overvoltage η.

Hence, if some foreign molecules are adsorbed at the metal surface at a distance closer than $2\rho_c$, when a section of an advancing step encounters them, it cannot advance further past the molecules, because in doing so, the radius of curvature of this portion of the step would have to be less than ρ_c. This, then, determines the highest surface concentration of adsorbed molecules which, under given conditions of overvoltage, still allows the uninhibited flow of steps, i.e.,

$$X_m = \frac{1}{(2\rho_c)^2} = \frac{10^{14}z^2F^2\eta^2}{4V^2\gamma^2} = \frac{10^{14}z^2F^2b^2i^2}{4V^2\gamma^2} \qquad (\eta < 10 \quad \text{mV}) \tag{63}$$

and b is the η–i slope at low overpotentials.

The next assumption of the model is that adsorption of the organic additive is strong enough that the mean lifetime of the molecules at the surface is sufficiently long that several steps moving across the surface encounter it, each one moving around and past it, as shown in Figure 37. Thus, the molecule becomes surrounded by layers of metal and eventually is incorporated into the body of the crystal. (Large molecules are more likely to become imbedded, since they would have much smaller vibration frequencies than individual atoms or ions and, hence, a correspondingly larger lifetime at the surface.) New molecules diffuse to the surface from the bulk of solution and become adsorbed. Thus, a steady state is attained when the rate of diffusion of the species becomes equal to the rate of incorporation into the metal. For the former, a spherical diffusion field can be assumed at the tip of the whisker. The maximum diffusion flux is then

$$(dn_i/dt)_{\text{diff}} = ND_iC_i/r \tag{64}$$

The rate at which the molecules are buried depends on the current density of metal deposition, i, and the surface area covered

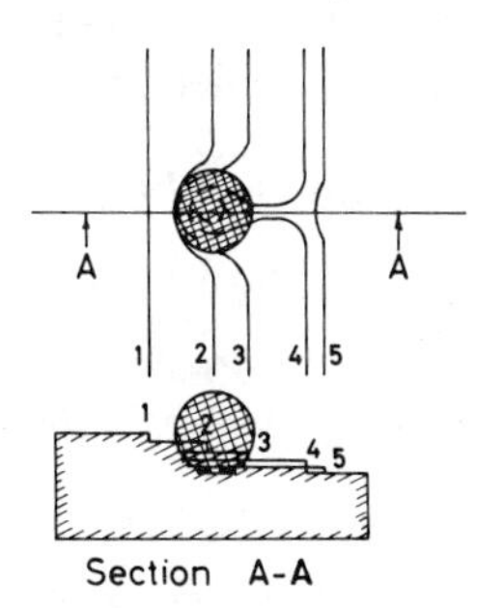

Figure 37. Surface steps (1, 2, 3, 4, 5) moving past an adsorbed molecule (from Price *et al.*[8]).

by each deposited metal atom, α, as well as on the height of the step, a. Thus,

$$(dn_i/dt)_{\text{inc}} = (Ni\alpha a/zFh)X_m \tag{65}$$

From the equality of the two rates and taking into account equation (63), one obtains

$$i_c = \left(4 \times 10^{-14} \frac{hM^2\gamma^2D_ib^2}{\alpha a\rho^2 zF} \frac{C_i}{r}\right)^{1/3} = A\left(\frac{C_i}{r}\right)^{1/3} \tag{66}$$

indicating that the critical current density of whisker growth should be directly proportional to the third root of additive concentration and inversely proportional to the same power of whisker radius. This relationship is seen in Figures 34 and 35 to follow closely the experimental results, the constant A being determined for the best fit of the experimental points.

The order-of-magnitude calculations of the constant A for a case of silver whiskers grown from 1 N $AgNO_3$ solution with 0.1 g liter^{-1} gelatin yield A of the order of 14, which is in good agreement with experiment ($b = 40$ A V^{-1} cm^{-2}; $D = 10^{-6}$ cm^2 sec^{-1}; $h = 3 \times 10^{-7}$ cm; $a = 3 \times 10^{-8}$ cm; $\alpha = 10^{-15}$ cm^2; $\gamma =$ 1000 ergs cm^{-2}; $r = 3 \times 10^{-4}$ cm).

These authors also calculated the minimum concentrations of metal ions in solution which could still give a sufficiently fast supply for the critical current density to be maintained. This is the concentration which, in the limiting current density condition, still gives i_c. In the spherical diffusion field, this is

$$C_m = i_c r/zFD_m \tag{67}$$

For a critical current density of 0.7 A cm^{-2}, indicated by the previous calculation, and taking $D_m = 10^{-5}$ cm^2 sec^{-1}, this comes out as 0.2 *N*, which is in accordance with the findings of Gorbunova *et al.*[7]

The theory could also be checked by calculating the minimum time of current interruption required to completely block the tip of the whisker and prevent further growth upon continuation of deposition. This could be estimated from the rate of diffusion of impurity molecules to the tip and the number of molecules m needed to block a cm^2 of surface. Thus,

$$t = mr/ND_iC_i \tag{68}$$

For an m value of 10^{13} cm^{-2}, the linear log t versus log C_i relation indicated by the equation (68) is seen in Figure 36 to follow closely the experimental points.

One should note that an alternative and more elaborate theory of whisker growth was offered by Krichmar,[46] who claimed a need for the new theory because of an apparent incapacity of the one discussed above to explain the observed independence of i_c on r at high r values.

The basic difference between the models of Krichmar and Price *et al.* is that the former assumes the deposition current to be controlled by the frequency of nucleation of new layers, and this is affected by the surface concentration of adsorbed molecules,[47] so that

$$i = i_0 \exp[-4r_c(NX_m)^{1/2}] \tag{69}$$

where i_0 is the current density in the absence of any foreign molecules. The surface concentration X_m is related to the volume concentration of the additive in the layer next to the surface, C_{is}, by an adsorption isotherm which at low C_{is} has a linear form, i.e.,

$$X_m = GC_{is} \tag{70}$$

With all the other relations the same as those of the theory of Price *et al.*, Krichmar obtains the relationship

$$i \ln \frac{i_0}{i} = \frac{4\gamma Mb}{\rho zF} \frac{NC_i G}{(iMGr/zFD_i h_i) + 1} \qquad (71)$$

At high i values, equation (68) is shown to reduce to

$$i = A'(C_i/r)^{1/3} \qquad (72)$$

and

$$A' = \left[\frac{16\gamma^2 NMD_i h_i b^2}{\rho zF}\right]^{1/3} \ln^{-2/3}\left[\frac{zFD_M C_M}{(16^2 NMD_i h_i b^2 r^2 C_i/\rho zF)^{1/3}}\right] \qquad (73)$$

At low i values, however, a relation independent of the whisker radius is obtained as

$$i = A'' C_i^{1/2} \qquad (74)$$

with

$$A'' = \frac{4\gamma Mb(GN)^{1/2}}{\rho zF} \ln^{-1}\left[\frac{zFD_M C_M}{4\gamma Mb(CNC_i)^{1/2} r/\rho zF}\right] \qquad (75)$$

More detailed and precise experiments are needed as to verify one or the other theory. In the region where both give the same i versus C_i/r relationship, Krichmar claims his constant A' to be close to the value found by Price *et al.** If so, experimental behavior in this region cannot be used for discriminating between the two theories.

Nevertheless, both theories agree in assigning the dominant role in whisker growth to the diffusion of foreign molecules from the bulk of solution to the surface and to their adsorption at the surface.

They also both suffer from the same shortcoming: neither of them offers good reasons for the appearance of whiskers, i.e., for the competition between adsorption of impurities and their burial in the metal at one crystal plane only. Hence, new experiments and thinking are needed on the problem of whisker initiation.

*This cannot be obtained by putting the appropriate data into equation (73). Hence, there appears to be a misprint in the original paper.

V. FORMATION OF POWDERED DEPOSITS

Powdery metal deposits differ from the ones described above mainly by the fact that they are very loosely connected with the substrate at which they are formed. Also, they tend to have a very high degree of dispersion, i.e., to consist of very small crystallites, often in the range of 100 Å. This usually results in the electrode turning black as soon as powder deposition starts.

Black silver powder was obtained using electrolysis by Priestley[50] as early as 1803. Extensive studies on a large number of metals (Cu, Ni, Co, Fe, Pb, Sn, Mn, Zn, Cd, Pd, W) have been carried out since then. Relatively recently, a detailed survey of this work was made by Ibl.[10] Hence, only basic features of the phenomenon of powder deposition will be reviewed below.

1. Common Features

The common features are:

(a) All metals which can be electrodeposited exhibit a tendency to appear in the form of powders at current densities larger than a certain critical value i_p. Some metals (Zn, Cd) also have a specific tendency to give powders at low current densities, lower than those needed for obtaining compact deposits. While this specific powder deposition may have different causes (deposition of colloidal metal dissolved elsewhere, secondary reduction by codeposited hydrogen, disproportionation of divalent ions, etc.), powder deposition at high current densities seems to be a direct consequence of the slow transport of depositing ions.

(b) At current densities equal to i_p and larger, an induction period is observed, during which a compact deposit is formed. The time of powder formation t_p can be observed visually as the electrode is seen to turn suddenly from a lustrous to a black appearance.

(c) Powder formation is enhanced by decreasing the concentration of depositing ions, increasing the concentration of indifferent electrolyte, increasing the viscosity of the solution, decreasing the temperature, and decreasing the velocity of motion of the solution.

(d) The particle size is strongly affected by conditions of deposition. Particle size increases with increasing concentration of depositing ions, decreasing current density, and increasing temperature and rate of stirring. Hence, all factors enhancing powder formation have an opposite effect on particle size.

2. Critical Current Density of Powder Formation and Limiting Current Density of Metal Deposition

All the above-mentioned facts concerning powder formation at high current densities (nonspecific powder deposition) clearly indicate this to be connected with the limitations in the transport of depositing ions. Kudra was first to observe quantitatively that, for other conditions constant, the product of the current density used and the square root of the time of powder formation is a constant quantity.[51,52] This has since been established on a large number of systems: $CuSO_4$, $CuCl_2$, $Cu(NO_3)_2$, $ZnSO_4$, $Zn(CH_3COO)_2$, $CdSO_4$, $CdCl_2$, $CdBr_2$, CdI_2, $Cd(CH_3COO)_2$, $AgNO_3$, $CoCl_2$, $CoBr_2$, $Co(CH_3COO)_2$, $CoSO_4$, $Pb(NO_3)_2$, $Pb(CH_3COO)_2$. A typical dependence of log t_p on log i is shown in Figure 38, and shows well-defined straight lines of slope -2.[10]

The product $it_p^{1/2}$, as shown in Figure 39, was found to be a linear function of the concentration of depositing species, with a slope somewhat dependent on the nature of both the cation concerned and the anion composing the electrolyte.

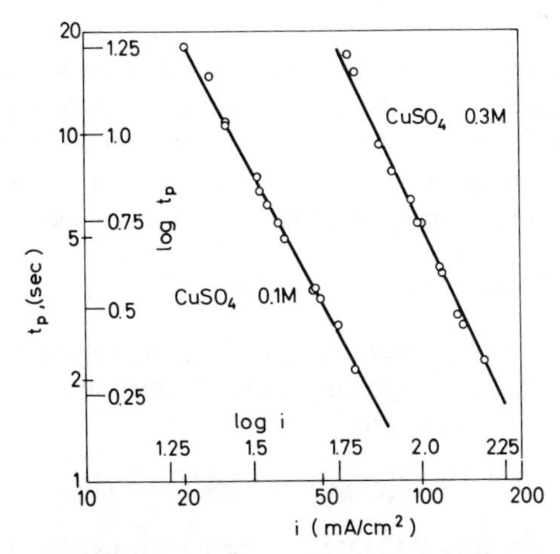

Figure 38. Dependence of time of powder formation t_p on the current density i in the electrolysis of copper sulfate solutions (from Ref. 10).

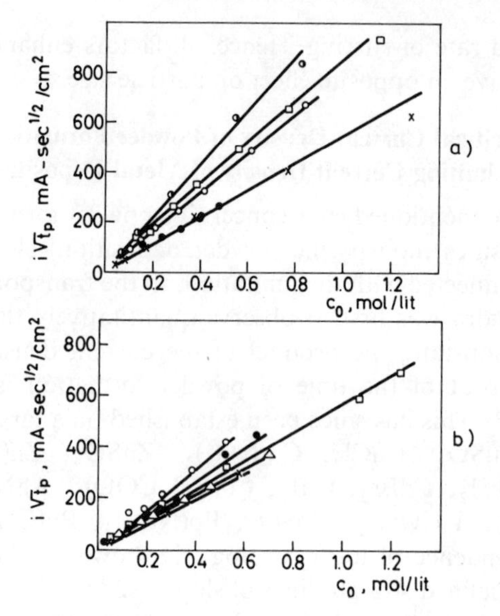

Figure 39. Dependence of $it_p^{1/2}$ on bulk concentration c_0 of electrolyzed solutions: (a) Δ, $Cu(NO_3)_2$; ○, $CuSO_4$; ×, $CdSO_4$; □, $CuCl_2$; ●, $Cd(CH_3COO)_2$. (b) Δ, $ZnSO_4$; ○, $Zn(CH_3COO)_2$; □, $CdCl_2$; ●, $AgNO_3$; ×, $AgNO_3$ in 25% NH_3.

Such dependences are characteristic of processes controlled by diffusion. Thus, in the absence of any convective motion of the electrolyte, the Sand equation relates the concentration of depositing ions at the electrode surface, C^s, to that in the bulk of solution, C°, and to the current density and time of deposition as

$$it^{1/2} = [zF/2(1 - n_c)](\pi D)^{1/2}(C^\circ - C^s) \tag{76}$$

where n_c is the transference number of the depositing cation. As C_s tends to zero, the chronopotentiometric transition time τ is reached and equation (76) reduces to

$$i\tau^{1/2} = [zF/2(1 - n_c)]\pi^{1/2}D^{1/2}C^\circ \tag{77}$$

The straight lines of Figure 39, passing through the origin, follow the same function as that defined by equation (77). Hence, the time of powder formation appears to coincide with the transition time, i.e., powdery deposits start forming as the reaction layer at the electrode is totally depleted of depositing ions.

To confirm more closely the applicability of equation (77), one could correct the product $it_p^{1/2}$ for the absolute values of the transference number and diffusion coefficients of two different metal ions, as well as for the concentration dependence of these quantities. As shown in Figure 40, the corrected points follow very closely the predicted common straight line over a considerable concentration range.

An additional check on the correlation between the conditions of powder formation and those of limiting current is obtained for the case of deposition at vertical electrodes in the presence of natural convection. The convective-diffusion limiting current density in this case is given by

$$i_L = [0.67zF/(1 - n_c)]D^{3/4}(g\alpha^*/\nu h)^{1/4}C^{\circ\,5/4} \tag{78}$$

where g is the gravitational constant, ν the kinematic viscosity, h the height of the electrode, and α^* is a complex factor, but in the case of the electrolysis of a pure binary electrolyte, is equal to the densification coefficient $(1/\rho)\,d\rho/dc$.

Hirakoso[53] studied the deposition of copper powder from $CuSO_4$ solutions containing H_2SO_4 under conditions close to those for which equation (78) should be applicable. A replot of his data on the dependence of i_p on concentrations on a log-log basis yielded a straight line with a slope of 1.21, which is very close to the expected value of 5/4.

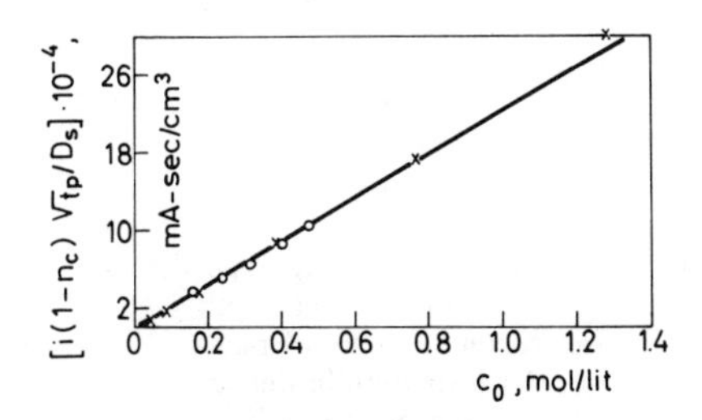

Figure 40. Dependence of $i(1 - n_c) \times (t_p/D_s)^{1/2}$ on bulk concentration of $CdSO_4$ (×) and $CuSO_4$ (○) (from Ref. 10).

To compare the absolute values of i_p with those of i_L under corresponding conditions, Ibl[54] rewrote equation (78) in the form used in chemical engineering, i.e.,

$$\mathrm{Nu} = 0.67(\mathrm{Sc} \times \mathrm{Gr})^{1/4} \quad (79)$$

where

$$\mathrm{Nu} = i_L(1 - n_c)h/zFDC^\circ \quad (80)$$

$$\mathrm{Sc} = \nu/D \quad (81)$$

$$\mathrm{Gr} = g\alpha C^\circ h^3/\nu^2 \quad (82)$$

The result of such a comparison is shown in Figure 41.

3. Interpretation of the Role of Transport Control in Powder Formation

The established equality between the critical current density of powder formation and the diffusion limiting current is, of course, of considerable practical interest, for it enables precise determination of the current density range and induction times at which powders are to be expected. A comprehensive theoretical explanation of this relation, i.e., the model of the mechanism of powder formation, is apparently not yet available.

The growth of the powder particles should probably be governed by the same laws as the amplification of surface roughness and the growth of dendrites. Yet, the reason for the appearance of an

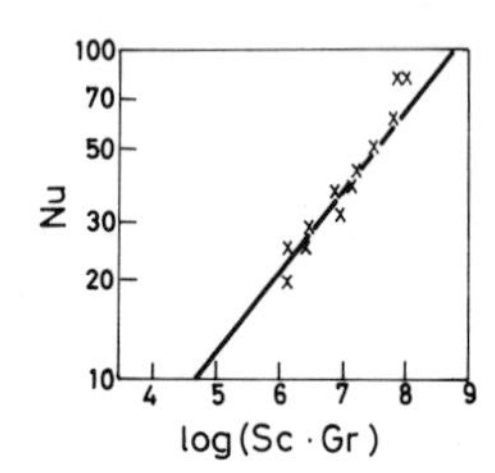

Figure 41. Comparison of limiting current with current density of powder formation in the deposition of copper from $CuSO_4$ solutions of various concentration (from Ibl[54]).

outgrowth so loosely connected to the substrate as is the powder particle has not been elucidated. For, it is this loose connection, sufficiently strong to allow relatively free flow of electrons and weak enough for the particle to fall off at a slightest provocation, which makes powders a category separate from the others.

Two possible bases for this difference have been proposed so far in a qualitative manner.

Powder particles are formed by nucleation and not by an extended growth of already existing irregularities (dislocations present in the substrate surface). The nucleation rate is known (cf., e.g., Fisher and Heiling[55]) to increase with decreasing concentration. Hence, under limiting current conditions, when the concentration of the depositing ions is virtually zero and the overpotential relatively high, a deposit of high dispersiveness should appear which does not reproduce the substrate surface.

It has indeed been found by Antanasiu and Calusaru[56] that powders do not form at the initial part of the limiting current plateau, but rather at the higher overpotential part, at which the tendency to a transition into another electrode process begins to develop. One should note that in model experiments described earlier on the amplification of surface roughness and the appearance of dendrites (see Ref. 21) it was also observed that these phenomena are pronounced at overpotentials of mixed activation and diffusion control. At overpotentials where diffusion is in practically complete control, quite irregular, treelike deposits are formed at peaks of the original substrate irregularities.

The high-nucleation-rate theory does account for the high dispersion of the deposited metal as well as for the change of dispersion with the factors outlined earlier. Thus, as a growing crystallite penetrates deeper into the diffusion field, further deposition at its tip occurs at an ever higher concentration. The rate of nucleation decreases and the crystal grows until the connection with the substrate is lost. This is the more pronounced the steeper is the concentration gradient in this diffusion field. The latter increases with increasing bulk concentration, rate of stirring, etc. Hence, a coarser deposit results.

However, this theory does not necessarily also explain the basic fact of loose connection between the new particle and the substrate. For, it is well known that extremely fine-grained but compact deposits can also be formed under certain conditions at which nucleation is favored.

Hence, in this connection, one should note the second theory proposed by Atanasiu and Calusaru.[57] They maintain that the energy barrier for quantum mechanical tunneling is widened with decreasing concentration of the discharging species. Hence, at a vanishingly small concentration, the electrons start meeting the ions at distances larger than the lattice parameter of the metal. Metal adatoms of a very high free-energy content should then wander in a layer somewhat more remote from the metal surface than in ordinary deposition and undergo three-dimensional nucleation. The nucleus should be attracted to the metal surface of any foreign particle, make electric contact, and continue to grow outward for reasons discussed earlier, until it becomes too heavy and falls off.

This is an attractive model, but other possibilities should not be overlooked. In particular, one should bear in mind the above-mentioned fact that powders start forming in the potential region in which some other electrochemical process starts developing with unknown effects on metal deposition.

VI. LEVELING

It has been known since the early work by Meyer,[58] Schmellenmeier,[59] and particularly Gardam[60] that in electrolytes containing specific substances as additives, a phenomenon opposite to the ones described so far can occur: a more rapid metal deposition at recessed than at elevated points of the surface, causing leveling of surface irregularities. The fact that this phenomenon is only observed at microprofiles not exceeding 100 μm in amplitude created a need to introduce the concept of "microthrowing power" as a category different from ordinary throwing power.[61] The latter is used in technical literature to describe the quality of electrolytes in plating on macroprofiles, at which a similar effect is never observed.

Detailed surveys of the literature on leveling are available.[62–65]

1. Basic Facts

In electrolytes with a leveling power, such as, e.g., Watts nickel solution containing coumarin, a cathodic current causes an exponential decrease of the amplitude of a surface irregularity, H_t, with time.

As shown in Figure 42, good linear dependence of log H_t on time was demonstrated by Krichmar and Pronskaya[66] by a direct microscopic observation of the change in the depth of a sinusoidally shaped profile when plated in such a bath. Leveling occurred in a wide range of profile wavelengths a, although a sharp increase of the time constant was found at $a = 100\ \mu m$.

The inverse of the time constant of the process was found to be almost linear with the coumarin concentration in the whole of the investigated concentration range up to 1 g liter^{-1}. It was found to be inversely proportional to the wavelength of the profile and linearly decreasing with increasing current density.

Similar leveling power is possessed by other additives such as 2-butyne-1,4-diol, quinoline ethiodide, thiourea, diethyl thiourea, etc. They all adsorb at an electrode surface and cause an increase in electrode polarization or a decrease of current density at a constant potential. The extent of the effect is shown in Figure 43, where the addition of 0.4 g liter^{-1} of 2-butyne-1,4-diol is seen to cause a shift in polarization of about 100 mV. Agitation is seen to produce a further increase in polarization, compared to some decrease at

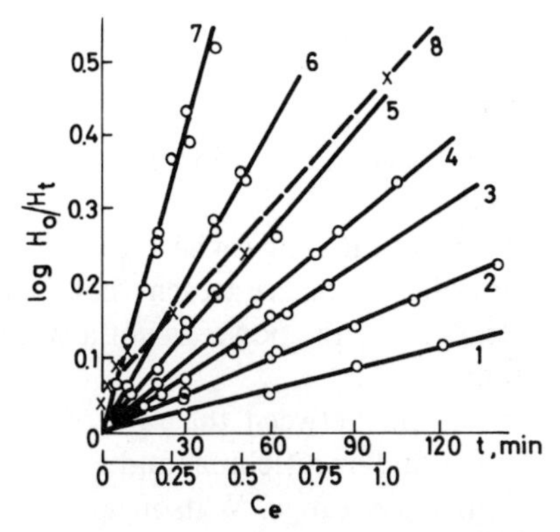

Figure 42. Dependence of $\log(H_0/H_t)$ on time for different coumarin concentrations in the electrolyte. $a = 100\ \mu m$, $i = 0.01\ A\ cm^{-2}$; curves: (1) 0.000; (2) 0.025; (3) −0.05; (4) 0.09; (5) 0.025; (6) 0.5; (7) 1.0 g liter^{-1}; (8) the dependence of $\log(H_0/H_t)$ at 10 min on coumarin concentration (from Krichmar and Pronskaya[66]).

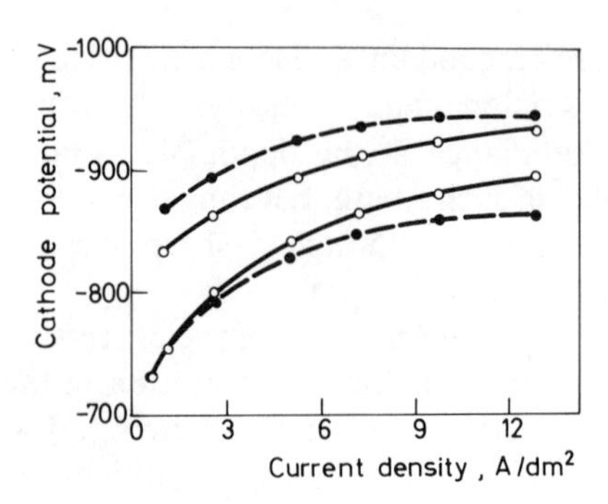

Figure 43. Polarization curves (versus s.c.e.) in nonagitated (open circles) and agitated (closed circles) Watt's nickel bath (50°, *p*H 3.0) in the absence of addition agents (lower pair) and in the presence of 0.4 g liter^{-1} of 2-butyne-1,4-diol (upper pair) (from Kardos and Foulke[64]).

higher current densities in the absence of such an additive. Experiments with a rotating disk electrode have shown that the inhibition is proportional to the diffusion rate of the additive, which obeys the $\omega^{1/2}$ dependence characteristic of that electrode. Thus, an equal increase in polarization, as shown in Figure 44, is produced with such pairs of concentrations and rotation speeds which give equal values of the produced $C\omega^{1/2}$.

It is in the concentration range where the effects of additive diffusion on electrode polarization are present that the leveling action takes place. However, the leveling power is not a simple function of additive concentration.

The linear relationship between the rate of leveling and the additive concentration found by Krichmar and Pronskaya[66] extends over a limited concentration range. Watson and Edwards[70] have demonstrated the more complex dependences shown in Fig. 45.

In the low-concentration range, some substances known as leveling agents can exhibit an inverse effect and act as activators for increasing surface roughness. On the other side, all agents are found to give maxima in the leveling power at some concentration. Such maxima were found by a number of other authors as well (cf. Voronko,[68] Kruglikov *et al.*,[69] and Rogers and Taylor[71]).

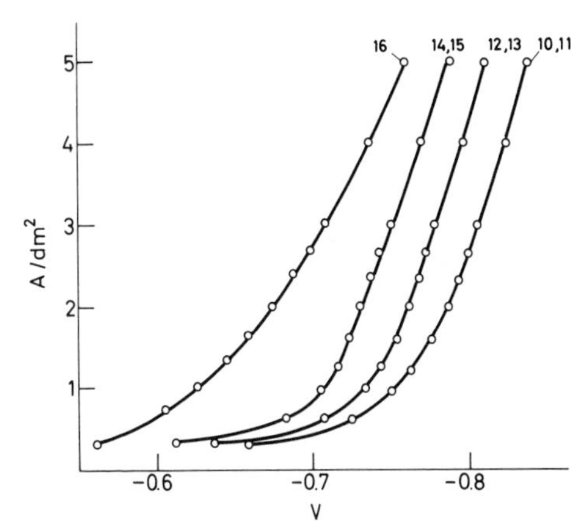

Figure 44. Current density versus cathode potential for nickel solution containing thiourea. Thiourea concentrations and rotation speeds: (10) 0.1 g liter^{-1} and 360 rpm; (11) 0.063 g liter^{-1} and 900 rpm; (12) 0.063 g liter^{-1} and 360 rpm; (13) 0.04 g liter^{-1} and 900 rpm; (14) 0.04 g liter^{-1} and 360 rpm; (15) 0.027 g liter^{-1} and 900 rpm; (16) without thiourea, 360 or 900 rpm (from Kruglikov *et al.*[69]).

A similar effect is produced by increasing the current density at a constant additive concentration (Figure 46).

Current interruption has a negative effect on the leveling. As shown in Figure 47, considerable effects are already exhibited at interruption frequencies around 1 Hz, particularly if the off periods are much longer than the on periods.

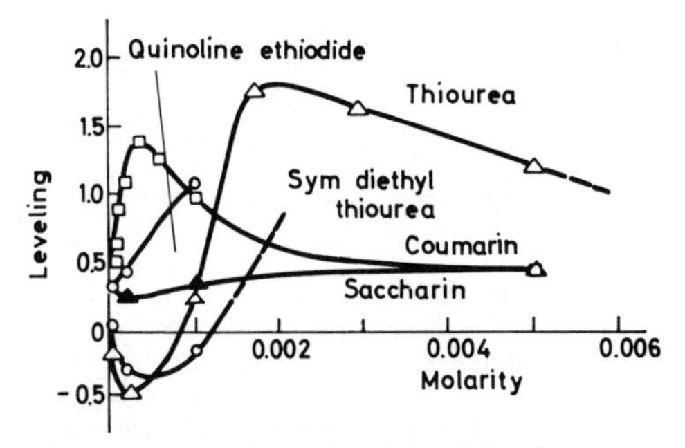

Figure 45. Experimental graphs of leveling versus addition-agent concentration (from Watson and Edwards[70]).

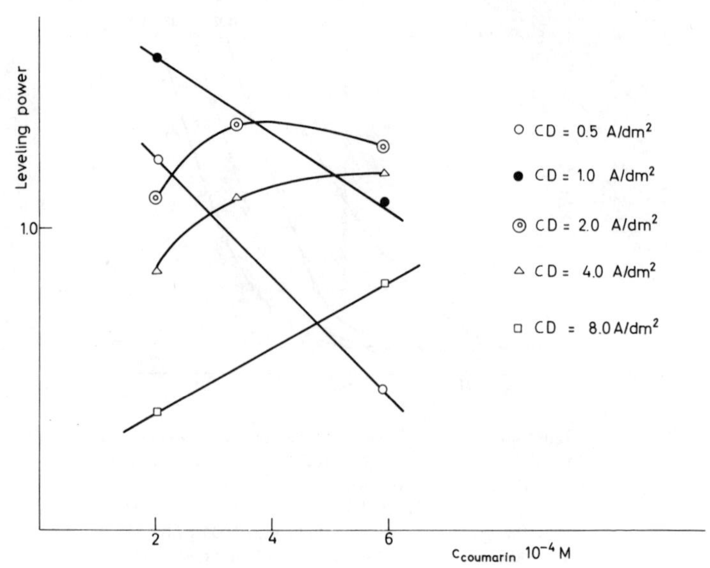

Figure 46. Leveling power of coumarin as a function of current density (from data of Kardos and Foulke[64]).

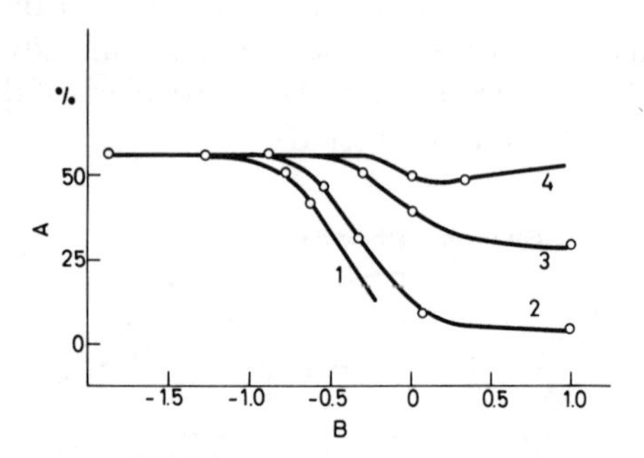

Figure 47. Percentage of leveling versus log of pulse frequency (Hz) in a Watt's nickel bath saturated with coumarin (30°C, *p*H 4.5, 4 A dm^{-2}, 430 rpm). Deposit thickness, 4 μm on flat surface (angle of groove, 60°; depth of groove, 50 μm). Cathodic time/interruption time: (1) 1/12; (2) 1/4; (3) 1/1; (4) 3/2 (From Kruglikov *et al.*[69]).

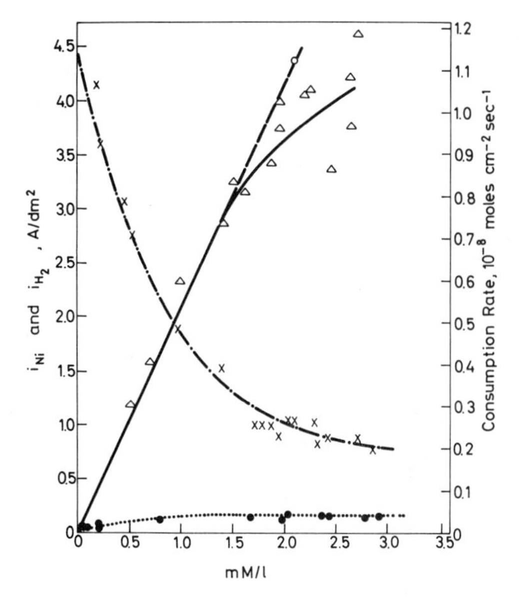

Figure 48. Rate of coumarin consumption (▲), partial current density of nickel deposition i_{Ni} (×), and hydrogen evolution rate i_{H_2} (●), versus coumarin concentration in a Watt's nickel bath at −960 mV (s.c.e.) at 980 rpm, 48.5°C, *p*H 4.0. Point *T* calculated from the Levich equation (From Rogers and Taylor[71]).

Finally, one should note that the leveling agent is consumed in the deposition process. The consumption may occur through incorporation of the agent in the deposit or through a chemical change, or both. In any case, the observed consumption rates show diffusion control over a wide range of conditions. Thus, in the experiments of Rogers and Taylor[71] with a rotating disk electrode in a Watts nickel bath, the rate of coumarin consumption is shown in Figure 48 to obey the Levich dependence over a considerable concentration range. Deviations are observed for $C > 2$ *m*mole liter^{-1}.

2. Model of Leveling

All the experimental evidence points to the conclusion that leveling takes place under conditions when the supply of a substance causing inhibition of the electrode process is under diffusion control. It was clear already to early investigators (cf. Gardam[72]) that the explanation should be sought in the local variations in the supply of the leveling agent over the surface profile. Peaks at the surface receive larger amounts of the additive than the recesses. This produces an increase in inhibition and a decrease in the local current density of deposition at the former relative to less-exposed parts of the surface. Thus, leveling is directly related to differences in surface concentration of the additive causing differences in the local current density of deposition.

Obviously, for this model to be operative, two conditions must be satisfied: (a) the additive must be consumed in some manner at the electrode, so that its continuous supply is needed so as to maintain a certain surface concentration, and (b) the diffusion layer must not follow the microprofile but must have a smoother outer boundary, so that variations in its thickness arise, which cause variations of the diffusion flux of the additive.

The first condition is fulfilled with all good leveling agents. Most of them undergo sufficiently strong adsorption to stay long enough at the metal surface to be surrounded by depositing atoms and be incorporated into the deposit. It is the balance between the rate of incorporation and that of diffusion of the substance from the bulk of solution which maintains a given surface concentration of the additive. The larger the diffusion flux, the higher is the steady-state surface concentration of the additive. Conversely, higher rates of metal deposition cause lowering of the latter.

There is an optimal range of additive concentration and current density of deposition at which the differences in inhibition of deposition between peaks and recesses, and hence the effect of leveling, are at maximum. At too low surface concentrations of the additive, i.e., low bulk concentration and high current density of deposition, the process is practically uninhibited and little difference in local rates of deposition can arise. This explains the decrease in the effect of leveling with increasing current density.

At somewhat higher bulk concentrations and lower current densities, linearity exists between the former and the surface concentration. This is the range of maximum difference in inhibition.

However, at still higher concentrations, adsorption–desorption equilibrium tends to be approached leading to a Langmuir-type relationship. Eventually, in spite of incorporation, saturation of the surface is reached and the surface concentration is no longer sensitive to local changes in additive diffusion flux. Hence, differences in inhibition vanish and leveling is lost. This is well demonstrated by Figure 45.

One should appreciate that some time is needed for the diffusion layer to develop to the extent that it separates from the surface microprofile and provides for local differences in additive diffusion flux. Hence, an induction time should be expected for the leveling effect to appear. This is demonstrated by the observed sensitivity of the process to current interruptions.

3. Quantitative Treatment

In attempting to make a quantitative treatment of the phenomenon of leveling, one encounters the problem of defining the leveling power. Watson and Edwards[70] suggested this to be given as the ratio $(i_r - i_p)/i_{av}$, where i_r, i_p, and i_{av} are the recess, peak, and average current densities, respectively. They found that this obeys a simple relation to the slopes of the potential–additive-concentration dependence, dE/dC_0, and of the potential–current–density dependence, dE/di, i.e.,

$$\frac{\Delta_i}{i} = \frac{i_r - i_p}{i_{av}} = \frac{KC^{\circ}_{ad}}{i_{av}} \frac{(dE/dC^{\circ}_{ad})_i}{i_{av}(dE/di)_{C^{\circ}_{ad}}} = K\frac{(dE/d \ln C^{\circ}_{ad})_i}{(dE/d \ln i)_{C^{\circ}_{ad}}} \tag{83}$$

Rogers and Taylor[71] suggested a similar definition.

The shortcoming of these definitions, however, is in that they depend on the particular part of the microprofile observed, i.e., on the amplitude of the surface irregularity at this point, since the difference Δi is a function of the latter. As such, it must also depend on the duration of deposition since the amplitude (and hence Δi) is also a function of time. It seems that a somewhat better characteristic of the leveling power can be found in the time constant of the process and this will be considered below.

Semiquantitative speculations on the effect of various factors on leveling power were made by the various workers who contributed to building up the model of the process. An attempt at a comprehensive quantitative consideration of the problem was made by Krichmar.[74]

The decrease in the amplitude of the surface irregularity H is determined by the difference of the current densities of deposition in the recess and at the peak:

$$-\frac{dH}{dt} = \frac{M}{2\rho zF}(i_r - i_p) \tag{84}$$

The current densities are functions of the effective diffusion layer thickness and the local concentration of the additive in the metal at the surface, i.e.,

$$i_r = i(\delta + H, C_{\text{ad}}^s - \Delta C_{\text{ad}}^s) \tag{85}$$

and

$$i_p = i(\delta - H, C_{\text{ad}}^s + \Delta C_{\text{ad}}^s) \tag{86}$$

where δ is the average diffusion layer thickness, C_{ad}^s is the average additive concentration, and ΔC_{ad}^s its deviation at the two points.

Using the first-order approximation, Krichmar[74] has shown that

$$-\frac{dH}{dt} = -\frac{M}{2\rho zF}\left[2\left(\frac{\partial i}{\partial \delta}\right)H + \left(\frac{\partial i}{\partial C_{\text{ad}}^s}\right)\Delta C_{\text{ad}}^s\right] \tag{87}$$

The partial derivatives in equation (87) could be assessed as follows: Taking the example of a sinusoidal microprofile with wavelength a, and for $a > H < \delta$, it was shown earlier that the difference in diffusion fluxes of metal ions at the two points is

$$\left(\frac{\partial c}{\partial y}\right)_r - \left(\frac{\partial c}{\partial y}\right)_p = -\frac{4\pi H\,\Delta C}{a\delta}\coth\frac{2\pi\delta}{a} \tag{88}$$

where ΔC is the average difference in metal ion concentrations at the surface and in the bulk of solution. If this is accepted as correct, then relating fluxes to current densities of deposition and the average flux $D\,\Delta c/\delta$ to the average density i, and using again the same mathematical procedure as above, we derive

$$i_r - i_p = -\frac{4\pi H_i}{a}\coth\left(\frac{2\pi\delta}{a}\right) = -2\left(\frac{\partial i}{\partial \delta}\right)H \tag{89}$$

from which the partial derivative is

$$\left(\frac{\partial i}{\partial \delta}\right) = \frac{2\pi i}{a}\coth\left(\frac{2\pi\delta}{a}\right) \tag{90}$$

[Note that H is missing in equation (88) in Krichmar's original derivation.]

To evaluate the other partial derivative, one has to assume some relationship between the current density of deposition and the surface concentration of the additive. In the absence of knowledge of the inhibition mechanism, Krichmar assumed an exponential relationship, which yields

$$\partial i/\partial C^s_{ad} = -K(\alpha F/RT)i \tag{91}$$

where K is an empirical constant.

This implies a direct or an indirect effect of the additive on the free energy of the transition state. If such a possibility is accepted, a linear dependence of that free energy on additive concentration is likely to be a good first approximation for small values of ΔC^s_{ad}.

The value of ΔC^s_{ad} can be obtained from the diffusion fluxes of the additive as

$$\Delta C^s_{ad} = (j_p - j_r)/V = (j_p - j_r)\rho zF/iM \tag{92}$$

where V is the volume of metal deposited per unit time.

The difference in diffusion fluxes obeys the same law as that for metal ions [equation (88)].

Hence, at $C^s_{ad} \to 0$,

$$C^s_{ad} = \frac{4\pi\rho zFHD_{ad}C^\circ_{ad}}{a\delta iM}\coth\left(\frac{2\pi\delta}{a}\right) \tag{93}$$

Introducing equations (90), (91), and (93) into (87), one obtains

$$-\frac{dH}{dt} = \left\{\frac{2\pi M}{zF\rho a}\left[\frac{K\alpha F^2 z\rho D_{ad}C^\circ_{ad}}{RT\delta M} - i\right]\coth\left(\frac{2\pi\delta}{a}\right)\right\}H$$

$$= \left\{\frac{A}{a}\left[B\frac{C^\circ_{ad}}{a} - i\right]\coth\left(\frac{2\pi\delta}{a}\right)\right\}H \tag{94}$$

Integration of this equation yields an exponential decrease of the amplitude with time,

$$H = H_0 e^{-t/\tau} \tag{95}$$

with the time constant

$$\tau = \frac{a}{A[B(C^\circ_{ad}/\delta) - i]}\coth\left(\frac{2\pi\delta}{a}\right) \tag{96}$$

Agreement between the dependences of τ on a, C°_{ad}, and i implied in equation (96) and those found experimentally is a good indication of the correctness of the theory. Data are lacking, however, on K, α, and δ for quantitative evaluation.

Besides the time constant of leveling, it is also of interest to evaluate the surface concentration of the addition agent and the rate of its consumption. The former can be obtained from the equality of the rate of diffusion of the additive to the surface and the rate of its consumption by incorporation or reduction (cf. Kardos[65]).

The steady-state rate of supply of the additive to the surface can be written as

$$j_{\text{ad}} = D(C^{\circ}_{\text{ad}} - C^{e}_{\text{ad}})/\delta_N \tag{97}$$

where C^{e}_{ad} is the concentration of the additive close to the electrode surface and δ_{N} is the local value of the Nernst diffusion layer thickness. The rate of consumption should be proportional to the surface concentration:

$$j'_{\text{ad}} = (K_i + K_r)C^{s}_{\text{ad}} \tag{98}$$

where K_i is the pseudo-rate constant of incorporation and K_r that for the reduction of the addition agent molecule to less adsorbable species which are released into the electrolyte.

K_i is a relatively simple function of current density. If h is the height of the metal layer needed to build in a molecule of the addition agent, then

$$K_i = iM/zF\rho h \tag{99}$$

The dependence of K_r on i must be of a somewhat more complex nature. Suppose the reduction is an activation-controlled process. Then,

$$K_r = K_r^{\circ} \exp\left(\frac{\alpha_c F}{RT} E\right) \tag{100}$$

Since metal deposition is of a similar nature at sufficiently high metal ion concentration $C_{\text{M}} \gg C^{\circ}_{\text{ad}}$, the i–E relation is likely to be

$$i = i^{\circ} \exp\left(\frac{\alpha c' F}{RT} E\right) \tag{101}$$

Hence, from (100) and (101),

$$K_r = [K_r^\circ/(i^\circ)^\alpha]i^\alpha \tag{102}$$

where $\alpha = \alpha_c/\alpha_c'$ is the ratio of the transfer coefficients for the two processes. K_r° and i° are the rate constant of reduction and the current density, respectively, at an electrode potential equal to the reference electrode potential (at which $E = 0$).

A relationship should exist between C_{ad}^e and C_{ad}^s. This can be obtained by a Langmuir-type treatment from the equality between the rate of adsorption and the sum of rates of desorption, incorporation, and reduction, i.e.,

$$K_a C_{\text{ad}}^e[1 - (C_{\text{ad}}^s/C_{\text{max}}^s)] = (K_d + K_i + K_r)C_{\text{ad}}^s \tag{103}$$

Using all the derived equations and for low surface coverage $(C_{\text{ad}}^s/C_{\text{max}}^s \ll 1)$, one obtains

$$C_{\text{ad}}^s = \frac{C_{\text{ad}}^\circ}{(K_d/K_a) + \{(1/K_a) + [1/(D/\delta_{\text{N}})]\}\{(M/zF\rho h) + [K_r^\circ/(i^\circ)^\alpha]i^{\alpha-1}\}i} \tag{104}$$

When the addition agent is strongly adsorbed, K_a is very large and, hence, the above relationship is reduced to

$$C_{\text{ad}}^s = \frac{DC_{\text{ad}}^\circ/\delta_N}{\{(M/zF\rho h) + [K_r^\circ/(i^\circ)^\alpha]i^{\alpha-1}\}i} \tag{105}$$

It is interesting to note that in all cases where $\alpha \to 1$, equation (92) indicates that it is the product of the surface concentration of the additive and the current density of deposition which is determined by the concentration of the addition agent and by the local value of the Nernst diffusion layer thickness.

Replacing equations (99), (102), and (104) back into (98), the rate of consumption of the addition agent is obtained as

$$j_{\text{ad}} = \frac{\{(M/zF\rho h) + [K_r^\circ/(i^\circ)^\alpha]i^{(\alpha-1)}\}C_{\text{ad}}^\circ i}{(K_d/K_a) + \{(M/zF\rho h) + [K_r^\circ/(i^\circ)^\alpha]i^{\alpha-1}\}\{(1/K_a) + [1/(D/\delta_{\text{N}})]\}i} \tag{106}$$

It is seen that for strongly adsorbed substances ($D/\delta_{\text{N}} \ll K_a \gg K_d$), the rate of consumption is equal to the limiting rate of diffusion, i.e.,

$$j_{\text{ad}} = DC_{\text{ad}}^\circ/\delta_{\text{N}} \tag{107}$$

The direct proportionality of the rate of incorporation to concentration of the additive and its inverse proportionality to the diffusion layer thickness (to $\omega^{1/2}$ in rotating disk experiments) as well as its virtual independence of the current density have been found for many cases.

The validity of the above type of consideration was checked by Krichmar,[75] who considered the consumption by incorporation only. He introduced the average value of δ_N from the theory of convective diffusion and applied the resulting equation to the results of Edwards[76] on the rate of incorporation of sulfur into nickel deposits from baths containing saccharin and *p*-toluene sulfamide. Good agreement between theory and experiment is claimed.

VII. ELECTROPOLISHING

1. Summary of Experimental Facts

The phenomenon of decreasing surface coarseness of metals upon anodic dissolution under certain conditions was first investigated by Jacquet[77,78] for copper in an orthophosphoric acid bath. A number of significant investigations have been carried out since.

In cases when polishing occurs, the current–voltage curve was found to exhibit a plateau characteristic of diffusion control of the dissolution process (Figure 49). Thus, Hoar and Rothwell[79] have shown this limiting current to exhibit a well-defined dependence on the flow rate of electrolyte past the anode, i.e., on the thickness of the hydrodynamic boundary layer.*

Some facts point to a complex nature of the phenomenon of electropolishing. Thus, it was already observed by Jacquet, and quantitatively recorded in some very recent investigations,[80] that within this plateau, a periodic change of potential occurs† at a virtually constant current density, i.e., parallel to the abscissa of Figure 49, whose time dependence is shown in Figure 50. In the case of

*Some controversy seems to exist concerning the character of this plateau. Thus, Shchigolev and Tomashov[81] claim that a potential jump, characteristic of sudden passivation, is obtained at a certain value of current density, rather than a continuous change caused by concentration polarization. Shchigolev[82] has shown this jump to occur at higher current density at artificially produced elevations than at a flat surface or at recesses (cf. Figure 51). However, the experiments of Hoar and Rothwell[79] seem unambiguous.

†A general theory of oscillations in diffusion controlled anodic reactions has been given by Wojtowicz and Conway.[131]

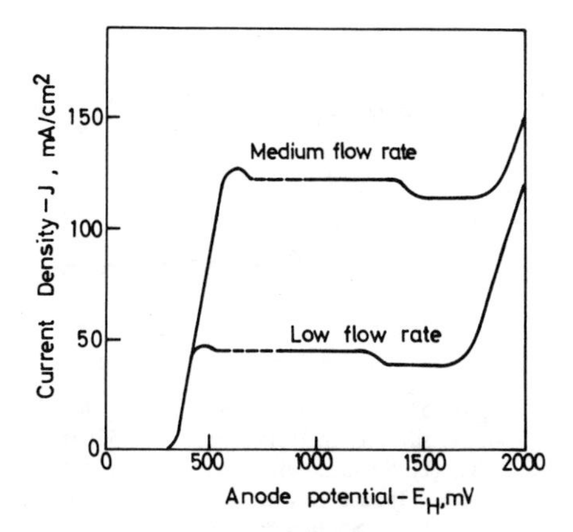

Figure 49. Schematic polarization curves of the copper–phosphoric acid electropolishing system and corresponding reversible potentials (from Hoar and Rothwell[79]).

99.99% electrolytic copper in the middle of the plateau potential region, at which maximum brilliance is obtained by polishing, these oscillations extend without any appreciable damping over a long period of time. As the current density is increased beyond the plateau value, they tend to vanish. Impure copper produces oscillations that are more intense but decay more rapidly with time.

It was also found relatively recently[83] that systems undergoing electropolishing exhibit a significant photoelectrochemical effect. As shown in Figure 52, this corresponds to the region of limiting current densities and also of the maximum polishing effect.

This points to the existence of a photosensitive semiconducting film at the surface and to a possible role of this film in the electro-

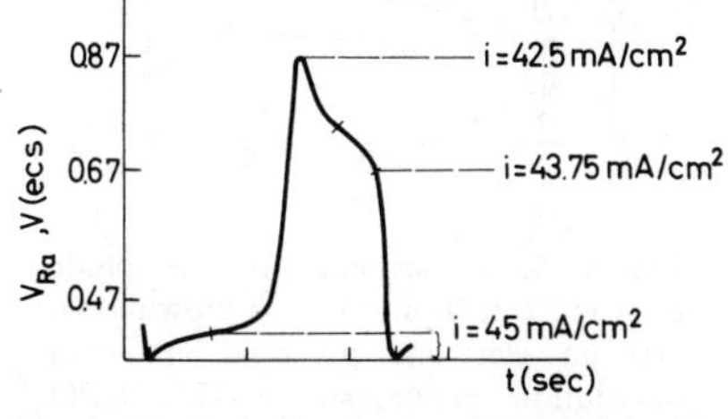

Figure 50. Periodic change of potential with time recorded during electropolishing (from Pointu[80]).

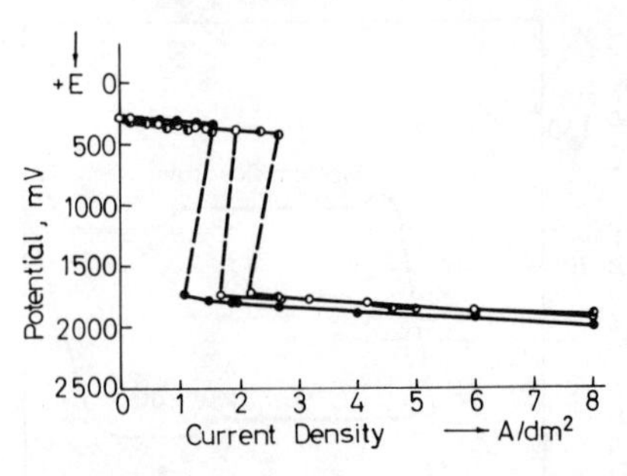

Figure 51. Anodic polarization curves for copper samples in H_3PO_4 (sp. gr., 1.55) at 25°C. (1) Surface with recesses; (2) flat surface; (3) surface with elevations (From Shchigolev.[82]).

polishing process. Subsequent measurements of the capacitance and resistance of the double layer as functions of potential (Figure 53) have shown[84] that this film must be a very thin and well-conducting one.

In spite of all this, considerable evidence has accumulated justifying the treatment of the electropolishing process as an essentially transport-controlled phenomenon.

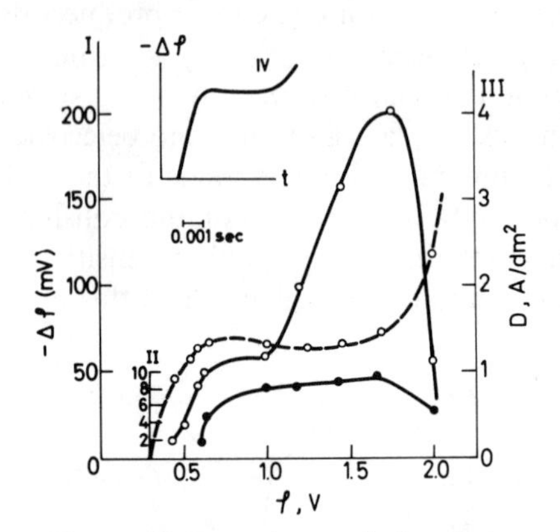

Figure 52. Dependence of the photoelectric effect (I) and its fast growing part (II) on electrode potential in anodic dissolution of copper in 72% H_3PO_4 solution. (III) polarization curve; (IV) the appearance of the initial stage of the effect (from Gretchukhina and Valeyev[83]).

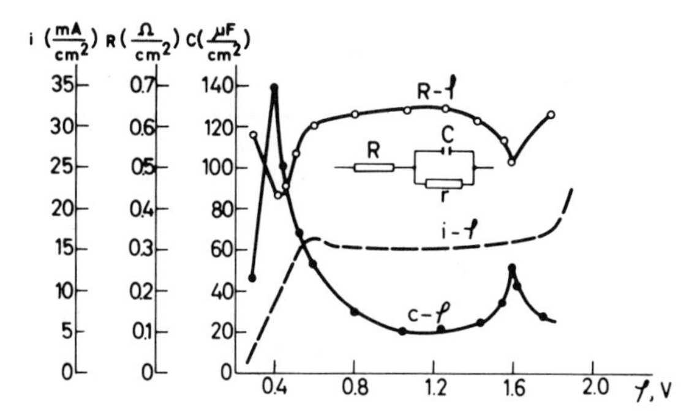

Figure 53 Potential dependence of the capacitance ($C - \varphi$) and resistance ($R - \varphi$) as well as the polarization curve ($i - \varphi$) at anodic dissolution of copper in 72% phospharic acid (from Valeyev and Petrov[84]).

Thus, in experiments at constant current densities i, transition times τ are observed and the product $i\tau^{1/2}$ was found to be constant,[85,86] and its value was shown to be reciprocal with the square root of the viscosity of the solution.

As shown in Table 1, the product $i\tau^{1/2}\eta^{1/2}$ is virtually independent of the concentration of dissolved copper within rather wide limits.* Yet, the Faradaic efficiency of metal dissolution during polishing is found to be 100%,[86] and hence polishing cannot be ascribed to the effect of any other process but the dissolution reaction. An insight into the smoothing efficiency of the electropolishing process can be obtained from the experiments of Edwards[86] on a model surface formed from a phonograph record negative being polished in a phosphoric acid bath. The height of the microgroove was measured, by observing its cross section under the microscope, as a function of the quantity of electricity passed, i.e., the amount of copper dissolved (Figure 54). The efficiency is found to be at a maximum in an unstirred bath and to decrease with stirring.

Chemical polishing was found to exhibit identical behavior to that of electrochemical polishing.

Krichmar has shown[87,88] that this decrease in height of surface irregularities is exponential with time. Very good linear relationships of log of recess depth versus time are shown in Figure 55, the slope depending on the concentration of phosphoric acid. When

*A converse conclusion has been reached, however, by some authors.[87]

Table 1
Dependence of $i\tau^{1/2}\eta^{1/2}$ on Concentration of Cu^{2+}

Cu^{2+} conc., g liter^{-1}	$i\tau^{1/2}$ A sec$^{1/2}$	η, sec (20 ml)$^{-1}$	$i\tau^{1/2}\eta^{1/2}$
0	0.174	54.9	1.29
5	0.168	57.9	1.28
10	0.164	61.7	1.29
15	0.159	65.9	1.29
20	0.154	70.8	1.29
25	0.151	76.0	1.32
30	0.146	80.5	1.31
35	0.142	85.1	1.31
40	0.136	89.4	1.29
45	0.132	94.6	1.28
50	0.129	98.6	1.28
55	0.124	103	1.26
60	0.120	109	1.25
65	0.117	113	1.24
70	0.114	119	1.25
75	0.108	128	1.22
80	0.102	133	1.18
85	0.100	142	1.19

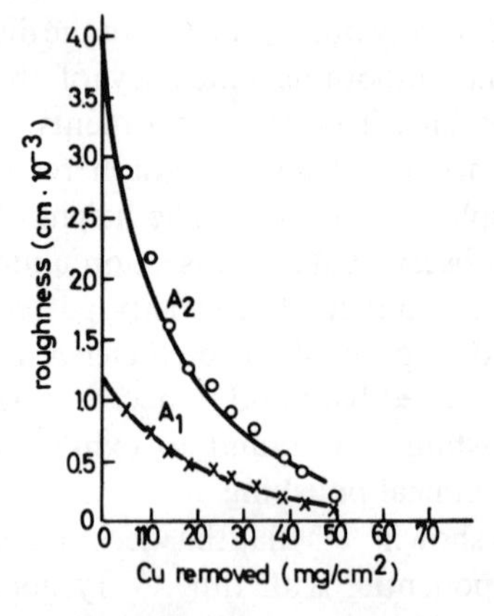

Figure 54. Smoothing efficiency on shielded anode. Negative surface. Dotted circle: maximum height; cross: average height (from Edwards[86]).

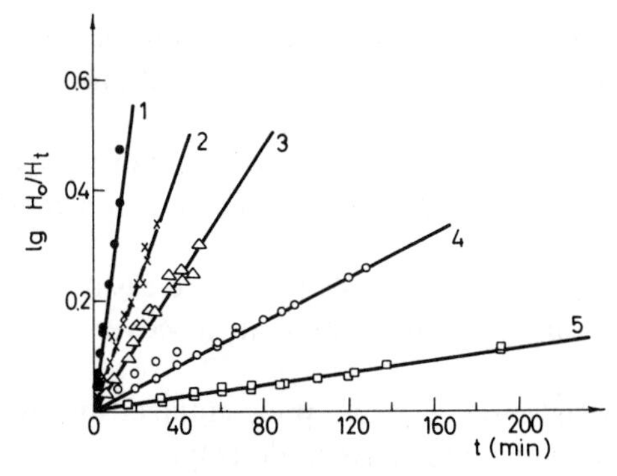

Figure 55. Dependence of log of depth of recess of the profile on bronze on time for different wavelengths *a* in 10 *M* H_3PO_4. Concentrations of H_3PO_4(moles l^{-1}) 2(1), 5(2), 7.5(3), 10(4), and 15(5). (from Krichmar[87]).

samples of well-defined sinusoidal profile were made, it could be shown that this slope (i.e., the exponential time constant) depends also on the wavelength of the profile, i.e., the distance between the peaks to be leveled.

2. Proposed Models of Electropolishing

Several possible phenomena can be put forth as the cause for the observed faster dissolution of the metal at elevated parts of the surface compared with those at the bottom of recesses:

(a) The electrical field distribution results in a higher concentration of current lines at convex than at concave surface irregularities (primary-current-distribution type of effect).[77,78]

(b) The surface free energy of metal atoms to be dissolved is higher at peaks of elevations than at bottoms of recesses (Kelvin-type effect).[89,90]

(c) The same is true of lattice defects and imperfections existing at the surface as a result of surface history (mechanical treatment, etc.).[91]

(d) The diffusion flux of products of the dissolution reaction away from the surface is larger at elevated points than at recesses since the effective diffusion layer is thinner at the former than at the latter.[85,92]

(e) Another reason for the increased diffusion flux is the convex diffusion field at the peaks, leading to conditions of spherical diffusion.[17]

(f) The same applies to the flux of complexing agent when the reaction product forms a complex with a component of the electropolishing solution.[86]

(g) Passive films may be formed more easily at recesses than at peaks, thus inhibiting dissolution at the covered portions of the surface.[89,93–96]

(h) In special cases, at peaks, preferential adsorption may occur of some species which are active in destroying protective films (e.g., ClO_4^- ions on aluminum, according to Darmois and Epelboin[97]).

A model proposed by Jacquet[77,78] upon the discovery of electropolishing ascribed the electropolishing effect to the first cause, i.e., to a primary-distribution type of effect on the microscale profile of the surface. The potential field in the vicinity of the surface with triangularly shaped irregularities (projections and recesses) is shown in Figure 56. Current lines should be perpendicular to the equipotential lines and the current intensity should be inversely proportional to the distance between the latter. Hence, it is seen that, due to difficulties in the transport of electricity, the current flowing to tips of projections should be larger than that to bottoms of recesses. Consequently, the faster dissolution of the former must lead to a

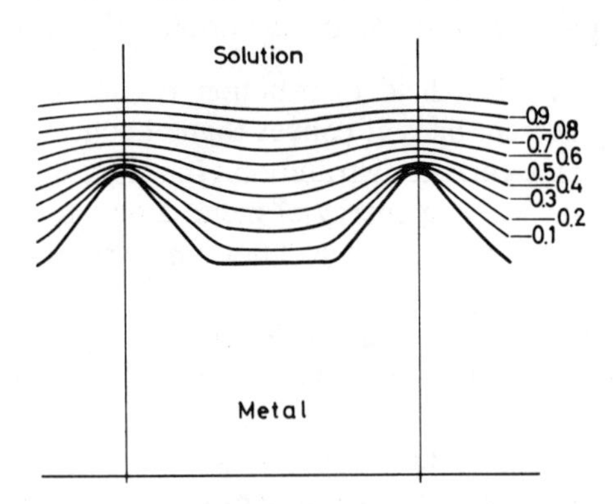

Figure 56. Potential field in the vicinity of triangularly shaped surface elevations (from Edwards[86]).

decrease in the amplitude of the surface profile. However, a simple calculation can show that under normal conditions of conductivity of the metal and the electrolyte throughout the system, the difference in current density cannot be nearly large enough to account for the observed rates of smoothing. Figure 57 represents an equivalent circuit corresponding to the model of the surface.

Hence, to account for the observed rates of smoothing, Jacquet had to assume that a film is formed by the reaction products at the dissolving surface with a considerably higher viscosity and electrical resistivity than the rest of solution. In such a case, R_A and R_B can become larger than R and determine the current. If the outer plane of the film is flattened by convective motion of the electrolyte to follow a line like that shown in Figure 57, (HH'), the resistances R_A and R_B can become significantly different and hence $\Delta i/i_A = (R_A/R_B) - 1$, sufficiently large for producing a considerable difference in the rates of metal dissolution at the two positions.

The model seems adequate for cases where polishing is connected with the appearance of insoluble films at the surface. However, if these films are compact enough to increase the resistivity for the passage of current, they are likely to be too strong for the slow, natural convective motion to affect their outer plane.

More adequate models seem to be those ascribing the phenomenon of electropolishing to some features of the secondary current

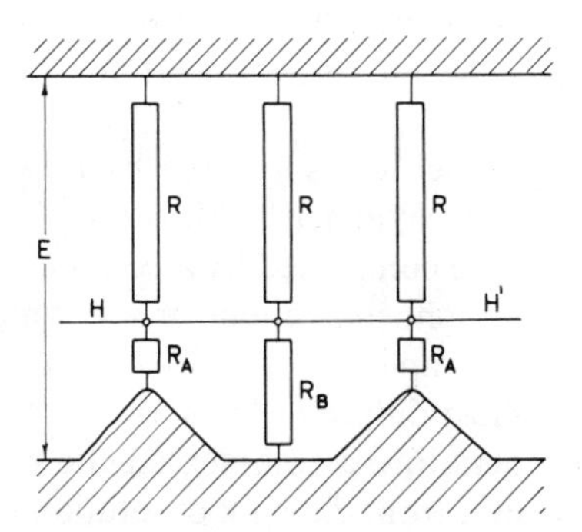

Figure 57. Equivalent circuit of a surface with triangularly shaped elevations.

distribution, i.e., some kind of diffusion control of the dissolution current.

A model based on the effect of slow diffusion of dissolution products away from the electrode is due to Elmore,[85] and a correct interpretation of its consequences was given by Krichmar.[92]

He first treated the smoothing of coarser surface irregularities at which the Kelvin effect has little influence. However, implicit in Krichmar's discussion is the assumption of a *totally reversible* electrode process. Hence, a constant overpotential over the surface implies a constant concentration different from the one in the bulk of the solution. In the steady state, the dissolution current density is equal to the diffusion flux and the latter is inversely proportional to the effective layer thickness, i.e., the distance between a given point at the surface and the hydrodynamic boundary layer. This makes the dissolution rates vary correspondingly over the surface profile. Moreover, the height of the elevations was shown to decrease exponentially with time which was in accordance with experiment (cf. Figure 55).

An additional effect is the increased free energy of the surface at points of very small radii of curvature ($<1\ \mu$m), as well as of lateral diffusion of components in a two-dimensional diffusion field. These were also given a quantitative treatment.

However, the Elmore–Krichmar model cannot under ordinary conditions result in an anodic limiting current. Krichmar[100] has shown that this can arise if the viscosity of the solution is exponentially dependent on the concentration of the reaction product. The latter relationship is found for copper ions in phosphoric acid solution.[101]

An alternative model was suggested by Edwards[86] and elaborated quantitatively by Wagner.[17] According to this model, the metal ions produced are complexed by a component of the electropolishing solution (e.g., phosphate ions or water molecules). Hence, for the reaction to be completed, not only must ions be formed, but complexing species must diffuse to the surface from the bulk of the solution to accept them into a complex reaction product. It is the diffusion of the acceptor from the bulk of solution which determines the overall rate of reaction. The diffusion field of the acceptor is the same as that for the ions formed and both should be of the same layout as the potential field shown in Figure 56, the equipotential lines now representing the lines of equal concentration of the reaction product. Hence, differences in acceptor fluxes at different points at the surface arise. The slower arrival of the acceptor at recessed parts should cause an increased concentration of free metal ions with the possible consequences (a) of increasing the cathodic partial current and thus reducing the net dissolution current, and (b) of producing such changes in the reaction layer as, e.g., the formation of oxide film by hydrolysis, thus making room for an additional phenomenon observed in electropolishing, namely that of brightening. This seems to be related to the dissolution of facets and other crystallites. Brightening seems to occur when the surface becomes covered by a protective film controlling the rate of dissolution and making the latter a random process, the energetic advantages of atoms at facets and dislocations being lost. Thus, one could conclude that the Edwards–Wagner model seems to provide a reasonable basis for developing a comprehensive theory of the electropolishing process.

3. Quantitative Treatment of Diffusion-Controlled Smoothing

Any surface profile can be represented by a Fourier series

$$y = \sum_{i=1}^{\infty} b_i \sin(2\pi x/a_i) \tag{108}$$

where y is the elevation, i.e., the distance from the "average surface plane." Hence, the problem can be analyzed by considering the simplest of all profiles which can be described by a single sine function of amplitude b and wavelength a. This is shown in Figure 7.

(*i*) *A Steady-State Rate of Dissolution*

Suppose that anodic dissolution assumes a reaction path which consists of three steps: (a) ionization of a metal adatom and its transfer through the double layer into solution; (b) complexing of this ion with an appropriate ligand (e.g., water molecule, phosphate ion, or hydroxyl ion); and (c) diffusion of the complex away from the electrode surface.

(*a*) *Slow ionization of metal adatoms.* If the first step were the rate-determining one, the rate of dissolution at any point would be given by a simple Butler–Volmer-type relation except that dependence of the surface free energy on the radius of curvature should be taken into account.

Two cases may arise, as discussed earlier [cf. equations (14a) and (14b)]. In the case of the dissolution of divalent metals (e.g., copper), it seems likely that only the anodic partial current should be affected by the Kelvin effect. Hence, equation (14b) should be applicable.

The difference in the rates of dissolution at the top and the bottom of the sine wave of Figure 7 is

$$\frac{dh}{dt} = \frac{dy(x_\lambda)}{dt} - \frac{dy(x_\nu)}{dt} = \frac{M}{\rho nF}[i(x_\lambda) - i(x_\nu)]$$

$$= \frac{Mi_0 f_a(\eta)}{\rho nF}\left[\exp\left(\frac{2\sigma V}{RTr}\right) - \exp\left(-\frac{2\sigma V}{RTr}\right)\right] \tag{109}$$

where r is the radius of curvature of the sine function at x_λ and x_ν. The effect is seen to depend on the rate at which the "average surface" dissolves, $dy(0)/dt = Mi_0 f_a(\eta)/\rho nF$. The relative difference in rates of dissolution, $(dh/dt)_n = (dh/dt)/[dy(0)/dt]$, is shown in Figure 58 as a function of the reduced radius of curvature $r_n = r/2\sigma V/RT$. For a typical set of data for σ, V, and T (cf. p. 210), a sharp dropping off of the effect should occur at a radius of $\sim 1\ \mu$m. Hence, this Evans–Krichmar-type mechanism can possibly account for the smoothing of very low surface coarseness at such current densities that cause no concentration polarization. However, it cannot give an account of the appearance of either the limiting current or of selective film formation at bottoms of surface recesses.

(*b*) *Inhibited complexing of the reaction product.* The reaction between a metal ion and a ligand is usually sufficiently fast, but *it is the insufficient supply of the latter which can cause the inhibition of this step and make it rate-determining.*

The anodic partial current is, of course, independent of the presence or absence of the ligand. But as the latter becomes scarce, the concentration of the free metal ions increases and as a result, the cathodic partial current is increased as well. In an ideal case, when diffusion of free metal ions away from the electrode is negligible, the difference between the two partial currents, i.e., the net dissolution current, must be proportional to the flux of the ligand. This case was considered in detail by Wagner,[17] who assumed a molecular diffusion mechanism of supply through a fixed hydrodynamic boundary layer of thickness δ, much larger than a and b.

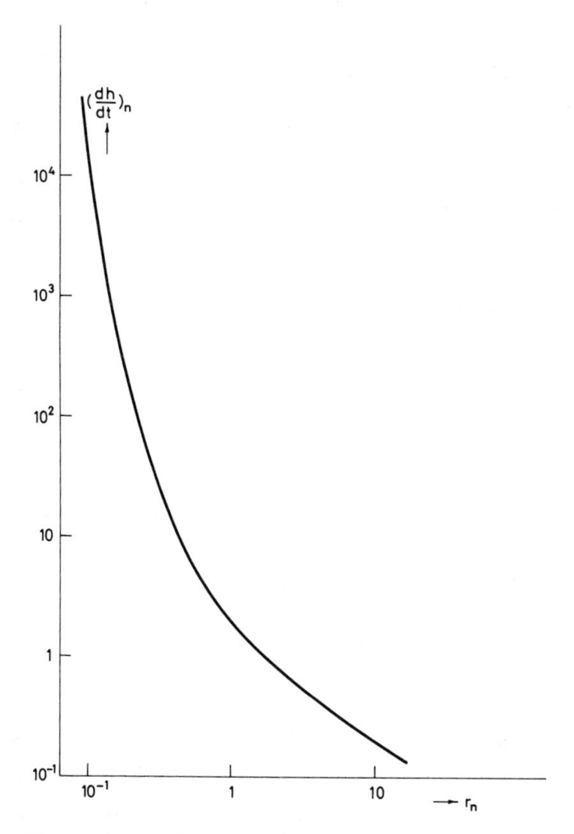

Figure 58. Relative difference in rates of dissolution as a function of the reduced radius of curvature.

Under steady-state conditions, the diffusion field is determined by Fick's second law,

$$\frac{\partial c}{dt} = D\left(\frac{\partial^2 c}{\partial x^2} + \frac{\partial^2 c}{\partial y^2}\right) = 0 \tag{110}$$

and the boundary conditions

$$c(x, b\sin(2\pi x/a), t) = 0 \tag{111}$$

$$c(x, b\sin(2\pi x/a), 0) = C^\circ \tag{112}$$

$$c(x, y \to \infty, t) = C^\circ \tag{113}$$

The solution of this equation as given by Wagner for the case of a flat profile ($a \gg b$) yields

$$c(x, y) = B[y - b\exp(-2\pi y/a)\sin(2\pi x/a)] \tag{114}$$

where B is a constant which is shown to be equal to the average concentration gradient at the anode $(\partial c/\partial y)_{av}$.

Differentiating equation (114) with respect to y, substituting for the electrode surface $y = b\sin(2\pi x/a)$, expanding the exponential function, and neglecting higher terms in (b/a), one obtains for the component of the flux of ligand perpendicular to the surface

$$\left(\frac{\partial c}{\partial y}\right)_{y=b\sin(2\pi x/a)} \simeq \left(\frac{\partial c}{\partial y}\right)_{av}\left(1 + \frac{2\pi b}{a}\sin\frac{2\pi x}{a}\right) \tag{115}$$

$$\text{if} \quad b \ll a \quad \text{and} \quad a \ll \delta$$

To obtain the decrease of surface coarseness with time, one should consider the change in position of the points x_v and x_λ at the surface in the y direction. Thus,

$$\frac{dh}{dt} = \frac{dy(x_\lambda)}{dt} - \frac{dy(x_v)}{dt} = DV\left[\left(\frac{\partial c}{\partial y}\right)_{x_\lambda} - \left(\frac{\partial c}{\partial y}\right)_{x_v}\right] \tag{116}$$

$$= DV\left(\frac{\partial c}{\partial y}\right)\frac{4\pi b}{a}$$

Considering that the steady-state limiting diffusion flux for the average (i.e., flat) surface is $(\partial c/\partial y) = C^\circ/\delta$ and that $h = 2b$, one obtains

$$dh/(h\,dt) = 2\pi DVC^\circ/a \tag{117}$$

Integrating equation (117), an exponential time dependence is obtained:

$$h = h_0 e^{-t/\tau} \tag{118}$$

where h_0 is the initial surface coarseness ($h_0 = 2b_0$) and

$$\tau = a\delta/2\pi DVC^\circ \tag{119}$$

Equation (119) shows that the electropolishing process is faster the smaller the diffusion layer thickness and the larger the bulk

Table 2
Time Constants for Electropolishing a Copper Surface with Grooves of 700 μm Wavelength in Phosphoric Acid Baths of Different Concentrations[98]

$C_{H_3PO_4}$, mole liter^{-1}	$[\log(h_0/h)]/t$, min^{-1}	τ, sec	C_{ligand}, mole liter^{-1}
5	7×10^{-3}	3.6×10^3	0.45
7.5	6.8×10^{-3}	3.8×10^3	0.42
10	3.5×10^{-3}	1.6×10^4	0.1
13	2.5×10^{-3}	2.2×10^4	0.07

concentration of the ligand. Also, it is faster the smaller the "wavelength" of the surface profile. The latter reflects the radius of curvature of elevations and recesses and the result indicates that "microroughness" will disappear more rapidly than "macroroughness," which is in accordance with experimental experience.

The exponential time dependence of the height of the elevations given by equation (118) is in agreement with the findings of Krichmar[87] (cf. Figure 55).

Table 2 lists the slopes of the $\log(h_0/h)$ versus t dependences as functions of phosphoric acid concentration for a copper disk with grooves of 100 μm wavelength.

For the set of values $a = 100\ \mu$m, $\delta = 100\ \mu$m, $V = 7$ cm^3 mole^{-1}, and $D = 10^{-5}$ cm^2 sec^{-1}, the derived time constants imply the ligand bulk concentrations as listed in the last column of Table 2. The structure of concentrated phosphoric acid solutions is not known sufficiently well enough to allow us to judge the validity of this speculation, but the result is too low for water to be considered as the ligand. The phosphate ion is more likely to be found in such concentrations.

The primary advantage of the Edwards–Wagner acceptor model of electropolishing is that a limiting current comes out as a straightforward result of the transport difficulties in the process.

(c) *Inhibited diffusion of reaction products away from the electrode.* Krichmar[99] has developed a comprehensive theory for this case by solving the diffusion equation (110) with the boundary conditions

$$c(x, b \sin(2\pi x/a), t) = C_0^* \qquad (120)$$

$$c(x, b \sin(2\pi x/a), 0) = 0 \qquad (121)$$

$$c(x, y \to \infty, t) = 0 \qquad (122)$$

where C^* is the local concentration of the reaction product at the electrode surface. At a given electrode potential, this is different from that at a flat surface, C^s, because of the effect of curvature on the free energy of metal adatoms, and (for $h_0 < a$) is given as

$$C^* = C^s\left(1 - \frac{8\sigma V\pi^2}{RT}\frac{b}{a^2}\sin\frac{2\pi x}{a}\right) \qquad (123)$$

The solution of this diffusion problem gives

$$C \simeq C^s\left[\frac{y}{\delta} - b\left(\frac{1}{\delta} + \frac{8\sigma V\pi^2}{RTa^2}\right)\frac{\sinh(2\pi y/a)}{\sinh(2\pi\delta/a)}\sin\frac{2\pi x}{a}\right] \qquad (124)$$

Differentiating equation (122), one obtains for the diffusion fluxes of the reaction products away from the surface at x_v and x_λ

$$\left(\frac{\partial c}{\partial y}\right)_{x_\lambda} = C^s\left\{\frac{1}{\delta} + \frac{2b_0\pi}{a}\left(\frac{1}{\delta} + \frac{8\sigma V\pi^2}{RTa^2}\right)\frac{\cosh[2\pi(\delta - b)/a]}{\sinh(2\delta/a)}\right\} \qquad (125)$$

$$\left(\frac{\partial c}{\partial y}\right)_{x_v} = C^s\left\{\frac{1}{\delta} - \frac{2b_0\pi}{a}\left(\frac{1}{\delta} + \frac{8\sigma V\pi^2}{RTa^2}\right)\frac{\sinh[2\pi(\delta + b)/a]}{\sinh(2\delta/a)}\right\} \qquad (126)$$

The rate of change of the surface coarseness h $(=2b)$ is obtained in the same manner as before, i.e.,

$$\begin{aligned}\frac{dh}{dt} &= \frac{dy(x_\lambda)}{dt} - \frac{dy(x_v)}{dt} = VD\left[\left(\frac{\partial c}{\partial y}\right)_{x\lambda} - \left(\frac{\partial c}{\partial y}\right)_{xv}\right] \\ &= -\frac{2V\pi DC^s h}{a}\left(\frac{1}{\delta} + \frac{8\sigma V\pi^2}{RTa^2}\right)\coth\frac{2\pi\delta}{a}\cosh\frac{\pi h}{a}\end{aligned} \qquad (127)$$

Upon integration, this again yields an exponential time dependence of the surface coarseness, given by equation (118) with

$$\tau = \left[\frac{2V\pi DC^s}{a}\left(\frac{1}{\delta} + \frac{8\sigma V\pi^2}{RTa^2}\right)\coth\frac{2\pi\delta}{a}\cosh\frac{\pi h}{a}\right]^{-1} \qquad (128)$$

or, for $\delta > a > h_0$,

$$\tau = \pi\delta^2/VDC^s \tag{129}$$

and for $a > \delta > h_0$,

$$\tau = \delta a/2V\pi DC^s \tag{130}$$

It is interesting to note that in the second case, the time constant is the same as that given by equation (119) except that the concentration of the ligand in the bulk of the solution is replaced by the surface concentration of the reaction product. In the two cases discussed, the effect of the radius of curvature i is seen to have vanished (no effect of surface tension). This is due to the assumption of a large wavelength of the surface profile. In another limiting case, when $\delta \gg a > h_0$, it is this factor that plays the major role. The time constant is obtained as

$$\tau = RTa^3/32V\sigma\pi^3DC^s \tag{131}$$

The surface concentration C^s could be estimated from the values of potential assuming a very fast electrode process (negligible activation overpotential). This, however, could be used only in the potential range before the appearance of the limiting current. Any calculation of C^s from potentials along the current plateau must give erroneous values since the theory in the form presented so far does not account for this phenomenon.

Krichmar[100] offered an explanation of the appearance of the anodic limiting current in terms of the effect of reaction products on the viscosity of the solution within the diffusion layer and hence on the value of the diffusion coefficient. If the latter is continuously decreasing with increasing concentration, this dependence can be represented by an exponential series

$$D = D^0 \sum_{k-1}^{\infty} a_k e^{KZC} \tag{132}$$

where D^0 is the diffusion coefficient in the bulk of solution and a and z are constants. The diffusion flux in the direction of change of D can be written as

$$j = -(D^0/\delta) \int_{C^s}^{C^\circ} \left(\sum_{k=1}^{\infty} a_k e^{-KZC} \right) dc \tag{133}$$

Integrating (109), one obtains

$$j = -(D^0/Z\delta) \sum_{k=1}^{\infty} (a_k/k)[\exp(-KZC^\circ) - \exp(-KZC^s)] \quad (134)$$

which, at high surface concentrations, i.e., for $C^s \gg C^0$, reduces to

$$j = -(D^0/Z\delta) \sum_{k=1}^{\infty} (a_k/k) \exp(-KZC^\circ) \quad (135)$$

The flux j is seen to be independent of the surface concentration C^s and hence of potential. Thus, it forms a basis for the appearance of a potential-independent limiting current.

If this is accepted, the model still exhibits some disadvantages compared to the Edwards–Wagner model. It gives no possibility for the change in the structure of the reaction layer at the electrode that is needed for uneven film formation in the recesses of the surface. If the latter is caused by the concentration of the reaction product exceeding a certain critical value, in the Krichmar model, this is attained at the same time over the whole surface. If anything, due to the Kelvin effect, the concentration can become higher at the tops of the elevations and film formation should be expected there first.

Hence, in conclusion, one could maintain that the inhibition of the third step in the dissolution reaction can have some smoothing effect on the dissolving surface, but it can hardly represent a basis for explaining all the phenomena observed in relation to electropolishing.

(ii) Surface Concentration of Reaction Products and Conditions for Film Formation

Whatever the structure of the surface film which is formed during electropolishing, its appearance must be connected with the concentration of the anodic dissolution products exceeding a certain value. The latter is determined by the solubility product of the film-forming substance and the concentration of the other component needed for the reaction.

At the beginning of the electropolishing procedure, as the electrode is submitted to anodic polarization, all of the overpotential is used for transmitting the activation-controlled current.

The faster the specific rate of the process, the larger the current. In the case of complex formation, the ligand is being used up and the free metal ions are produced in excess. The net anodic current is determined by the relation

$$i = nF\vec{K}\exp[(\alpha_a F/RT)E] - nF\overleftarrow{K}C_i\exp[(-\alpha_c F/RT)E] \quad (136)$$

and in a steady-state situation, this is equal to the flux of the ligand. For the same electrode potential E over the entire surface, the current can be adjusted to the changing flux only by an appropriate variation in the concentration of the ionic reaction product C_i. Hence, since the flux is smaller at recesses than at tops of elevations, C_i must be correspondingly higher in the former than at the latter. In the limiting current region, the concentration of the ligand tends to zero, while the concentration of the free ions assumes definite values varying with the surface profile. Hence, if the pH situation at the surface is such that metal ions tend to form hydroxides or oxides, they will do this more readily at recesses than at elevations. The difference in concentrations can be evaluated as follows: The difference in dissolution current densities at the top and bottom of the sinusoidal profile shown in Figure 7, Δi, can be obtained from any one of the equations (109), (116), or (127). When the two current densities are expressed using equation (136) one obtains, after rearranging,

$$(C_i)_{x_\lambda} - (C_i)_{x_\nu} = \Delta i\{\exp[+(\alpha_c F/RT)E]/nFK\} \quad (137)$$

Considering that the critical concentration is given by

$$(C_i)_{\text{crit}} = S/C_f \quad (138)$$

where S is the solubility product and C_f is the concentration of the other component of the film-forming substance, the film is forming selectively in the recesses in the potential region in which

$$(C_i)_{x_\lambda} > (C_i)_{\text{crit}} > (C_i)_{x_\nu} \quad (139)$$

and a maximum electropolishing effect should be expected.

VIII. DEPOSITION OF METALS AT A PERIODICALLY CHANGING RATE

In plating practice, it has been known for a relatively long time that the application of a periodically changing current leads to improvements in the quality of the deposits obtained (cf. Ref. 101). Three types of current have found use: (a) reversing currents; (b) sinusoidal ac superimposed on a direct cathodic deposition current; (c) pulsating currents.

1. Reversing Currents

This type of deposition current is represented schematically in Figure 59. It is characterized by the cathodic (depositing) current density I_c and the anodic (dissolving) current density I_a, as well as by times of flow of the current in the cathodic and anodic direction, t_c and t_a respectively. Naturally, $t_c > t_a$, and $t_c + t_a = T$ represents the full period of the reversing-current wave.

A number of workers[104–112] have investigated the potentialities of reversing current in plating practice. They have shown that compared with normal dc deposition, the operating current densities can in some cases be considerably increased (thus reducing the plating time). More-ductile deposits can be obtained, with less internal stress and often having a better appearance. A decreased average grain size results in somewhat increased hardness.

A detailed survey of the effect of reversing current on various processes of plating and dissolution is given by Bakhvalov.[101] Its use in charging secondary chemical energy sources is discussed by Wales[116] and Arouete.[117]

It is interesting to note that positive effects are not always achieved with reversing current. Thus, while in plating copper from

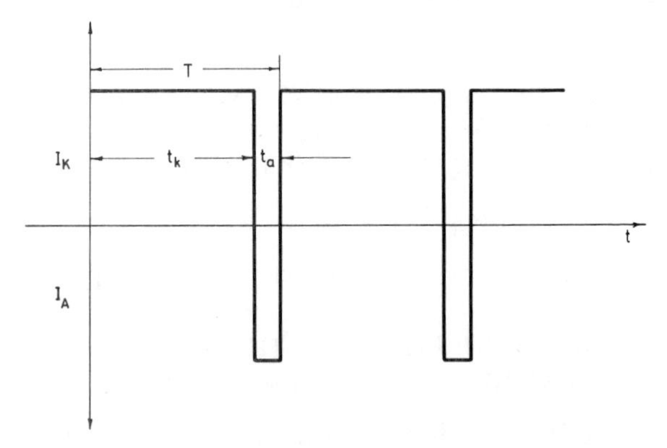

Figure 59. Time dependence of current in reversing-current electrolysis.

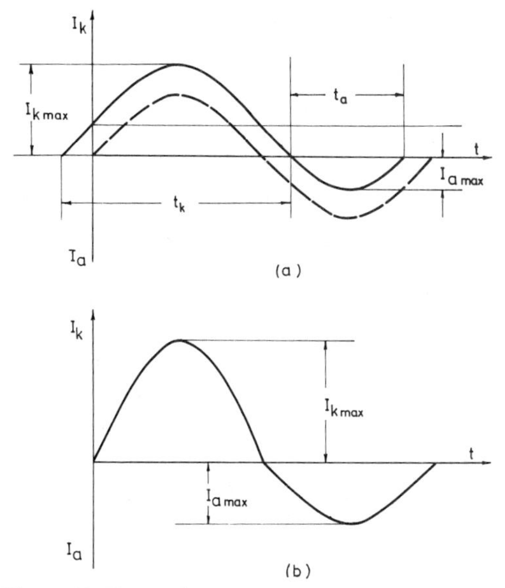

Figure 60. Shape of (a) asymmetric sinusoidal and (b) nonsinusoidal current waves.

cyanide bath the current density of the reversing current can be 2–3 times larger than the dc current, leading to the same quality of the deposit,[101,102,114] this is not so in using the acid sulfate baths.[113]

It appears that little effect is obtained when deposition is activation-controlled, while positive effects are found in operating conditions causing a diffusional control of the deposition process.

It appears that the most important characteristic of a reversing current is the ratio t_c/t_a. Thus, Vene and Nikolayeva[115] claim that, irrespective of the overall current density or of the length of the reversing-current wave, the best results are obtained at $t_c/t_a = 7$. At higher ratios, the effect of reversing the current rapidly decreases, while at lower ratios, the current efficiency is considerably reduced and the porosity of the deposit is increased.

However, Bibikov[118] has shown that in zinc deposition from acid electrolytes, at any value of t_c/t_a, the current efficiency decreases with increasing current density. Also, the upper limit of the applicable current densities decreases with increasing duration of the reversing-current wave T.

2. DC with Superimposed AC

This kind of deposition current is shown in Figure 60. Two types are employed. The first type (Figure 60a) is obtained by imposing a sinusoidal current wave upon continuous dc. The resulting current is termed asymmetric sinusoidal current. Asymmetric nonsinusoidal current is obtained when a sinusoidal current wave is superimposed on half-rectified ac of the same phase and frequency (Figure 60b).

DC with superimposed ac is essentially similar to the reversing current. The latter can be treated mathematically as a Fourier series of the former. Hence, similar effects are to be expected.

Indeed, Romanov[122] and Wales[123] reported on beneficial effects of asymmetric currents on charging silver–zinc batteries. Romanov[119] and Bek and Kudryavtsev[124] found an improvement in the quality of zinc deposited from alkaline electrolytes. Bek *et al.*[126] obtained bright copper deposits using this type of current, while Papereka and Avramenko[127,128] found that such a deposit can have less internal stress compared to the one obtained by dc.

These effects, however, seem to be observable only under certain well-defined conditions. Thus, Romanov[119] investigated the yield of dendrites in the deposition of zinc from alkaline zincate solution at 4 mA cm^{-2} overall current density during 14 hr of

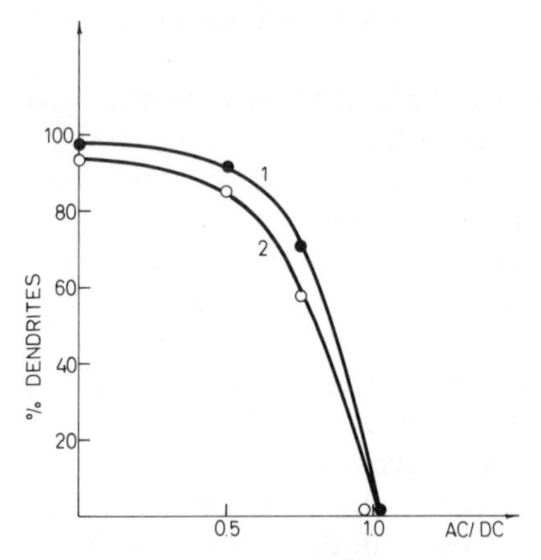

Figure 61. Yield of dendritic deposit as a function of the ratio i_{ac}/i_{dc} in the deposition of zinc from alkaline zincate solution (from Romanov[119]).

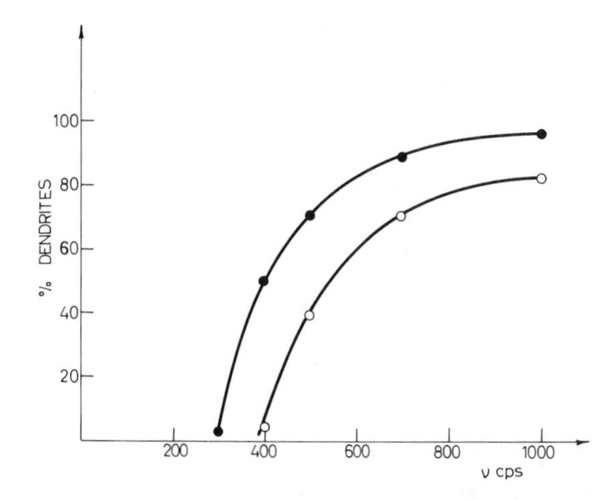

Figure 62. Yield of dendritic deposit as a function of frequency of the pulsating current in the deposition of zinc from alkaline zincate solution (from Romanov[119]).

electrolysis at 20°C as a function of the ratio of the ac to dc component, and of the frequency of the applied ac. As seen in Figure 61, appreciable effects start appearing at values of the ratio $i_{ac}/i_{dc} = 0.6$. Note that this is the value at which the superposition leads to the appearance of anodic overall current in some time interval t_a, which also increases with increasing i_{ac}/i_{dc}. A major effect, the virtual absence of dendrites, is already obtained at $i_{ac}/i_{dc} = 1$. A further increase of the ratio has no effect on the quality of the deposit, while reducing to some extent the current efficiency. The frequency dependence (shown in Figure 62) reflects the decrease in the full period of the wave and, hence, also of the period of relaxation of the system from deposition. It is seen that beyond 300–400 Hz, i.e., below relaxation time intervals of the order of 1 msec, there is a fairly sharp decrease of the effect. Similar results were obtained by Bek and Kudryavtsev.[124]

It has been suggested[129] that the appearance of sponge or dendrites is due to a slow supply of depositing ions to the cathode and that providing some relaxation time in the system, in the form of current interruption or reversal, helps this supply to be more efficient.

3. Pulsating Currents

Pulsating current consists of a periodic repetition of square or sinusoidal pulses, i.e., is similar in shape to the current shown in Figure 59 except for the absence of the anodic component. The system is suddenly submitted to current during the time interval t_k and then left to relax for a period t_0.

A number of investigators have tested the effects of such a deposition pattern on the quality of deposit when, e.g., charging silver-zinc cells,[15,16,119] or a silver oxide electrode,[122] and depositing zinc from alkaline zincate solutions,[117] or copper from acid copper sulfate baths.[120]

Positive effects are generally found. Romanov[15] was the first to carry out an extensive quantiative study by observing the appearance of dendritic zinc deposits from alkaline zincate solutions under the influence of varying parameters of the pulsating current and comparing this with the appearance of dendrites in dc electrolysis of the same systems. The results of that comparison are given in Figure 63. The shape of the current wave had practically no influence on the result, i.e., square current pulses gave an almost identical

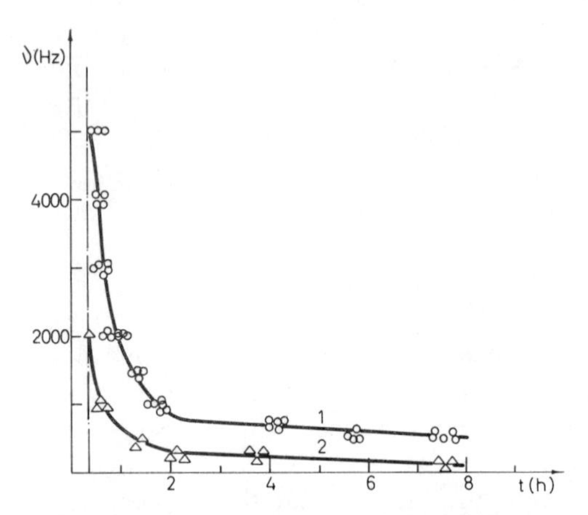

Figure 63. Time of appearance of visible dendritic zinc deposit as a function of frequency of the pulsating current (half-rectified ac). Current density, 0.5 A dm^{-2}; temperature, 20°C. Curves: (1) nickel anode; (2) zinc anode; (3) time of appearance under dc conditions (from Romanov[15]).

frequency dependence of the time of appearance as did the half-rectified ac.

It is evident that a sharp increase in the time of appearance is obtained at frequencies below 500 Hz. However, it is also obvious that, at least as far as the square-wave current is concerned, at very low frequencies, dc conditions are approached and hence a return to very short times of appearance is to be expected. Romanov apparently did not reach that condition in his experiments.

The use of controlled pulsating current, as in the experiments of Romanov, has two shortcomings if the phenomena are to be contrasted with a rational model:

(a) The overall current which is controlled is actually an integral of local current densities, which can follow an unpredictable distribution pattern. In effect, the controlled overall current merely determines an average electrode potential, which in turn is partly responsible for the local current densities producing dendrites or compact deposit.

(b) The controlled current is used both for double-layer charging and the deposition process. The capacitance current of the periodic charging and discharging of the double layer produces a smearing of the Faradaic current wave. Hence, as the frequency increases, the Faradaic current wave flattens and approaches dc even though the overall current appears as a pulsating one.

Hence, controlled-potential results have a better chance to yield to theoretical interpretation. These were obtained in experiments by Despić and Popov,[130] who deposited copper from copper sulfate solution first onto a phonograph record negative and then also onto a wire electrode and made microscopic observations of the electrode cross section dipped into wax and polished. The morphology of the deposit was found (Figure 64) to depend very strongly on the frequency of the sinusoidally pulsating potential (half-rectified ac potential wave). As seen in Figure 65, the average amplitude of the surface irregularities after a constant time of deposition decreased strongly with increasing frequency, so that at about 10,000 Hz, a virtually compact deposit was obtained at an equivalent dc current density which would have produced a fully dendritic deposit.

Note that these experiments are in apparent contradiction with those of Romanov, in which dendritic growth is stimulated by increasing frequency.

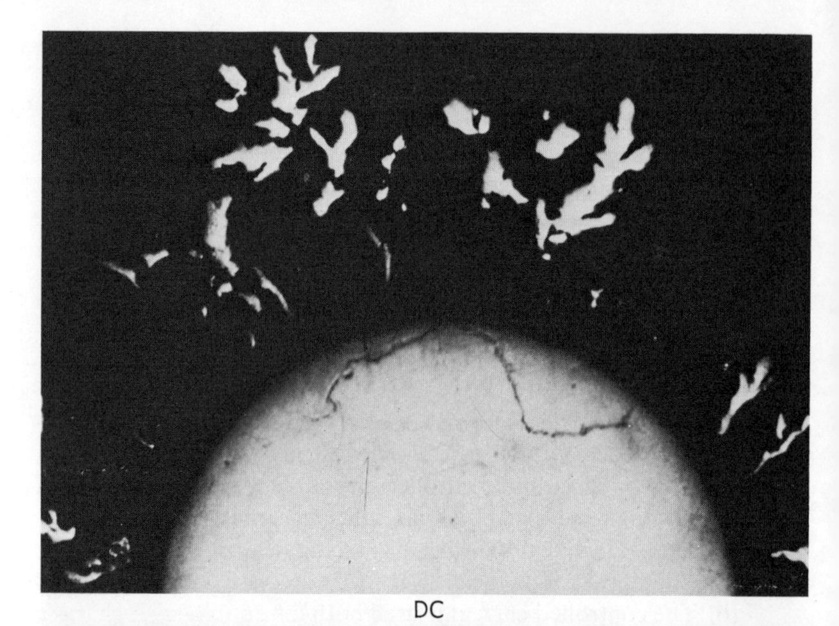

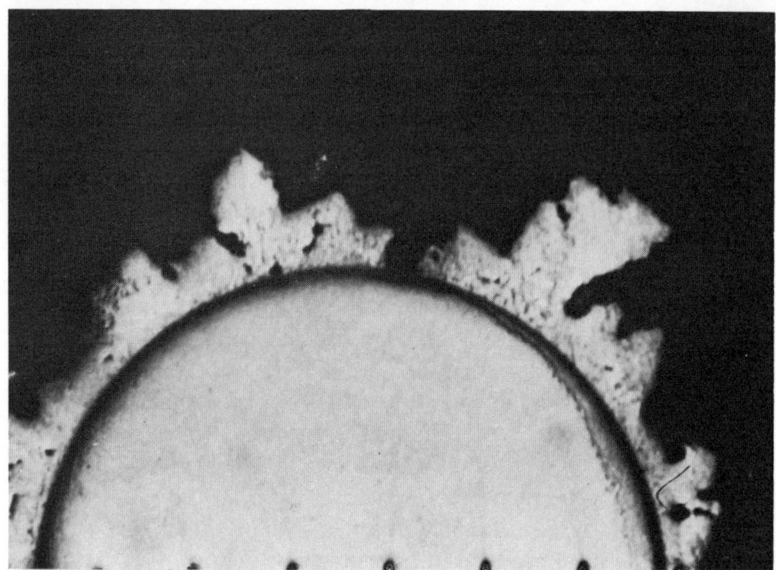

Figure 64. Micrographs of cross sections of deposits obtained with sinusoidally pulsating potential of equal amplitude (−100 mV) and varying frequency (from Despić and Popov[21]).

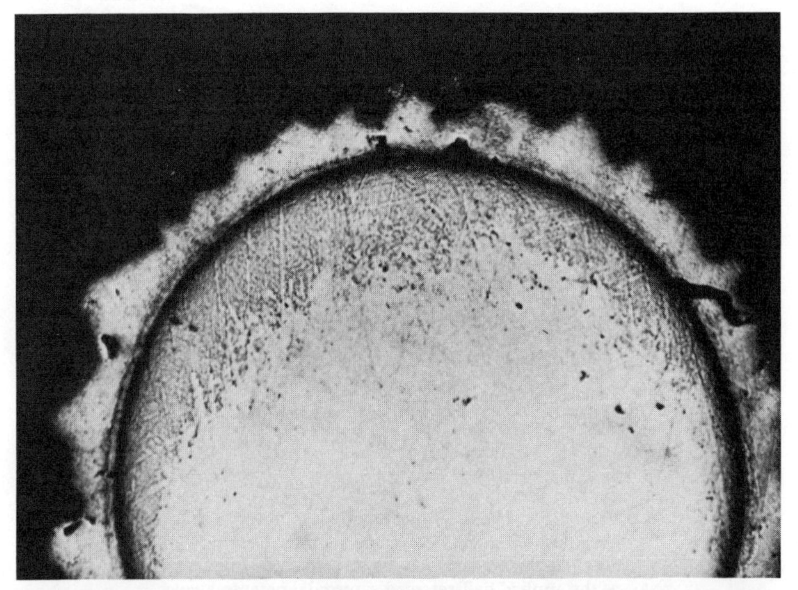

$f = 10^3$ Hz

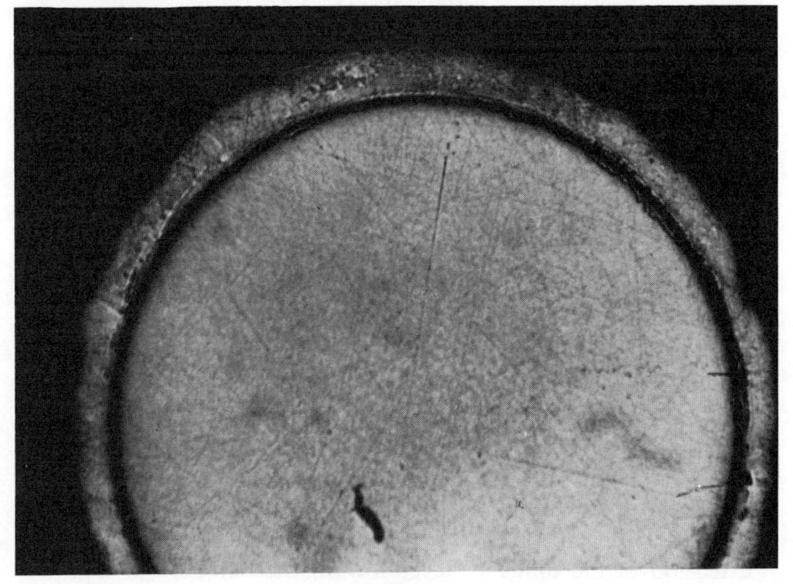

$f = 10^4$ Hz

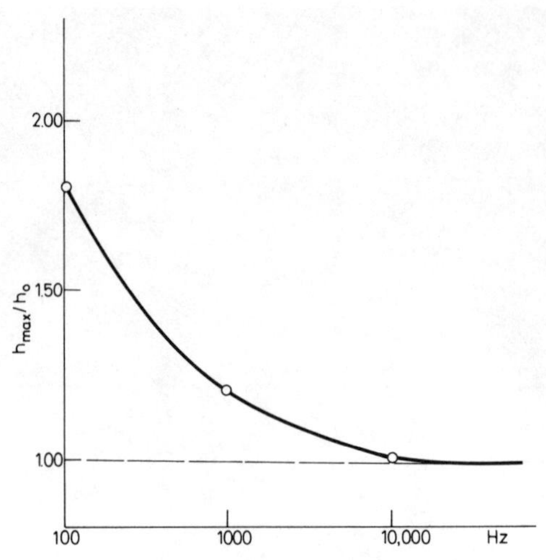

Figure 65. The average amplitude of the surface irregularities after 1 hr of deposition of zinc from alkaline zincate solution, as a function of frequency of the applied half-rectified sinusoidal potential wave (amplitude, −100 mV) (from Despić and Popov[130]).

4. Theory of the Effect of Pulsating Electrolysis on the Morphology of the Deposit

This is based on the concept that a periodic relaxation of the diffusion field prevents the development of phenomena seen to appear under dc conditions of electrolysis (as discussed in preceding sections).[130] Such effects are known to be due to inhibited diffusion of the depositing substance through diffusion layer which is steadily extending into the solution and separating from the surface microprofile so as to be of different thickness at different points. It is clear that any method which would prevent the diffusion layer from becoming so thick as to stop following closely the microprofile would also prevent amplification of the surface roughness and its consequences. The experiments with pulsating current and potential resulting in a reduction of roughness amplification were taken as an indication that it may be a feature of the pulsating deposition rate to limit the extension of the diffusion layer toward the bulk of the solution. Therefore, the mathematics of the diffusion problem under such conditions were considered and the results analyzed.

(*i*) *Case of a Pulsating Potential Imposed on a Totally Reversible Electrode*

Fick's second law was solved[130] for the set of boundary conditions which reflect the case when an electrode at which a very fast electron exchange takes place is submitted to a series of square-wave potential pulses, changing the electrode potential from its reversible value to a given, more negative value and back. Since the process is very fast, the electrode reacts in such a manner that the concentration of depositing ions at the electrode surface follows closely the potential pulse and, hence, varies from the bulk value C_0 to a value close to zero and back, according to the equation

$$C(0, t) = C_0 \exp\left[\frac{+zFE(t)}{RT}\right] = f(t) \tag{140}$$

Negligible values of $C(0, t)$ are achieved if, e.g., for copper, potential pulses larger than 60 mV are used. Other boundary conditions are

$$C(x, 0) = C^\circ \tag{141}$$

$$C(x, t)_{x\to\infty} = C^\circ \tag{142}$$

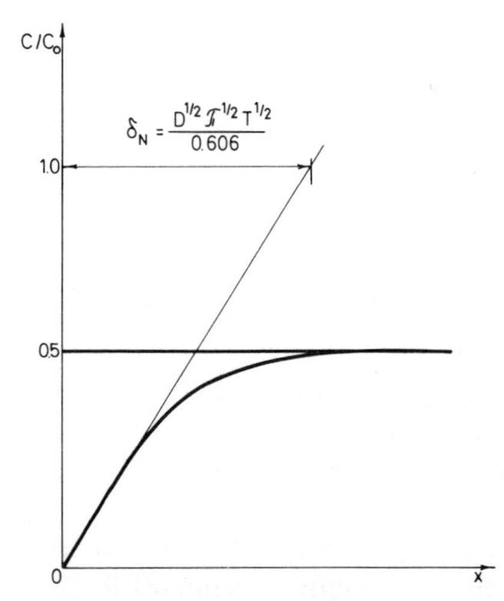

Figure 66. The relative concentration of the depositing substance as a function of distance from the electrode surface after a very large number of pulses of duration T.

The solution yields the concentration of the depositing substance at the end of the "on" and "off" periods as a function of time in the nth pulse of duration T as

$$C_{\text{on}} = C(x, (2n+1)T) = \frac{2C_0}{\pi^{1/2}} \left[\int_0^{\frac{x}{2D^{1/2}[(2n+1)T]^{1/2}}} \exp -u^2 \, du + \int_{\frac{x}{2D^{1/2}(2nT)^{1/2}}}^{\frac{x}{2D^{1/2}[(2n-1)T]^{1/2}}} \exp -u^2 \, du + \cdots + \int_{\frac{x}{2D^{1/2}(2T)^{1/2}}}^{\frac{x}{2D^{1/2}T_2^{1/2}}} \exp -u^2 \, du \right] \quad (143)$$

$$C_{\text{off}} = C(x, 2nT) = \frac{2C_0}{\pi^{1/2}} \left[\int_0^{\frac{x}{2D^{1/2}(2nT)^{1/2}}} \exp -u^2 \, du + \int_{\frac{x}{2D^{1/2}[(2n-1)T]^{1/2}}}^{\frac{x}{2D^{1/2}[(2n-2)T]^{1/2}}} \exp -u^2 \, du + \cdots + \int_{\frac{x}{2D^{1/2}T^{1/2}}}^{\infty} \exp -u^2 \, du \right] \quad (144)$$

The concentration distribution obtained is shown in Figure 66. This indicates two important conclusions:

(a) After a very large number of pulses, the concentration away from electrode surface assumes a value of $0.5C_0$, and, hence, an average concentration polarization is never larger than about 10 mV, i.e., on the average, the limiting current conditions are never achieved.

(b) A virtual limiting current is reached during each potential pulse (implied in the assumed boundary condition with $E > -60$ mV). This is proportional to the concentration gradient at the electrode surface, and is infinite at the beginning of the potential pulse and equal to the one shown in Figure 66 at its end.

This same concentration gradient determines the Nernst diffusion layer thickness δ_N, which can be derived from equation (143) as

$$\begin{aligned}\delta_N &= \frac{C_0}{(\partial C/\partial x)_{x=0}} \\ &= \frac{D^{1/2}\pi^{1/2}T^{1/2}}{[(2n+1)^{-1/2} - (2n)^{-1/2} + (2n-1)^{-1/2} - \cdots + 1]} \\ &= \frac{D^{1/2}\pi^{1/2}T^{1/2}}{a} \qquad (145)\end{aligned}$$

Computer analysis has shown that the sum in the denominator in the first line exhibits an asymptotic tendency to the value given in the second line with $a = 0.606$. Hence, the diffusion layer thickness after the nth pulse is not much different from that at the end of the first pulse. Thus, the latter can be taken as a good enough criterion of the diffusion situation with a pulsating potential.

This result shows that, indeed, the diffusion layer boundary under a pulsating potential is held at the electrode surface fixed to a finite distance not changing with time. Note that this obviously represents a method of fixing the diffusion layer thickness additional to the well-known methods based on defined hydrodynamic boundary layers.

The above analysis has shown that the diffusion layer thickness at an ideally reversible electrode is proportional to the square root of the pulse duration, i.e., varies inversely with the square root of pulse frequency.

Hence, with increasing frequency, the diffusion layer should more closely follow the surface microprofile and at some frequency, a completely even deposition can be expected in spite of the fact that it is done under limiting current conditions. This explains the observed effects of the frequency of a pulsating potential on the morphology of the deposit (cf. Figures 64 and 65).

If it is assumed that noticeable amplification of surface roughness will start taking place at a diffusion layer thickness equal to the radius of curvature of a surface irregularity (for, then, a spherical diffusion supply will start being appreciably different from one due to linear diffusion), one can calculate the limiting radius of curvature of irregularities which should be amplified as a function of the frequency of the applied pulsating potential, i.e.,

$$r_{\text{ampl}} \leq \delta_{\text{N}} \tag{146}$$

This is shown in Figure 67. Thus, at a given frequency, surface irregularities should become amplified until their radii reach the limiting value. After this, amplification should cease and, further on, a practically even deposition should take place. The calculation indicates that in the frequency range above about 10,000 Hz, the irregularities should stop amplifying as their radii reach the level of about 1 μm. This is in accordance with experiments which, at these frequencies, produce what appears as a smooth surface.

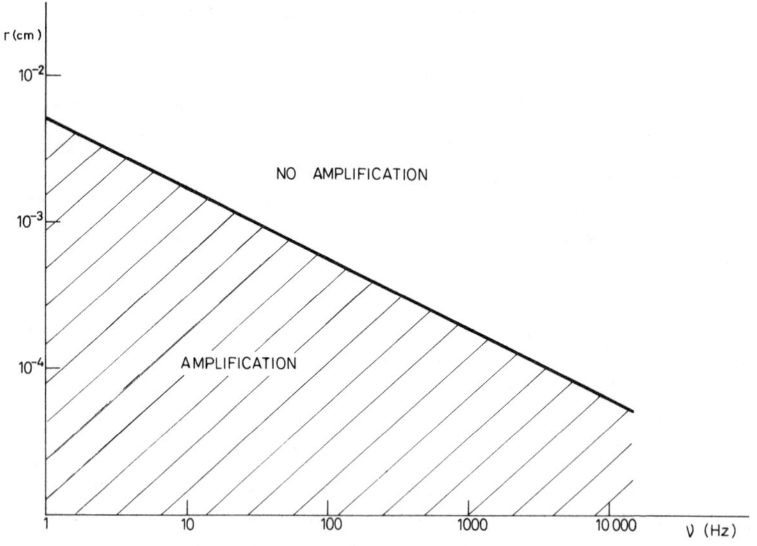

Figure 67. Radii of curvature of tips of surface irregularities which are amplified at different frequencies (from Despić and Popov[130]).

(*ii*) *Case of a Pulsating Potential Imposed on an Electrode with Low Exchange Current Density*

To obtain an insight into the development of the diffusion layer in such a case, one should solve Fick's second law using a boundary condition which is based on the fact that the diffusion flux at the electrode surface is proportional to the current density and the latter is related to the potential by the Butler–Volmer equation. Thus,

$$\left(\frac{\partial c}{\partial x}\right)_{x=0} = \frac{i}{zFD} = \frac{i_0}{zFD}\left\{\frac{C(0,t)}{C_0}\exp\left[-\frac{\alpha_c F}{RT}E(t)\right] - \exp\left[\frac{\alpha_a F}{RT}E(t)\right]\right\} \tag{147}$$

As a first approximation, one can maintain this flux to be constant within a pulse period since major changes, from a certain value to zero and back, are induced by stepwise changing of the potential in the "on" and "off" periods, respectively, while the change of $C(0, t)/C_0$ is a second-order effect except at an overwhelming concentration polarization.

Using other boundary conditions as before, one obtains for the concentration profiles in the on and off periods

$$\begin{aligned} C_{\text{on}} = C(0, (2n+1)T) = C_0 + aA\pi^{-1/2}\{&(t-T)^{1/2} \\ &- t^{1/2} + (t-3T)^{1/2} \\ &+ (t-2T)^{1/2} + \cdots + [t-(2n+1)T]^{1/2} \\ &- (t-2nT)^{1/2}\} \end{aligned} \tag{148}$$

$$\begin{aligned} C_{\text{off}} = C(0, (2n+2)T) = C_0 + aAT^{1/2}\pi - 1[&(2n+1)^{1/2} \\ &- (2n+2)^{1/2} + (2n-1)^{1/2} \\ &- (2n)^{1/2} + \cdots - (2T)^{1/2} + T^{1/2} \end{aligned} \tag{149}$$

The diffusion layer thickness at the nth potential pulse can be calculated in the same manner as before. It is given by

$$\delta_{\text{N}} = 2\frac{D^{1/2}}{\pi^{1/2}}\sum_{m=0}^{m=2\nu t}(-1)^m\left(t - \frac{m}{2\nu}\right)^{1/2} = \frac{D^{1/2}t^{1/2}}{\pi^{1/2}}, \qquad t \to \infty \tag{150}$$

Equation (150) indicates that, contrary to the case of a reversible process, δ_{N} does depend on time. The diffusion layer is extending in much the same way as in the case of dc deposition. Hence, no effect of pulsating potential in the sense of preventing amplification of surface roughness can be expected. Since this is in effect a situation of controlled current, one could conclude that this is a possible reason why in the controlled current experiments of Romanov[15] no conditions for complete absence of dendritic growth could be found. However, since equation (150) does not predict any frequency dependence, the findings of Romanov should be ascribed primarily to the effect discussed earlier (see p. 291), i.e., the smearing effect of capacitance current.

IX. GENERALIZED MODEL OF TRANSPORT-CONTROLLED PROCESSES AND APPROACHES TO QUANTITATIVE ANALYSIS OF ITS CONSEQUENCES

The above review demonstrated the basic similarity in the mechanisms of all the phenomena described. In all cases, it has been established that diffusion of some species taking part in the process plays a dominant role in shaping the observed phenomenon. Three phases of development are common to all these processes:

(a) There is an initial period in which the process leads to a depletion of the solution at and near the electrode surface, in the species essential for effecting the desired phenomenon, to the point where the conditions for the initiation of the phenomenon are established.

(b) There is a period of development of the diffusion layer, from a close following of the surface microprofile to some thickness at which the surface irregularities are buried deep in the layer and at which it can be said to follow smoothly the macrogeometry of the electrode.

(c) There is a period in which the diffusion layer attains a constant thickness due to some external factor such as convection of the solution.

It is interesting to note that all quantitative theories of the phenomena described, except the last one, (periodically changing current) have been related so far to the final period only.

1. Qualitative Consideration of the Development of the Diffusion Layer

(*i*) *The Initial Period*

This has been noted experimentally only in the case of powder deposition. This is probably due to the fact that the formation of powders requires considerable concentration polarization to be attained before the loose connection between the deposit and the substrate is effected. However, the inevitable existence of this phase in the process makes approximate all calculations of initiation times based on steady-state concepts. Any exact quantitative theory must obviously obtain the period of reaction layer depletion as a natural result of using the appropriate boundary conditions (see below).

(*ii*) *The Diffusion-Layer Development*

At the onset of the deposition or dissolution process, the diffusion layer follows closely the electrode surface and at any point, a linear diffusion law can be applied to a good approximation. Since this contains no parameters of surface geometry, the diffusion flux and hence the surface concentration of the relevant species is the same over the whole surface. The process proceeds at an even rate.

As the diffusion layer develops to a thickness comparable to the local amplitude of the surface profile, its outer boundary* starts separating from the latter. There are two causes of the onset of an uneven supply of diffusing species to the surface, and hence, of an uneven rate of the process: (i) differences in the thickness between different points imply different diffusion fluxes for linear diffusion; (ii) rounding off of the diffusion layer at elevations leads to a transition from linear diffusion to some diffusion regime (e.g., spherical, cylindrical, paraboloidal) at which the lateral component of the diffusion flux starts playing an appreciable role. A converse situation develops at surface recesses.

As the process proceeds, the outer boundary of the diffusion layer is becoming smooth and the relative difference between its thickness at elevations and at recesses is decreasing.

*See footnote on p. 203.

(*iii*) *The Steady State of the Process*

At some point of this development, convection becomes appreciable and limits further extension of the diffusion layer. A quasi-steady state is attained, since the metal surface, i.e., the inner diffusion layer boundary, continues to change. However, for the purpose of the theory, in very short time intervals, this could be considered as a true steady-state situation.

One should note that all the phenomena considered depend on *differences* in the supply of the relevant species at different points of the surface. The above model allows two conclusions:

1. The difference in supply between, e.g., the peak of an elevation and the bottom of a recess, is increasing as the diffusion layer starts separating from the microprofile, passes through a maximum, and decreases as the diffusion layer becomes thicker.
2. Since the diffusion layer thickness at which the diffusion layer separates from the surface profile depends on the amplitude of the profile, so do the maximum difference in supply and the phenomenon observed as its consequence. Thus, smaller surface irregularities should pass through a maximum rate of amplification or leveling sooner than the larger ones. Also, for any diffusion layer thickness there will be a corresponding amplitude of the surface at which amplification or leveling will be most pronounced. Because of the exponential time dependence of these phenomena and the fact that this effect of the amplitude is reflected in the time constant of this dependence, this conclusion is equivalent to the observation that for any fixed diffusion layer thickness, there is a narrow range of surface irregularities that will undergo the given processes.

2. Quantitative Aspects

The task of the theory is to determine the time dependence of the phenomena concerned as a function of the different process parameters. We can take as being representative of all the phenomena, except powder formation, the relative motion between two points at the surface, perpendicular to the surface. As a matter of convention, the theory has so far always considered as the two points the peak of an elevation and the bottom of a recess. Hence, the motion is expressed as the change in local amplitude of the surface profile.

A system of equations can be set up covering all cases considered. The first is a rate equation. In its most general form it can be written as

$$\frac{dh}{dt} = \frac{M}{\rho} \frac{d(\Delta N)}{dt} = \frac{M}{\rho n F} \Delta i =$$

$$= \frac{M}{\rho n F} \left\{ \left[\vec{K}(1 - \theta_a) \prod_i C_i^{\nu_i} \exp\left(\frac{\alpha_a F}{RT} \Delta\varphi \right) - \overleftarrow{K}(1 - \theta_a) \right. \right.$$

$$\left. \times \prod C_j^{\nu_j} \exp\left(-\frac{\alpha_c F}{RT} \Delta\varphi \right) \right]_p$$

$$- \left[\vec{K}(1 - \theta_a) \prod_i C_i^{\nu_i} \exp\left(\frac{\alpha_a F}{RT} \Delta\varphi \right) - K(1 - \theta_a) \right.$$

$$\left. \left. \times \prod C_j^{\nu_j} \exp\left(-\frac{\alpha_c F}{RT} \Delta\varphi \right) \right]_r \right\} \tag{151}$$

where the first term refers to the process at the peak of the elevation and the second to the process at the bottom of the recess; θ_a is the value of the surface coverage by all adsorbing species foreign to the electrochemical process, C_i and C_j are concentrations of all species participating in some way in the anodic and cathodic directions of the electrode process, respectively, close to the electrode surface. Other symbols have the meanings usually assigned to them in electrochemical literature.*

For all practical purposes, equation (151) can be considerably simplified. The electrode potential can be considered as virtually constant over the surface of all metallic electrodes. The rate constants are usually independent of position and so are concentrations of all but one species, the one relevant to the observed phenomenon.

*The absolute inner potential difference $\Delta\varphi$ can be replaced by the electrode potential E relative to some standard reference electrode without any change in the equation except in the meaning of the rate constants. The latter are then defined as the specific rates of the two processes, anodic and cathodic, at the reference electrode potential.

Hence, equation (151) can be rewritten as

$$\frac{dh}{dt} = \frac{M}{\rho nF}\left\{\left[\vec{K}' \exp\left(\frac{\alpha_a F}{RT}\Delta\varphi\right)\right][(1-\theta_a)_p C_p^{\nu_{ap}} - (1-\theta_a)_r C_r^{\nu_{ar}}] - \left[\overleftarrow{K}' \exp\left(-\frac{\alpha_c F}{RT}\Delta\varphi\right)\right][(1-\theta_a)_p C_p^{\nu_{cp}} - (1-\theta)_r C_r^{\nu_{cr}}]\right\} \quad (152)$$

where ν_a or ν_c are equal to zero if the species is participating in the cathodic or the anodic process, respectively. They both may be zero if the phenomenon is caused by the difference in θ_a.

The surface coverage θ_a is related to the concentration of the adsorbing species in the immediate vicinity of the surface ($x = 0$) by some adsorption isotherm. In the most general case, this can be written as

$$\theta_a/(1-\theta_a) = f(C_a, \Delta\varphi) \quad (153)$$

The local concentrations of the adsorbing species C_a and the species participating in the electrode process C are determined by the diffusion law, i.e.,

$$(\partial c/\partial t) = D(\partial^2 c/\partial x^2) + \partial^2 c/\partial z^2 + \partial^2 c/\partial y^2 \quad (154)$$

and the appropriate boundary conditions. The conditions applicable to all cases are

$$C(x, z, f(x, z), 0) = C^\circ, \quad C(x, z, y, t)_{\substack{y\to\infty \\ \text{or} \\ y=\delta}} = C^\circ \quad (155)$$

It is the introduction of the boundary condition reflecting the perturbation of the system and the geometry of the surface, as given by $f(x, z)$, which differs in different cases and which calls for special methods of treating the particular problem.

3. Some Approaches to Quantitative Solutions

Different approaches to solving the above system in special cases can be grouped into the following general methods:

(*a*) *Finding exact explicit solutions.* This can be done only for rather simplified models such as those used by Despić *et al.*[2] for the amplification of surface roughness, Barton and Bockris[19] and Hamilton[33] for dendrite growth, Wagner[17] and Krichmar[99] for electropolishing, etc., which all assume a constant diffusion layer thickness, much larger than the height of surface irregularities, and a rather simple third boundary condition which either allows the application of Fisk's first law, or enables an exact solution to be obtained for the diffusion field around a well-defined surface profile, e.g., sinusoidal.

This approach, has revealed important properties of the investigated phenomena. They were all shown to exhibit an exponential time dependence in the initial stage of development. The time constant was shown to be the essential criterion for quantitative consideration of the phenomenon.

Equations (10), (24), (32), (96), (119), and (128) exhibit considerable similarity. Three cases should be distinguished, as pointed out in the discussion of equation (128), depending on the properties of the microprofile of the substrate for deposition or dissolution:

(A) Surface roughness consisting of very sharp peaks (very small radius of curvature or wavelength of the profile) leads to the surface energy effect becoming dominant [cf. equation (131)]. The time constant is inversely proportional not only to diffusional parameters (diffusion coefficient and concentration), but also to the surface tension at the metal–electrolyte boundary at the given potential.

(B) At profiles with wavelength smaller than δ, but large enough for the Kelvin effect to become negligible, the phenomena are governed by the direct supply of material through the diffusion layer (the lateral component of supply being negligible). The time constant then depends basically on the square of the diffusion layer thickness as well as on other diffusional parameters [see equations (24) and (130)].

(C) At profiles with large wavelengths, the lateral supply at elevations becomes equal in importance to the direct one (spherical diffusion conditions). The time constant is determined by the product of the diffusion layer thickness and wavelength [see equations (119), (130)] or radius of curvature [see equation (32)], depending on how the profile is defined. Hence, it is seen that the time constant is not a simple characteristic of the system concerned, but also depends on the morphology of the substrate.

Different surface elements can be expected to undergo any one of the processes described, at different rates, and with different times of appearance, and it is only at model surfaces with well-defined morphologies that a single expression can be adequate.

(*b*) *The first-order approximation.* This method is typified by the work by Krichmar[74] on the theory of leveling. It can be applied to cases where the differences between the quantities of interest are small enough that functions describing them can be considered to be linear in the vicinity of the expected variations. Thus, if the current density is a function of the effective diffusion layer thickness and of the local concentration of the species involved, for two points at the surface, one can write equations (85) and (86) where δ and C^s_{ad} are the average values. The difference between the two functions is obviously a function of the small parameters H and C^s_{ad} only. This can be totally differentiated to give equation (87). If the parameters are indeed very small, which is the case if the two points are not too different in elevation from an average surface plane, the partial derivatives $\partial i/\partial H$ and $\partial i/\partial \Delta C^s_{\text{ad}}$ can be considered constant. Hence, if they can be evaluated independently, the mathematical problem can be solved. Krichmar has shown one possible way of doing this, and each particular problem seems to be approachable in a similar way. The method has the shortcoming, however, of being applicable to initial stages of development of the phenomenon only, in the region where the partial derivatives are sufficiently close to constant values.

(*c*) *Use of computer calculations.* Analog simulation of the diffusion field around a well-defined surface profile consisting of triangularly shaped elevations and recesses was made by V. O. Kardos.[73] This author has demonstrated that in the important special case of steady-state deposition under complete diffusional control of the electrode process, and provided parameters are well matched, a complete analogy exists between the spatial arrangement of the lines of flow of the matter in the process of diffusion and of the electric current lines in the primary current distribution. This is due to the full analogy between the electric field and the diffusional field, arising from the formal identity between the Laplace and the Fick equations and the boundary conditions that usually hold in such cases. Hence, complete diffusion control has the same effect on deposition as the primary current distribution. They both tend to produce variations in local deposition rates on both the macroprofile and the microprofile of the surface. However, the diffusional effects are much more strongly pronounced because of the much smaller relative values of the diffusion coefficient than of the specific electric conductivity.

However, the analogy only makes it possible to use the model to evaluate the effect of transport on the morphology of the deposit after the outer diffusion layer boundary has become smooth and parallel to the average electrode surface. In most of these cases, explicit analytic solutions were shown to be possible.

Digital simulation of the events at a surface perturbed by a current or potential pulse seems to offer better insight.[130]

The penetration of the diffusion layer into a motionless electrolyte from the moment of application of the pulse can be followed for a two-dimensional model surface like the one shown in Figure 68 if a numerical calculation of the concentration time dependence is carried out for solving the diffusion equation by the method of finite differences. The field observed is a square of dimensions y_0 by x_0. The bottom boundary represents the metal surface at a constant potential different from the reversible one, the upper boundary represents the bulk of solution at concentration C_0, while the sides are characterized by the absence of exchange of matter, i.e., $(\partial c/\partial x)_{x=0} = (\partial c/\partial x)_{x=x_0} = 0$. Instead of the Nernst diffusion layer boundary, the program was made to give the isoconcentration lines of $C = \frac{1}{2}C_0$ at different time intervals. These are shown in Figure 68 at times equal to $0.1y_0^2/D$, $5y_0^2/D$, and

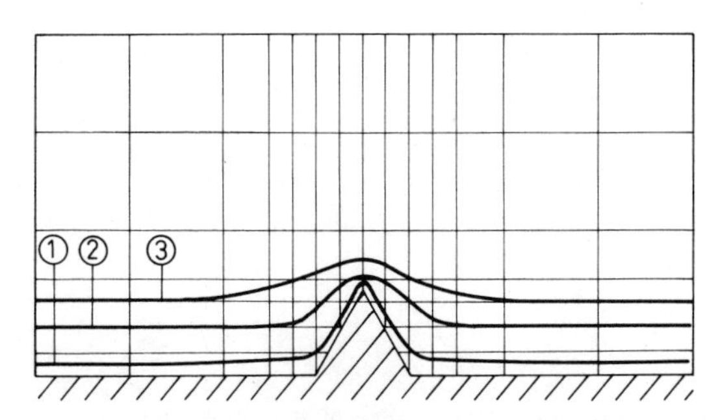

Figure 68. Calculated isoconcentration lines ($c = 0.5C_0$) at different times after the beginning of deposition: (1) $0.1\ y^2/D$ sec; (2) $5y^2/D$ sec; (3) $10y^2/D$ sec.

$10y_0^2/D$ seconds after the application of the potential pulse. The normalization factor y_0^2/D would be equal to 10^{-1} for a typical situation of $y_0 = 10$ mμ and the diffusion coefficient $D = 10^{-5}$.

The aim of this work was to show the time dependence of the amplification factor of the surface roughness, which was seen to have to exhibit a maximum at times at which the diffusion layer starts separating from the surface profile (see p. 302).

The amplification factor was derived as [see equations (22) and (21)]

$$\frac{dh}{hdt} = \frac{1}{h}\left[\frac{dy(0)}{dt} - \frac{dy(x_0)}{dt}\right] = \frac{M}{h\rho nF}[i(0) - i(x_0)]$$

$$= \frac{MD}{\rho nFh}\left[\left(\frac{\partial c}{\partial y}\right)_{x=0} - \left(\frac{\partial c}{\partial y}\right)_{x=x_0}\right] \quad (156)$$

and is shown in Figure 69 as a function of normalized time. The maximum is found at y_0^2/D, which for the typical case given above amounts to 10^{-1} sec.

The same treatment can be used for the analysis of the effect of pulsating potential. Figure 70 shows that the isoconcentration line

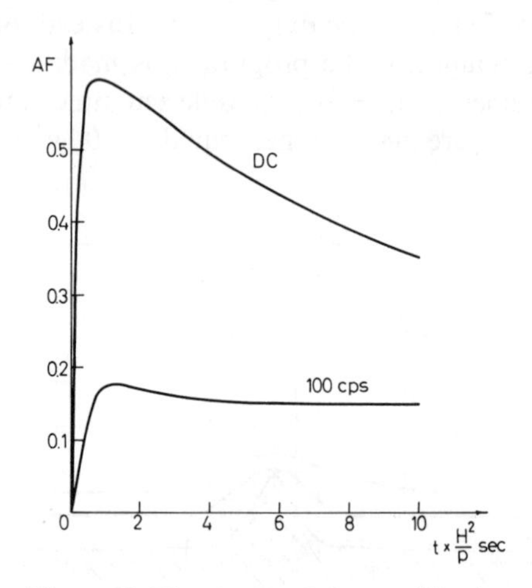

Figure 69. The change of the amplification factor with time of electrodeposition with dc and with a square-wave overpotential pulsating at 100 cps.

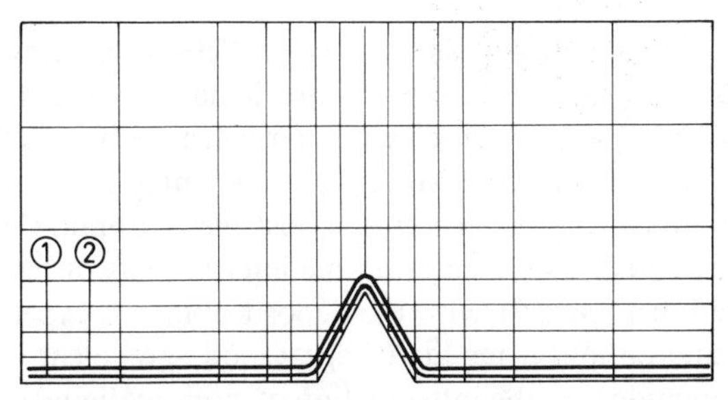

Figure 70. Calculated isoconcentration lines ($c = 0.5C_0$) at different times after the beginning of deposition with a square-wave overpotential pulsating at 100 cps. At 0.01 sec (1) and after 0.5 sec.

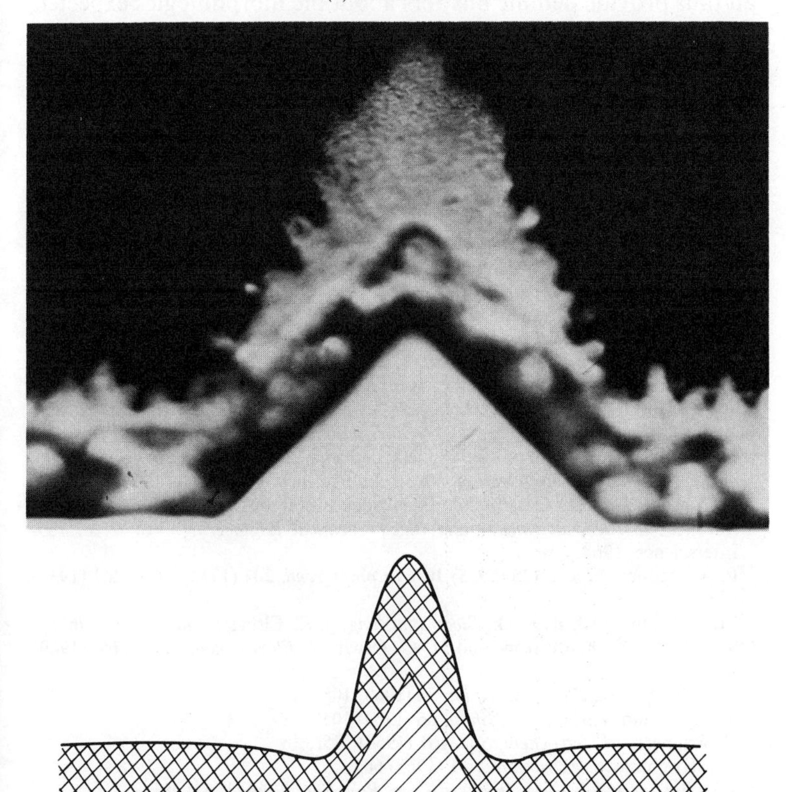

Figure 71. Comparison of the calculated and experimental metal deposits obtained with pulsating square-wave overpotential (600 mV) at similar frequencies.

of $C = \frac{1}{2}C_0$ for potential pulses of $100y_0^2/D$ cps stops moving away at times larger than $5y_0^2/D$, i.e., before it has separated from the surface profile. The derivations given earlier (p. 297) have thus been quantitatively confirmed. Moreover, by integrating fluxes at different points at the surface, it was possible to obtain the appearance of the deposit after a certain time interval, e.g., of $10y_0^2/D$. This is compared in Figure 71 with the deposit obtained experimentally with the potential pulsating at 200 cps. Good agreement is evident.

We conclude that the introduction of modern digital computers can allow the mathematical difficulties connected with solving Fick's second law for unconventional surface profiles, to be overcome, and can thus provide definite answers about the morphologies expected with the different phenomena that have been discussed in this chapter.

REFERENCES

[1]N. Ibl and K. Schadegg, *J. Electrochem. Soc.* **114** (1967) 54.
[2]A. R. Despić, J. Diggle, and J. O'M. Bockris, *J. Electrochem. Soc.* **115** (1968) 507.
[3]A. R. Despić and K. I. Popov, unpublished results.
[4]G. Wranglen, *Electrochim. Acta* **2** (1960) 130.
[5]D. D. Saratovkin, *Dendritic Crystalization*, Consultants Bureau, New York, 1959.
[6]L. Graf and W. Weser, *Electrochim. Acta* **2** (1960) 145.
[7]K. M. Gorbunova and A. J. Zhukova, *Zh. Fiz. Khim.* **23** (1949) 605.
[8]P. B. Price, D. A. Vermilyea, and M. B. Webb, *Acta Met.* **6** (1958) 524.
[9]W. H. Dettner and J. Elze, *Handbuch der galvanotechnik*, p. 882, Carl Hanser Verlag, München, 1964.
[10]N. Ibl, The formation of powdered metal deposits, in *Advances in Electrochemistry and Electrochemical Engineering*, Vol. 2, Interscience, New York, 1962.
[11]O. Kardos and D. G. Foulke, Electrodeposition on small-scale profiles, in *Advances in Electrochemistry and Electrochemical Engineering*, Vol. 2, p. 185, Interscience, 1962.
[12]P. A. Jacquet, *Nature* **135** (1935) 1076; *Compt. rend.* **201** (1935) 1973; **202** (1936) 403.
[13]M. A. Brimi and J. R. Luck, *Electrofinishing*, p. 72, Elsevier, New York, 1965.
[14]S. Arouete, K. F. Blurton, and H. G. Oswin, *J. Electrochem. Soc.* **116**, (1969) 166.
[15]V. V. Romanov, *Zh. Prikl. Khim.* **36** (1963) 1057.
[16]V. V. Romanov, *Zh. Prikl. Khim.* **36** (1963) 1050.
[17]C. Wagner, *J. Electrochem. Soc.* **101** (1954) 225.
[18]S. I. Krichmar, *Elektrokhimiya* **1** (1965) 609.
[19]J. L. Barton and J. O'M. Bockris, *Proc. Roy. Soc.* **A268** (1962) 485.
[20]J. W. Diggle, A. R. Despić, and J. O'M. Bockris, *J. Electrochem. Soc.* **116** (1969) 1503.
[21]A. R. Despić and K. I. Popov, *Bull. Soc. Chim. Belgrade*, **36** (1971) in press.

[22]A. Papapetrou, *Z. Krist.* **92** (1935) 89.
[23]F. Weinberg and B. Chalmers, *Can. J. Phys.* **29** (1951) 382; **30**, (1952) 488.
[24]G. Wranglen, *Trans. Roy. Inst. Techn. Stocholm.*, **1955**, No. 94.
[25]Ling Jang, Chien-Yeh Chien, and R. G. Hudson, *J. Electrochem. Soc.* **106** (1959) 632.
[26]W. B. Hilling and D. Turnbull, *J. Chem. Phys.* **24** (1956) 914.
[27]F. Ogburn, C. Bechtoldt, J. B. Morris, and A. de Koranyi, *J. Electrochem. Soc.* **112** (1965) 575.
[28]C. J. Bechtold, F. Ogburn, and J. Smit, *J. Electrochem. Soc.* **115** (1968) 813.
[29]S. Tajima and M. Ogata, *Electrochim. Acta* **15** (1970) 61.
[30]S. Tajima and M. Ogata, *Electrochim. Acta* **13** (1968) 1845.
[31]J. Smit, F. Ogburn, and C. J. Bechtold, *J. Electrochem. Soc.* **115** (1968) 371.
[32]T. B. Reddy, *J. Electrochem. Soc.* **113** (1966) 117.
[33]D. R. Hamilton, *Electrochim. Acta* **8** (1963) 731.
[34]J. O'M. Bockris and B. E. Conway, in *Modern Aspects of Electrochemistry, No. 3*, Chapter 4, p. 224, Ed., J. O'M. Bockris, Butterworths, London, 1964.
[35]G. P. Ivantsov, *Dokl. Akad. Nauk. SSSR* **58** (1947) 567.
[36]J. W. Faust, Jr. and H. F. John, *J. Electrochem. Soc.* **110** (1963) 109.
[37]J. W. Faust, Jr., *J. Electrochem. Soc.* **114** (1967) 1311.
[38]J. B. Kushner, *Metal Progr.* **81** (1962) 88.
[39]H. Fisher, *Elektrolytishe Abscheidung und Electrokristallisation von Metallen*, Springer Verlag, Berlin, 1954.
[40]J. W. Faust, Jr. and H. F. John, *J. Electrochem. Soc.* **108** (1961) 855.
[41]N. A. Pangarov, *Phys. Stat. Sol.* **20** (1967) 371.
[42]N. A. Pangarov and V. Velinov, *Electrochim. Acta* **13** (1968) 1909.
[43]N. A. Pangarov, *Electrochim. Acta* **9** (1964) 721.
[44]K. M. Gorbunova and P. D. Dankov, *Zh. Fiz. Khim.* **23** (1949) 616.
[45]L. Graf and W. Morgenstern, *Z. Naturforsch.* **10a** (1955) 345.
[46]S. I. Krichmar, *Dokl. Akad. Nauk SSSR* **151** (1963) 616.
[47]S. I. Krichmar, *Zh. Fiz. Khim.* **34** (1962) 663.
[48]P. A. van der Meulen and H. V. Lindstron, *Trans. Electrochem. Soc.* **103** (1956) 390.
[49]T. C. J. Ovenston, C. A. Parker, and A. E. Robinson, *Trans. Electrochem. Soc.* **104** (1957) 607.
[50]J. Priestley, *Gilbert's Analen* **12** (1803) 466.
[51]O. Kudra and M. E. Lerner, *Ukrain. khim. Zh.* **17** (1951) 890.
[52]O. Kudra and E. Gitman, *Elektroliticheskoe poluchenie metallicheskiekh poroshkov*, p. 43, Izd. Akad. Nauk UkrSSR, Kiev, 1952.
[53]K. Hirakoso, *Denkikogaku Kyokoishi* **3** (1935) 7; *C.A.* **29** (1935) 5749*u*.
[54]N. Ibl, *Helv. Chim. Acta* **37** (1954) 1149.
[55]H. Fisher and H. F. Heiling, *Trans. Inst. Metal Finishing, 4th Internat. Conference, London*, pp. 31, 74 (1954); H. F. Heiling, *Metals*, **8** (1954) 438.
[56]J. Atanasiu and A. Calusaru, *Studii si cercetari de metalurgie* **2** (1957) 337; *C.A.* **52** (1958) 13470*h*.
[57]A. Calusaru, Thesis, Inst. polytechnic, Bucharest (1957).
[58]W. R. Meyer, *Proc. Am. Electropl. Soc.* **23** (1935) 116, 135; **24** (1936) 135.
[59]H. Schmellenmeier, *Korrosion u. Metallsch.* **21** (1945) 9.
[60]G. E. Gardam, *J. Electrodep. Tech. Soc.* **22** (1947) 155.
[61]C. E. Reinhard, *Proc. Am. Electropl. Soc.* **37** (1950) 171.
[62]E. B. Leffler and H. Leidheiser, *Plating* **44** (1957) 388.
[63]A. T. Vagramyan and Z. A. Solovyova, *Methody Issledovaniya Electroosazhdeniya Metallov*, Moscow, 1960.

[64]O. Kardos and D. G. Foulke, Electrodeposition on small-scale profiles, in *Advances in Electrochemistry and Electrochemical Engineering*, Vol. 2, Interscience, New York, 1962.
[65]O. Kardos, *Proc. "Surface 66,"* p. 62.
[66]S. I. Krichmar and A. Ya. Pronskaya, *Zh. Fiz. Khim.* **39** (1965) 741.
[67]S. A. Watson and J. Edwards, *Trans. Inst. Metal Finishing* **34** (1957) 167.
[68]A. A. Voronko, *Zh. Prikl. Khim.* **35** (1962) 2802.
[69]S. S. Kruglikov, and N. T. Kudriavtsev, G. F. Vorobiova, A. Ya. Antonov, *Electrochim. Acta* **10** (1965) 253.
[70]S. A. Watson and J. Edwards, *Trans. Inst. Metal Finishing* **34** (1957) 167.
[71]G. T. Rogers and K. I. Taylor, *Electrochim. Acta* **8** (1963) 887.
[72]G. E. Gardam, in *Proc. 3rd Intern. Electrodeposition Conf., London* (1947), p. 203.
[73]O. Kardos, *Proc. Am. Electropl. Soc.* **43** (1956) 181.
[73a]S. A. Watson, *Trans. Inst. Metal Finishing* **37** (part 4) (1960), p. 144.
[74]S. I. Krichmar, *Zh. fiz. khim.* **39** (1965) 602; *Elektrokhimiya* **1** (1965) 858.
[75]S. I. Krichmar, *Elektrokhimiya* **2** (1966) 726.
[76]J. Edwards, *Trans. Inst. Metal. Finishing* **39** (1962) 52.
[77]P. A. Jacquet, *Bull. Soc. chim. France* **3** (1936) 705.
[78]P. A. Jacquet, *Trans. Electrochem. Soc.* **69** (1936) 629.
[79]T. P. Hoar and G. P. Rothwell, *Electrochim. Acta* **9** (1964) 135.
[80]B. Pointu, *Electrochim. Acta* **14** (1969) 1207.
[81]P. V. Shchigolev and N. D. Tomashov, *Dokl. Akad. Nauk SSSR* **100** (1955) 327.
[82]P. V. Shchigolev, *Electrolytic and Chemical Polishing of Metals*, Akad. Nauk SSSR, Moscow, 1959 (in Russian).
[83]T. V. Grechukhina and A. Sh. Valeyev, *Elektrokhimiya* **3** (1967) 1080.
[84]A. Sh. Valeyev and G. I. Petrov, *Elektrokhimiya* **3** (1967) 624.
[85]W. C. Elmore, *J. Appl. Phys.* **10** (1939) 724; **11** (1940) 797.
[86]J. Edwards, *J. Electrochem. Soc.* **100** (1953) 189C; **100** (1953) 223C.
[87]S. I. Krichmar, *Zh. Priklad. Khim.* **37** (1964) 2244.
[88]S. I. Krichmar and A. Ya. Pronskaya, *Elektrokhimiya* **2** (1966) 69.
[89]U. R. Evans, *Trans. Electrochem. Soc.* **69** (1936) 652.
[90]S. I. Krichmar, *Dokl. Akad. Nauk. SSSR* **101** (1955) 297.
[91]G. S. Vozdvizhenskii, *Dokl. Akad. Nauk. SSSR* **59** (1948) 1587.
[92]S. I. Krichmar, *Dokl. Akad. Nauk SSSR* **100** (1955) 481.
[93]T. P. Hoar and J. A. S. Mowat, *Nature* **165** (1950) 64.
[94]T. P. Hoar and T. W. Farthing, *Nature* **169** (1952) 324.
[95]K. Huber, *Chimia* **4** (1950) 54; *Z. Electrochem.* **55** (1951) 165.
[96]W. Machu and A. Ragheb, *Z. Metallk.* **47** (1956) 176.
[97]E. Darmois and M. I. Epelboin, *Bull. Soc. Franc. électriciens* **4** (1954) 344.
[98]S. I. Krichmar, *Zh. Fiz. Khim.* **39** (1965) 1373.
[99]S. I. Krichmar, *Zh. Fiz. Khim.* **37** (1963) 2397.
[100]S. I. Krichmar, *Elektrokhimiya* **2** (1966) 1103.
[101]G. T. Bakhvalov, *New Technology of Electrodeposition of Metals*, Izd. Met., 1966 (in Russian).
[102]N. N. Bibikov, *Electrodeposition of Metals by AC*, Mashgiz, Moscow–Leningrad, 1961 (in Russian).
[103]G. T. Bakhvalov, *Tr. 2-oi Vses. Konf. Akad. Nauk po Teoret. i Prikl. Khimii*, p. 202, SSSR, 1949.
[104]G. T. Bakhvalov, *Sb. Nauchn. Tr. MITCMiZ im. Kalinina*, p. 222, Metalurgizdat, 1950.
[105]A. I. Levin *et al.*, *Sb. Statei UPI*, W43, p. 99 (1953).

[106]G. W. Gernsted, *Steel* **120** (1947) 100; **120** (1947) 134; *Steel Processing* **33** (1947) 479; **33** (1947) 498.
[107]French patent No. 932039, 10/III, 1948.
[108]G. W. Gernsted, *Plating* **35** (1948) 708.
[109]USA patent No. 2470755, 24/V, 1949.
[110]USA patent No. 2451340, 12/X, 1948.
[111]USA patent No. 2451341, 12/X, 1948.
[112]USA patent No. 2636850, 28/V, 1953.
[113]V. I. Chernenko *et al.*, *Elektrokhimiya* **4** (1968) 519.
[114]D. Rossi, *Galvanotechnics* **1953**, 793.
[115]Yu. Ya. Vene and S. A. Nikolaeva, *Zh. Fiz. Khim.* **39** (1955) 811.
[116]C. P. Wales, *J. Electrochem. Soc.* **115** (1968) 985.
[117]S. Arouete *et al.*, *J. Electrochem. Soc.* **116** (1969) 166.
[118]N. N. Bibikov, *Tekhnologya Transportnovo machinostroeniya*, No. 3, VPTI, 1956.
[119]V. V. Romanov, *Zh. Prikl. Khim.* **34** (1961) 2692.
[120]A. P. Popkov, *Zh. Prikl. Khim.* **39** (1966) 1747.
[121]C. P. Wales, *J. Electrochem. Soc.* **113** (1966) 757.
[122]V. V. Romanov, *Vestn. Elektropromishlenosty* **9** (1960); V. V. Romanov, *Zh. Fiz. Khim.* **34** (1961) 2692.
[123]C. P. Wales, *J. Electrochem. Soc.* **115** (1968) 680.
[124]R. Yu. Bek and N. T. Kudriavtsev, *Zh. Prikl. Khim.* **34** (1961) 2013.
[125]R. Yu. Bek and N. T. Kudriavtsev, *Zh. Prikl. Khim.* **34** (1961) 2020.
[126]R. Yu. Bek, D. Yu. Gamburg, and N. T. Kudriavtsev, *Zh. Fiz. Khim.* **35** (1962) 2244.
[127]M. Ya. Popereka and V. I. Avramenko, *Zh. Fiz. Khim.* **39** (1965) 1875.
[128]M. Ya. Popereka and V. I. Avramenko, *Elektrokhimiya* **1** (1965) 894.
[129]V. V. Stender and M. D. Zholuev, *Zh. Prikl. Khim.* **32** (1959) 1296.
[130]A. R. Despić and K. I. Popov, *J. Appl. Electrochem.* **1** (1971) 39.
[131]J. Wojtowicz and B. E. Conway, *J. Chem. Phys.* **52** (1970) 1407.

5

Mechanisms of Stepwise Electrode Processes on Amalgams

V. V. Losev

Karpov Institute of Physical Chemistry
Moscow, USSR

I. INTRODUCTION

The purpose of this review is to present a complete kinetic analysis of the anodic dissolution and cathodic deposition of metals (with particular reference to amalgams) for cases where multivalent ions are involved and the charge transfer is a stepwise process and low-valence intermediates are formed; the criteria for the stepwise mechanism that may be developed from such an analysis will also be presented. A description of stepwise electrochemical reactions of various classes has been given by Simonova and Rotinyan.[1]

The mechanism of anodic dissolution and cathodic deposition of metals has received considerably less attention than that given to some complex electrochemical redox reactions. This is due mainly to the difficulties of carrying out the experiments necessary for an elucidation of the mechanism of the charge-transfer step in these processes, and, for many metals, the rate of the electrochemical reaction is so high that in steady-state polarization measurements, the kinetics of the overall process are determined by steps such as bulk and surface diffusion, and incorporation of metal particles into the crystal lattice.

Because of the greater reversibility of metal discharge-ionization processes, e.g., in comparison with that of the process of hydrogen evolution, it is often possible to study, over a given range of potentials, the kinetics not only of the forward (for example, anodic) process, but also of the corresponding reverse process, and to measure the exchange current directly. This possibility is particularly important for correctly determining the mechanism of complex multistep processes. Since at equilibrium, and within a narrow range of deviations from equilibrium conditions, the forward and reverse reactions should proceed through the same steps, the kinetics and the mechanistic conclusions must be consistent for both the forward and reverse directions of the electrochemical reaction.[2,3]

In the case of solid metal electrodes, processes of diffusion of the discharged particle to the point of crystal growth and subsequent incorporation into the crystalline lattice occur after the primary electron transfer, and the kinetic analysis is then complex. Also, the true and apparent surface areas of solid metals usually differ and can be time-dependent, and the metal surface is often energetically nonhomogeneous, leading to a dependence of the rate constant of a given process on the exposure of crystal planes.

These difficulties can be largely overcome if *amalgams* are used instead of solid metal electrodes. In the case of amalgams, the cleanliness of the surface can also be readily ensured and its passivation more easily prevented. Furthermore, an amalgam electrode has a smooth, homogeneous surface. The high overvoltage of hydrogen on mercury and amalgams can considerably facilitate the study of electrodic processes in those cases where the hydrogen overvoltage on the corresponding solid metal is low and where the accompanying evolution of hydrogen would hence complicate the results of measurements on metal deposition at potentials both near and away from the equilibrium potential. An important added advantage of the use of amalgams over that of solid electrodes lies in the possibility of investigating the dependence of the exchange current and the rate of the anodic process on the concentration of metal in the amalgam. This makes it possible to determine the anodic reaction order and thus to obtain additional kinetic characteristics of the electrode process. Finally, the processes of metal discharge and ionization on amalgam electrodes occur without surface diffusion and lattice building steps.

For an amalgam electrode M(Hg) in contact with a solution containing simple ions M^{n+} of the same metal M, the overall electrode process involving multiple electron transfer is

$$M \rightleftarrows M^{n+} + ne \qquad (1)$$

and usually coincides with the formal electrochemical reaction. Reaction (1) will provide a basis for the theoretical analysis which follows.

II. BRIEF HISTORICAL SURVEY

Anodic dissolution and cathodic deposition of metals are accompanied by the transfer of metal particles across the phase boundary. In this process, the particles transferred to the metal phase to a certain extent retain their ionic nature. Because of this feature of the overall reaction (1), many investigators have tended to regard the reaction as a direct transfer of an ion across the phase boundary, i.e., a simple one-step process which does not involve the formation of intermediate ions or compounds of low valence.

Vetter, for example, who has greatly contributed[2,3] to the development of the theory of stepwise electrode processes, considers the theory inapplicable to reactions of type (1), since in such reactions, it is not the electron, but the ion which is regarded as undergoing transfer to or from the surface of the metal into the solution.[2]

A similar view was held by Grahame,[4] who suggested that because in such reactions, electrons do not leave the metal phase, the usual view that one-electron transfer steps are most probable[5–7] cannot therefore be applied. Some investigators[8a,b,9–16] have implicitly considered simultaneous transfer of charges across the phase boundary in the case of reaction (1).

The formation of low-valence intermediates in reactions of type (1) was first recognized by Heyrovsky,[17] who observed that such reactions at a mercury cathode are greatly retarded by surface-active substances (unlike processes involving M^+ ions). On the supposition that one-electron processes always proceed rapidly, Heyrovsky concluded that multielectron transfer processes take place according, for example, to the scheme

$$Zn^{++} + e \rightarrow Zn^+$$

$$2Zn^+ \rightarrow Zn + Zn^{++}$$

where the irreversibility of the overall process is considered to be determined by the disproportionation step which is retarded by the surface-active substances.[17]

Heyrovsky's theory was criticized by Loshkarev and Krjukova,[18] who showed that one-electron electrochemical processes can also be retarded by surface-active substances. Frumkin[19,20] considered the mechanism proposed by Heyrovsky improbable, pointing out that unstable ions such as Zn^+ would more readily undergo direct reduction at an electrode than enter into the disproportionation reaction (cf. Refs. 21 and 22). In general, however, the idea of formation of low-valence intermediates requires full consideration.

One of the first studies in which the stepwise nature of reactions of type (1) was proposed dealt with the anodic dissolution of iron in alkaline solutions, which proceeds according to the following overall reaction:

$$\mathrm{Fe} + 2\mathrm{OH}^- \rightleftarrows \mathrm{Fe(OH)_2} + 2\mathrm{e}$$

Kabanov and Leikis[23] showed that the stationary anodic polarization curve for this reaction has an unusually low slope ($\sim$0.040 V); their results could be represented approximately by

$$i = k[\mathrm{OH}^-]^{3/2} \exp(3F\phi/2RT) \tag{2}$$

where ϕ is the measured electrode potential and i the current density. In explaining these results, Kabanov *et al.*[24] assumed that the first one-electron step proceeds under quasiequilibrium conditions (cf. Ref. 25) leading to the formation of an intermediate surface compound $\mathrm{Fe(OH)_{ads}}$,

$$\mathrm{Fe} + \mathrm{OH}^- \rightleftarrows \mathrm{Fe(OH)_{ads}} + \mathrm{e} \tag{3}$$

followed by a second, rate-controlling step,

$$\mathrm{Fe(OH)_{ads}} + \mathrm{OH}^- \rightleftarrows \mathrm{FeO_{ads}} + \mathrm{H_2O} + \mathrm{e} \tag{4}$$

and then by a fast chemical reaction involving $\mathrm{FeO_{ads}}$ resulting in the eventual formation of a $\mathrm{Fe(OH)_2}$ precipitate.

The rate equation corresponding to this mechanism is

$$\begin{aligned} i_a &= k_a[\mathrm{Fe(OH)_{ads}}][\mathrm{OH}^-] \exp\left(\frac{\beta_{2,a}F\phi}{RT}\right) \\ &= k_a'[\mathrm{OH}^-]^2 \exp\left[\frac{(1 + \beta_{2,a})F\phi}{RT}\right] \end{aligned} \tag{5}$$

which accords with the experimental data if $\beta_{2,a}$, the symmetry factor for the anodic electron transfer in step (4), is ~ 0.5.

Theoretical calculation of the rates of stepwise electrode processes, for the purpose of comparing them with the rate of the corresponding one-step process, is difficult. Indeed, for multielectron redox electrode processes, the quantum-mechanical analysis does not provide an unambiguous answer whether the process proceeds stepwise or as a simple one-step reaction, although the stepwise path is more probable.[26a,26b] When the intermediate low-valence ions are sufficiently stable, and their equilibrium concentration is known, an approximate calculation can be made of the relative free energies of the transition states for various steps of the overall reactions, as carried out by Conway and Bockris[26b] for the reaction $Cu \rightleftarrows Cu^{++} + 2e$. It was found that for the one-step discharge of Cu^{++} ions, the activation energy is much greater than that for a stepwise process involving the formation of Cu^{+} ions. In the case of the reaction $Ni \rightleftarrows Ni^{++} + 2e$, such calculations are more difficult since stable Ni^{+} ions do not exist in solution. In such cases, and in general, experimental criteria for stepwise mechanisms are required.

III. PROCESSES WITH A SINGLE LIMITING STEP

Complex electrode processes in which the valence of reacting species varies by more than unity can be subdivided into two main classes:

(a) Reactions in which the initial and final products are in solution and the charge transfer across the phase boundary is brought about by electrons only, e.g., redox reactions such as

$$M^{p+} \rightleftarrows M^{q+} + (q - p)e \tag{6}$$

(b) Reactions accompanied by a transition of reacting particles from one phase to another with participation of ions in the charge transfer.

The processes (1) examined in this study fall in class (b). However, when these reactions occur in several successive electron-transfer steps, only the first of these steps (in the anodic direction) is accompanied by the transfer of "ions" across the phase boundary, while the remaining steps are simply redox reactions involving electron transfer of the type (6). The rate equation will first be derived with reference to an overall electrochemical process proceeding through successive single-electron steps. In this case, the rate equations for the first step $M \rightleftarrows M^+ + e$, in which the formal transfer takes place, not of the electron, but of the ion across the boundary surface, are identical for both ionic and electron transfer. Thus, for the purpose of simplification, we shall also formally consider this step as a redox electron-transfer step. Hence, all conclusions are applicable not only to amalgam and solid metal electrode processes, but also to "redox" reactions.

In the present section, the kinetics[27,28] as well as the experimental data for stepwise processes with a single limiting step on amalgam electrodes will be examined in order to develop methods for determining the kinetic parameters and to define the criteria for stepwise mechanisms.

1. Rate Equations with Allowance for the ψ_1 Potential

The following assumptions will be made:

(a) Overall anodic and cathodic processes as well as individual single-electron steps are first-order reactions with respect to the corresponding reacting species. Only simple cations take part in the electrode process.

(b) The overall electrode process consists only of electron-transfer steps; that is, coupled chemical steps are absent.

(c) Specific adsorption of reactants, intermediates, and products does not occur, and the bulk (c_i) and surface (Γ_i) concentrations of unstable intermediates are low so that $\Gamma_i \propto c_i$[28a] and the adsorption of intermediates will not affect the kinetics.[29]

(d) Parallel and side electrochemical reactions are absent.

(e) Diffusion limitation is absent.*

The overall charge-transfer reaction at an amalgam electrode can be described by equation (1) and the reaction is assumed to be a series of single-electron steps of the type

$$M^{(l-1)+} \rightleftarrows M^{l+} + e \tag{7}$$

It is assumed that the rates of the mth step,

$$M^{(m-1)+} \underset{k_{k,m}}{\overset{k_{a,m}}{\rightleftarrows}} M^{m+} + e \tag{8}$$

*If diffusion of intermediates from the electrode and their chemical reactions with solution components are taken into account, the rate equations become considerably more complicated.[29a,29b]

where $1 \leq m \leq n$, is so much lower than the rate constants of the remaining steps that they can be considered to be at equilibrium even at appreciable polarizations. Let the stoichiometric number[8a] $\nu = 1$; then,[30a] in the usual way, for anodic and cathodic partial current densities,

$$i_{a,m} = k_{a,m}[M^{(m-1)+}]_s \exp\left[\frac{\beta_{a,m}F(\phi - \psi_1)}{RT}\right] \tag{9}$$

and

$$i_{k,m} = k_{k,m}[\mathrm{M}^{m+}]_s \exp\left[-\frac{\beta_{k,m}F(\phi - \psi_1)}{RT}\right] \tag{10}$$

respectively; in (9) and (10), ψ_1 is the double-layer potential at the distance of closest approach of nonspecifically absorbed ions from the surface relative to the potential in the bulk of the solution; $\beta_{k,m}$ and $\beta_{a,m}$ are true cathodic and anodic symmetry coefficients of the mth step, which obey the usual condition

$$\beta_{k,m} + \beta_{a,m} = 1 \tag{11}$$

with $\beta_{k,m}, \beta_{a,m} > 0$.

Then, relating surface and bulk ion concentrations,[30b]

$$[M^{(m-1)+}]_s = [\mathrm{M}^{(m-1)+}] \exp[-(m - 1)F\psi_1/RT] \tag{12}$$

so that equations (9) and (10) give

$$i_{a,m} = k_{a,m}[\mathrm{M}^{(m-1)+}] \exp(\beta_{a,m}F\phi/RT) \times \exp[-(m - 1 + \beta_{a,m})F\psi_1/RT] \tag{13}$$

and

$$i_{k,m} = k_{k,m}[\mathrm{M}^{m+}] \exp(-\beta_{k,m}F\phi/RT) \exp[-(m - \beta_{k,m})F\psi_1/RT] \tag{14}$$

Under equilibrium conditions, $i_{a,m} = i_{k,m} = i_{0,m}$ and

$$i_{0,m} = k_{a,m}[\mathrm{M}^{(m-1)+}] \exp[\beta_{a,m}F\phi_e/RT] \times \exp[-(m - 1 + \beta_{a,m})F\psi_1/RT]$$

From this equation, the dependence of the exchange current of the mth step on the concentrations of ions $\mathrm{M}^{(m-1)+}$ and M^{m+} is readily obtained by expressing the equilibrium potential ϕ_e in terms of concentrations with the aid of the Nernst equation for step (8), taking into account condition (11); i.e.,

$$i_{0,m} = i^{\circ}_{0,m}[\mathrm{M}^{(m-1)+}]^{\beta_{k,m}}[\mathrm{M}^{m+}]^{\beta_{a,m}} \times \exp[-(m - 1 + \beta_{a,m})F\psi_1/RT] \quad (15)$$

where $i^{\circ}_{0,m}$ is the standard exchange current for step (8).

In relating the rate $i_{a,m}$ of the mth anodic step to the overall anodic partial current i_a, it is to be noted that in the steady state,

$$i_a = ni_{a,m} \quad (16)$$

and similarly

$$i_k = ni_{k,m} \quad (17)$$

The concentrations of the intermediates in equations (13) and (14) are determined by the usual quasiequilibrium conditions for all steps except the mth step; i.e.,

$$[\mathrm{M}^{(m-1)+}] = K_{(m-1)}[\mathrm{M}]\exp[(m-1)F\phi/RT] \quad (18)$$

$$[\mathrm{M}^{m+}] = K_{(m)}[\mathrm{M}^{n+}]\exp[-(n-m)F\phi/RT] \quad (19)$$

where [M] is the concentration of the metal in the amalgam and $K_{(m)}$ and $K_{(m-1)}$ are constants. Equations (16) and (18), with equation (13), then yield the expression for the anodic partial current

$$i_a = nk_{a,m}K_{(m-1)}[\mathrm{M}]\exp(\alpha_a F\phi/RT)\exp(-\alpha_a F\psi_1/RT) \quad (20)$$

where α_a is now the transfer coefficient of the anodic process, determined from the slope of the curve of ϕ versus $\ln i_a$ with the aid of equation (24) (see below); (here, changes in ψ_1 as a function of the potential are disregarded). α_a is related to the symmetry coefficient* $\beta_{a,m}$ through the equation

$$\alpha_a = m - 1 + \beta_{a,m} \quad (21)$$

Similarly,

$$i_k = nk_{k,m}K_{(m)}[\mathrm{M}^{n+}]\exp(-\alpha_c F\phi/RT) \times \exp[-(n - \alpha_k)F\psi_1/RT] \quad (22)$$

*The terms symmetry coefficient (β) and transfer coefficient (α) are distinguished here in the usual way; the former refers only to the activation step in the rate-controlling electron-transfer process, while the latter refers to the factor governing the overall potential dependence of the rate of the reaction, i.e., taking account of prior steps in potential-dependent quasiequilibrium.

where α_K is the cathodic transfer coefficient related to $\beta_{k,m}$ of equation (14) by

$$\alpha_k = n - m + \beta_{k,m} \tag{23}$$

For multielectron transfer, several equilibria are involved with relations similar to equations (18) and (19).

For representation of experimental data, equations (20) and (22) can be put in the more convenient forms

$$\phi = -\frac{RT}{\alpha_a F}\ln(nk_{a,m}K_{(m-1)}) - \frac{RT}{\alpha_a F}\ln[\mathrm{M}] + \frac{RT}{\alpha_a F}\ln i_a + \psi_1 \tag{24}$$

$$\phi = \frac{RT}{\alpha_k F}\ln(nk_{k,m}K_{(m)}) + \frac{RT}{\alpha_k F}\ln[\mathrm{M}^{n+}] - \frac{RT}{\alpha_k F}\ln i_k - \left(\frac{n-\alpha_k}{\alpha_k}\right)\psi_1 \tag{25}$$

and for the exchange current density, i.e., where $\phi = \phi_e$,

$$\begin{aligned} i_0 &= nk_{a,m}K_{(m-1)}[\mathrm{M}]\exp(\alpha_a F\phi_e/RT)\exp(-\alpha_a F\psi_1/RT) \\ &= nk_{k,m}K_{(m)}[\mathrm{M}^{n+}]\exp(-\alpha_k F\phi_e/RT) \\ &\quad \times \exp[-(n-\alpha_k)F\psi_1/RT] \end{aligned} \tag{26a}$$

Also, from equations (16) and (17), the exchange current of the overall process is related to that of the limiting step by the equation

$$i_0 = ni_{0,m} \tag{26b}$$

By solving equation (26a) for ϕ_e,

$$\phi_e = \frac{RT}{(\alpha_k + \alpha_a)F}\ln\frac{k_{a,m}K_{(m-1)}}{k_{k,m}K_{(m)}} + \frac{RT}{(\alpha_k + \alpha_a)F}\ln\frac{[\mathrm{M}^{n+}]}{[\mathrm{M}]} - \left[\frac{n-(\alpha_a+\alpha_k)}{\alpha_a+\alpha_k}\right]\psi_1 \tag{27}$$

is obtained. Comparison of equation (27) with the Nernst equation for process (1) yields

$$\alpha_a + \alpha_k = n \tag{28}$$

which also follows from equations (21) and (23), using equation (11).

Also, in terms of the overpotential $\Delta\phi\ (= \phi - \phi_e)$,

$$i_a = i_0\exp(\alpha_a F\Delta\phi/RT) \tag{29}$$

$$i_k = i_0 \exp(-\alpha_k F \Delta\phi / RT) \tag{30}$$

so that,[31] with equation (28),

$$i_a / i_k = \exp(nF\Delta\phi / RT) \tag{31}$$

which enables the curve of $\Delta\phi$ versus $\ln i_k$ to be evaluated if that of $\Delta\phi$ versus $\ln i_a$ is known.

The dependence of the exchange current on the concentrations of metal in the amalgam and in the solution, as well as the dependence on the ψ_1 potential, can be determined[28,32,33] by rewriting ϕ_e in equation (26a) in terms of these concentrations with the aid of the Nernst equation and by using equation (28), i.e.,

$$i_0 = nk_{a,m}K_{(m-1)}[\exp(-\alpha_a F\psi_1/RT)][\mathrm{M}]^{\alpha_k/n}[\mathrm{M}^{n+}]^{\alpha_a/n} \tag{32}$$

which allows double-layer effects[30a] to be calculated. These are determined by the sign of the amalgam surface charge q_{M}. When $q_{\mathrm{M}} < 0$ holds, a decrease in ionic strength brings about an increase in the absolute value of negative ψ_1 potential, thus accelerating the exchange; for positive q_{M}, the opposite effect arises.

The net current density i (cf. Ref. 34a, b, c) should be equal to the difference in the rates of the coupled partial processes for more complex cases when i_a and i_k correspond to different electrode processes. Here,

$$\begin{aligned} i = i_a - i_k = {} & nk_{a,m}K_{(m-1)}[\mathrm{M}]\exp(\alpha_a F\phi/RT)\exp(-\alpha_a F\psi_1/RT) \\ & - nk_{k,m}K_{(m)}[\mathrm{M}^{n+}]\exp(-\alpha_k F\phi/RT)\exp[-(n-\alpha_k)F\psi_1/RT] \end{aligned} \tag{33}$$

which determines the stationary polarization curve. More conveniently,

$$i = i_a - i_k = i_0[\exp(\alpha_a F\Delta\phi/RT) - \exp(-\alpha_k F\Delta\phi/RT)] \tag{34}$$

where $i > 0$ as $\Delta\phi > 0$ for anodic polarization and $\Delta\phi < 0$ for cathodic polarization.

The above relations can be used for determining, on the basis of polarization measurements, the kinetic parameters of process (1), namely the exchange current i_0, the transfer coefficients α_a and α_k, and the reaction orders.

2. Rate Equations in the Case of an Excess of Supporting Electrolyte

In the general case, ψ_1 varies with ϕ as well as with $[\mathrm{M}^{n+}]$. Therefore, for the cathodic process [see equation (25)], both $d\phi/d(\ln i)$ and

$d(\ln i)/d(\ln[M^{n+}])$ determined far from ϕ_e will not be constant. Hence, determination of the kinetic parameters should be carried out with an excess of inert electrolyte at constant ionic strength μ. For such conditions, equations (20), (22), and (33), and equation (26a) for the exchange current, become

$$i = i_a - i_k = k_a[\mathrm{M}]\exp(\alpha_a F\phi/RT) - k_k[\mathrm{M}^{n+}] \times \exp(-\alpha_k F\phi/RT) \quad (35)$$

and

$$i_0 = k_a[\mathrm{M}]\exp(\alpha_a F\phi_e/RT) = k_k[\mathrm{M}^{n+}]\exp(-\alpha_k F\phi_e/RT) \quad (36)$$

or

$$\phi_e = -(RT/\alpha_a F)\ln k_a - (RT/\alpha_a F)\ln[\mathrm{M}] + (RT/\alpha_a F)\ln i_0 = (RT/\alpha_k F)\ln k_k + (RT/\alpha_k F)\ln[\mathrm{M}^{n+}] - (RT/\alpha_k F)\ln i_0 \quad (37)$$

where

$$k_a = nk_{a,m}K_{(m-1)}\exp(-\alpha_a F\psi_1/RT)$$

$$k_k = nk_{k,m}K_{(m)}\exp[-(n-\alpha_k)F\psi_1/RT]$$

Also, equation (32) becomes*

$$i_0 = i_0^\circ[M]^{\alpha_k/n}[M^{n+}]^{\alpha_a/n} \quad (38)$$

where i_0° is a standard exchange current.[11,12]

When $\Delta\phi \gg 0$, equations (24) and (25) become, respectively,

$$\phi = -(RT/\alpha_a F)\ln k_a - (RT/\alpha_a F)\ln[\mathrm{M}] + (RT/\alpha_a F)\ln i_a \quad (39a)$$

and

$$\phi = (RT/\alpha_k F)\ln k_k + (RT/\alpha_k F)\ln[\mathrm{M}^{n+}] - (RT/\alpha_k F)\ln i_k \quad (39b)$$

and equation (34) yields, accordingly,

$$\Delta\phi = -(RT/\alpha_a F)\ln i_0 + (RT/\alpha_a F)\ln i_a \quad (40)$$

and

$$\Delta\phi = (RT/\alpha_k F)\ln i_0 - (RT/\alpha_k F)\ln i_k \quad (41)$$

*Equation (38) differs from the usual expressions describing i_0 as $f[\mathrm{M}]$ in the amalgam and in the solution[11,12] in that such expressions are obtained by assuming the "simultaneous" transfer of all the n electrons when the electrode process takes place; this corresponds to the condition $\alpha_a + \alpha_k = 1$, in which case the experimentally determined transfer coefficients coincide with the symmetry coefficients. Here, on the other hand, it is assumed that a stepwise mechanism takes place and therefore that $\alpha_a + \alpha_k = n$.

The transfer coefficients α_a and α_k can be determined from the Tafel slopes b_a and b_k in the usual way, and

$$\alpha_a/\alpha_k = b_k/b_a \tag{42}$$

and from equation (28), for decadic logarithms,

$$(1/b_a) + (1/b_k) = nF/2.3RT \tag{43}$$

A general conclusion with regard to the values of the Tafel slopes of the polarization curves for successive processes in a stepwise mechanism with a single rate-limiting step is[8b] that these slopes should decrease with the increase in the number of equilibrium steps preceding the limiting step. Thus, for the anodic process, the maximal value of the Tafel slope is $b_a = 2.3RT/\alpha_{a,1}F$ (that is, 116 mV for $\alpha_{a,1} = 0.5$) when the first step ($m = 1$) is limiting; with an increase in the number of the preceding equilibrium steps, b_a progressively decreases,[8b] reaching the limiting value $b_a = 2.3RT/(n - 1 + \alpha_{a,m})F$ for the limiting step of transfer of the last electron ($m = n$).

In the usual way, extrapolation of curves of $\Delta\phi$ versus $\ln i_a$ or $\Delta\phi$ versus $\ln i_k$ to $\Delta\phi = 0$ gives i_0, which may also be obtained for small $\Delta\phi$ from the transfer resistance,[2]

$$R_d = \left(\frac{\partial(\Delta\phi)}{\partial i}\right)_{i=0} = \frac{RT}{(\alpha_k + \alpha_a)F}\frac{1}{i_0} = \frac{RT}{nF}\frac{1}{i_0} \tag{44}$$

As follows from equation (34), values of exchange current obtained by extrapolating anodic and cathodic polarization curves $i_{0,\text{ext}}$ should coincide, and agree with the value of the exchange current $i_{0,\text{ex}}$ determined from exchange measurements near ϕ_e, for example, by means of the radiotracer or impedance methods or by calculation from equation (44).

For the simple single-step mechanism, there is similar agreement between $i_{0,\text{ext}}$ obtained from ϕ versus $\ln i$ curves and i_0 calculated from R_d. This can be readily proved by examining equation (47) (given below) for this mechanism. Thus, by comparing the extrapolated exchange currents (as well as the exchange currents calculated from R_d), the stepwise mechanism (for the case of the single limiting step analyzed here) cannot be distinguished from the simple, single-step mechanism.

Unlike the rates of simple, homogeneous exchange reactions, which are proportional to the concentration of exchanging particles raised to the first power,[35] the rate of electrochemical heterogeneous exchange [equation (38)] depends on this concentration raised to a fractional power (since $\alpha_k < n$ and $\alpha_a < n$). Thus, a simultaneous tenfold increase in the concentration of M and M^{n+} leads only to a tenfold increase in the exchange current, whereas the rate of a corresponding homogeneous exchange reaction, under the same conditions, would increase a hundredfold. Basically, this arises because a change in the concentration of one of the reacting components is accompanied by a Nernstian change in the electrode potential, which also determines the rate, and the exchange process occurs in two independent electrochemical steps.

Equation (38) is useful for the evaluation of transfer coefficients in cases where Tafel regions of the polarization curves are difficult to obtain, owing, for example, to the high rate of the electrode process; thus,

$$\left\{\frac{\partial(\ln i_0)}{\partial(\ln[\mathrm{M}])}\right\}_{[\mathrm{M}^{n+}]} = \frac{\alpha_k}{n} \quad \text{or} \quad \left\{\frac{\partial(\ln i_0)}{\partial(\ln[\mathrm{M}^{n+}])}\right\}_{[\mathrm{M}]} = \frac{\alpha_a}{n} \tag{45}$$

Similarly, by considering the dependence of i_0 or ϕ_e [compare equations (39a) and (39b) with (37)],

$$\left[\frac{\partial\phi_e}{\partial(\lg i_0)}\right]_{[\mathrm{M}]} = \frac{RT}{\alpha_a F} \quad \text{and} \quad \left[\frac{\partial\phi_e}{\partial(\lg i_0)}\right]_{[\mathrm{M}^{n+}]} = \frac{RT}{\alpha_k F} \tag{46}$$

It should be pointed out that for conditions where a stepwise mechanism applies (i.e., as assumed here) on the basis of experimental data, e.g., from the Tafel slope of the polarization curve [equations (39a)–(41)] and the dependence of i_0 on [M], [M^{n+}], and ϕ_e [equations (45) and (46)], only the transfer coefficients α_k and α_a can be determined. The true symmetry coefficients $\beta_{k,m}$ and $\beta_{a,m}$ characterizing the shape of the energy barrier for the rate-limiting step can be determined only by considering a definite limiting step. The transfer coefficient is the sum of the corresponding symmetry coefficient and the number of electrons participating in the quasi-equilibrium steps preceding the limiting step [equations (21) and (23)]. Such an explanation of the physical meaning of the transfer coefficient was first made by Frumkin[25] in his kinetic analysis of cathodic evolution of hydrogen.

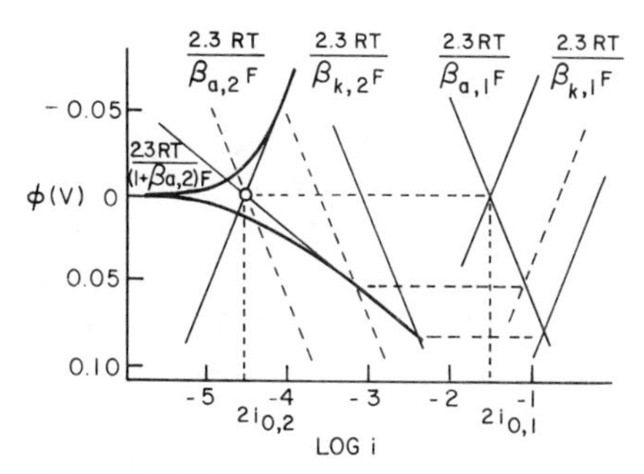

Figure 1. Polarization curves for a stepwise process.

In the simple single-step mechanism, for which the equation of the polarization curve [cf. equation (34)] has the usual form

$$i = i_a - i_k = i_0[\exp(\beta_{a,1} nF\,\Delta\phi/RT) - \exp(-\beta_{k,1} nF\,\Delta\phi/RT)] \quad (47)$$

the true symmetry coefficients $\beta_{a,1}$ and $\beta_{k,1}$ are obtained directly from the Tafel slopes and n without any additional assumptions (see, for example, Ref. 2, §51), i.e.,

$$\alpha_k = n\beta_{k,1} = 2.3RT/b_kF \qquad \text{and} \qquad \alpha_a = n\beta_{a,1} = 2.3RT/b_aF \quad (48)$$

with $\beta_{k,1} + \beta_{a,1} = 1$, and $n = 1$ usually.

The fundamental difference between the physical meanings of the transfer coefficients for the stepwise and for the simple single-step mechanism can be illustrated by the following example.* Let $n = 2$ at 20°C and the experimentally determined slopes $b_a = 39$ mV and $b_k = 116$ mV (Figure 1). In the case of the single-step mechanism, the symmetry coefficients, according to equations (48), will be $\beta_{k,1} = 0.25$ and $\beta_{a,1} = 0.75$, while in the case of the successive electron-transfer mechanism, the transfer coefficients are $\alpha_k = 0.5$ and $\alpha_a = 1.5$. Since $\alpha_a > \alpha_k$, the second step is normally the rate-limiting one and the first step is almost at equilibrium. The symmetry coefficients $\beta_{k,2} = \beta_{a,2} = 0.5$, obtained with the aid of equations (21) and (23), correspond to the rate-limiting step (Figure 1). Thus, in the first case, a sharply asymmetric energy barrier would correspond

*The concepts used here with regard to the physical significance of the transfer coefficients determined from experimental data were developed for two- and three-step electrode processes[36–38] and then for a process consisting of n steps.[27,28] Analagous ideas were expressed by Lovrecek[39] and Mohilner.[40]

to the given experimental data, whereas in the case of a stepwise mechanism, this barrier would be symmetric.

Unlike the symmetry coefficients, the physical meaning and the absolute values of the exchange current obtained by extrapolating ϕ versus ln i curves to ϕ_e or by calculation from R_d are the same for both mechanisms. This can be readily proved by comparing the equations for the polarization curves of the stepwise [equation (34)] and the simple process [equation (47)] as well as by taking into account the fact that the relation (44) obtained from them is the same for both mechanisms.

In deriving the main equation for the polarization curve, equation (33), it was assumed that the anodic and cathodic processes were first-order reactions with regard to the concentration of metal in the amalgam and in the solution. To compare this equation and the derived relations (equations (35) and others) with the experimental data for the purpose of testing the validity of the above assumption, it is necessary to determine the dependence of the rate of a given process (for example, anodic) on the amalgam concentration at constant $[M^{n+}]$ and ϕ.

As in the case of chemical reactions, a study of the orders of the electrochemical reactions in a given component must be carried out with the concentrations of the remaining components of the solution being kept constant.[2] It should be noted that the determination of electrochemical reaction orders is as necessary for elucidating the mechanism of metal discharge-ionization processes as is the determination of other kinetic parameters. The orders must be determined preferably (cf. Ref. 8b and 40a) at constant *electrode potential*, rather than at constant $\Delta\phi$, the overpotential.

When linear Tafel regions are difficult to obtain, α_a or α_k can be evaluated from equation (37) or (38). For determination of the anodic reaction order with respect to amalgam concentration according to equation (37), it is necessary to obtain several curves (ϕ_e versus ln i_0) at different concentrations of the amalgam and then determine the rate of the anodic process at a constant potential from the intersection of these curves with the constant-potential line.

3. Principal Features of Stepwise Mechanisms

It is of interest now to examine what the principal features are of stepwise mechanisms that can be formulated on the basis of electrode kinetics for processes with a single limiting step; as shown above, the transfer coefficients are related by equation (28), $\alpha_a + \alpha_k = n$. By using equations (48), it can be shown that for the simple, single-step mechanism, this sum has the same magnitude. Thus, equation (28) does not allow a distinction to be made between a stepwise mechanism having a single limiting step and a simple, single-step mechanism. Similarly, the above-mentioned comparison of the magnitudes of $i_{0,\mathrm{ext}}$ values determined from experimental data with the aid of equations for different mechanisms also does not enable differentiation between these two cases to be made. For both mechanisms, the i_0 values obtained from extrapolated i_a and i_k curves should agree with one another and with the i_0 value measured at ϕ_e. It can be readily shown that the dependence of the ratio i_a/i_k on the potential [equation (31)] is the same for both cases.

The stepwise mechanism can be proved only by comparing the values of the transfer coefficients α_k and α_a obtained from independent measurements. From equations (21) and (23), it follows that the ratio of the transfer coefficients, which equals that of the anodic and cathodic slopes [equation (42)], depends on the numerical value of the order of the limiting step m; thus,*

$$\alpha_a/\alpha_k = b_k/b_a = (m - 1 + \beta_{a,m})/(n - m + \beta_{k,m}) \tag{49}$$

For example, if the rate-controlling process is the first step ($m = 1$), and taking into account equation (11), the ratio obtained from equation (49) is $\alpha_a/\alpha_k \leq 1/(n - 1)$. If the limiting step is that involving the transfer of the last electron ($m - n$), it is seen that $\alpha_a/\alpha_k \geq n - 1$. Thus, for these limiting cases, there is a large difference between α_k and α_a. These two cases are of great practical interest, since electrode processes on amalgam electrodes are usually studied with bivalent and trivalent metals.

The dependence of the ratio α_a/α_k on the numerical order of the step is given in Table 1 for various values of m and n using equation (49).

It follows from Table 1 that in nearly every case, the transfer coefficients, and thus the anodic and cathodic slopes [equation (42)], should differ considerably for bivalent and trivalent metals in a stepwise mechanism with a single rate-limiting step. For the simple

*Cf. the discussion originated by Gileadi, in Ref. 8b.

Table 1
Values of the Ratio α_a/α_k^*

n	m			
	1	2	3	4
2	1/3	3	—	—
3	1/5	1	5	—
4	1/7	3/5	5/3	7

*$\beta_{k,m} = \beta_{a,m} = 0.5$ is assumed.

one-step processes the ratio α_a/α_k is usually close to unity.* Therefore, a considerable deviation of this ratio from unity is an indication of a stepwise mechanism and the absolute value of this ratio ($\alpha_a/\alpha_k > 1$ or $\alpha_a/\alpha_k < 1$) can indicate which step is the limiting one. For the example cited above ($\alpha_k = 0.5$ and $\alpha_a = 1.5$ with $n = 2$, Figure 1), $\alpha_a/\alpha_k = 3$ and hence $m = 2$, i.e., the rate-limiting step is the transfer of the second electron. In the particular case where $n = 3$ and the limiting step is the transfer of the second electron ($m = 2$), the above characteristic feature is not applicable since $\alpha_a/\alpha_k = 1$.[1]

In general, within the framework of the stepwise mechanism, the possibility should also be considered that "simultaneous" transfer of several (in the simplest case, two) electrons takes place in the limiting step.[1]† Thus, the rate equations and the relation between the transfer coefficients for a three-electron overall process $M \rightleftarrows M^{3+} + 3e$ with a fast first step may now be obtained:

$$M \underset{k_{k,1}}{\overset{k_{a,1}}{\rightleftarrows}} M^{3+} + e \tag{50}$$

*Values of transfer coefficients close to 0.5 (i.e., $\alpha_a/\alpha_k \doteqdot 1$), were obtained, for example, for the following processes: $Ti^{3+} \rightleftarrows Ti^{4+} + e$[41], $Eu^{2+} \rightleftarrows Eu^{3+} + e$[42], $V^{2+} \rightleftarrows V^{3+} + e$[43], $Cr^{2+} \rightleftarrows Cr^{3+} + e$ on mercury,[43a] and $Fe^{2+} \rightleftarrows Fe^{3+} + e$ on platinum.[43b] For these processes $n = 1$, i.e., the transfer coefficients coincide with the symmetry coefficients; hence the values of symmetry coefficients are also close to 0.5 (the energy barrier of the limiting step is symmetric).

†A stepwise process involving "simultaneous" transfer of two electrons in a rate-limiting step can apparently occur under conditions when the intermediates are extremely unstable and their formation requires a very high activation energy, so that such a simultaneous transfer of electrons is, energywise, more advantageous than the combination of two successive one-electron transfers.[26a]

with the limiting step

$$M^{+} \underset{k_{k,2}}{\overset{k_{a,2}}{\rightleftarrows}} M^{3+} + 2e \tag{51}$$

In excess of supporting electrolyte and at a constant μ, the rate of the step (51) is given by

$$i_{a,2} = k_{a,2}[M^{+}]\exp(2\beta_{a,2}F\phi/RT) \tag{52}$$

$$i_{k,2} = k_{k,2}[M^{3+}]\exp(-2\beta_{k,2}F\phi/RT) \tag{53}$$

where $\beta_{k,2} + \beta_{a,2} = 1$. After substituting [M] for [M^{+}] with the aid of the equilibrium condition for the first step, the partial rates of the overall process are (in the absence of ψ_1 effects)

$$\begin{aligned} i_a = \tfrac{3}{2}i_{a,2} &= \tfrac{3}{2}k_{a,2}K_{(1)}[M]\exp[(1 + 2\beta_{a,2})F\phi/RT] \\ &= \tfrac{3}{2}k_{a,2}K_{(1)}[M]\exp(\alpha_a F\phi/RT) \end{aligned} \tag{54}$$

$$\begin{aligned} i_k = \tfrac{3}{2}i_{k,2} &= \tfrac{3}{2}k_{k,2}[M^{3+}]\exp(-2\beta_{k,2}F\phi/RT) \\ &= \tfrac{3}{2}k_{k,2}[M^{3+}]\exp(-\alpha_k F\phi/RT) \end{aligned} \tag{55}$$

These equations yield

$$\alpha_k = 2\beta_{k,2} \tag{56}$$

$$\alpha_a = 1 + 2\beta_{a,2} \tag{57}$$

the sum of apparent transfer coefficients ($\alpha_a + \alpha_k = 3$) and their ratio

$$\alpha_a/\alpha_k = (1 + 2\beta_{a,2})/2\beta_{k,2} \tag{58}$$

Thus when $\beta_{k,2} = \beta_{a,2} = 0.5$, $\alpha_a/\alpha_k = 2$ and from the definitions of the Tafel slopes of the anodic and cathodic curves, $b_a = 2.3RT/2F$ and $n_k = 2.3RT/F$, i.e., 29 and 58 mV at 20°C, respectively.

The second possible variant of the mechanism under consideration is

$$M \underset{k_{k,1}}{\overset{k_{a,1}}{\rightleftarrows}} M^{++} + 2e \qquad \text{(the rate-limiting step)} \tag{59}$$

$$M^{++} \underset{k_{k,2}}{\overset{k_{a,2}}{\rightleftarrows}} M^{3+} + e \tag{60}$$

and, in an analogous manner, the following results are obtained:

$$\alpha_k = 1 + 2\beta_{k,1} \tag{61}$$

$$\alpha_a = 2\beta_{a,1} \tag{62}$$

$$\alpha_a/\alpha_k = 2\beta_{a,1}/(1 + 2\beta_{k,1}) \tag{63}$$

so that with $\beta_{a,1} = \beta_{k,1} = 0.5$, $\alpha_a/\alpha_k = 0.5$. As can be seen from Table 1, the α_a/α_k has entirely different values for various rate-limiting one-electron steps. Therefore, by establishing whether the ratio α_a/α_k is close to 2 or to 0.5, it is possible to determine whether the reaction proceeds according to the mechanism described by equations (50)–(51) or by equations (59)–(60).

This analysis shows that a comparison of the transfer coefficients calculated from independent measurements helps to establish the stepwise mechanism and makes it possible to determine the rate-limiting step.*

It is of considerable importance in the proof of a stepwise mechanism with a single limiting step to ascertain whether the assumption made at the beginning of Section III is correct, namely, that equilibrium is maintained in all the charge-transfer steps except the rate-limiting step. As will be shown in Section IV, the nonidentical dependence of the rates of different successive charge-transfer steps on the potential leads to the appearance of a break in the vicinity of the equilibrium potential (at comparable rates of such steps) on the curve of the partial current versus the potential. This break corresponds to the transition from one limiting step to the other (this will be referred to as the "break" criterion).[121a] Thus, the retention of linearity of this curve in accordance with equation (39a) in the entire region around the equilibrium potential serves as direct proof that the rate of one of the successive steps is significantly smaller than that of the other steps. These steps can hence be taken to be approximately at equilibrium.[36,37] Thus, experimental demonstration of the absence of a "break," together with the indication of a stepwise mechanism based on the relation between the transfer coefficients and determination of the reaction orders makes it possible fully to substantiate a stepwise mechanism having a single limiting step.

*The considerable difference between the independently measured transfer coefficients of a complex electrode process ($\alpha_a/\alpha_k = b_k/b_a \sim 4$) for the condition $\alpha_k + \alpha_a = 2$ [equation (28)] was first observed by Esin and Loshkarev[9] for the reaction $Sn^{2+} \rightleftarrows Sn^{4+} + 2e$. The comparison of the independently measured transfer coefficients for forward and reverse processes was used for establishing the stepwise mechanism and the limiting step of electrochemical reactions on amalgam electrodes.[36–38] Later, the same approach was used by Bockris and his colleagues,[44,45] Gerischer and Perez-Fernandez,[46] and Lovrecek and Markovac.[39,47]

Thus, for elucidating the mechanism of a complex metal deposition and dissolution process, the possibility of isolating one of the coupled processes (for example, the anodic process) and studying the dependence of its rate on the potential, the concentration of reacting species, and on other factors within a broad range of potentials, including the equilibrium potential, is of great significance. Such a possibility is especially important for electrode processes having relatively high rate constants, where well-known difficulties arise in steady-state measurements.

If, for example, the effect of the reverse process can be avoided by means of direct measurements of the partial rate of the forward process in the region of ϕ_e, e.g., by radioisotope studies, the Tafel region corresponding to equation (39a) or (39b) can be obtained at relatively low current densities, and thus the distorting effect of concentration polarization can be avoided. By plotting the partial current curve for a given (e.g., the anodic) process in the region of ϕ_e as well as under conditions where the reverse (cathodic) process is predominant, it is possible to verify experimentally the principle of the independence of the electrode partial reactions[30a] for process (7). This principle is used in deriving the steady-state equation (33) for the polarization curve. Finally, these measurements make it possible to compare the transfer coefficients of the direct and reverse processes (determined from the Tafel slopes at the same potential and, in particular, at the equilibrium potential) when the sum of these coefficients definitely conforms to equation (28). As noted above, the possibility of ascertaining whether these conditions are met precisely at a constant potential is of great importance for the elucidation of the mechanism of a complex electrode process.

For substantiation of a stepwise mechanism, it is important that i_0 for process (1) can be directly determined together with the usual polarization data. Thus, from the dependence of i_0 on the concentrations [equation (38)], the kinetic parameters for process (1) can be evaluated under conditions for which it is otherwise difficult to obtain linear Tafel regions in ordinary or partial polarization curves owing, for example, to concentration polarization. In addition (as will be shown in Section IV), by comparing the exchange current determined by an independent method with that determined from polarization measurements carried out simultaneously (for example, by extrapolating linear Tafel regions to ϕ_e), it is possible to formulate important criteria for stepwise mechanisms.

The problem of isolating and investigating the kinetics of one of the coupled processes over a wide range of potentials, and particularly in the region of ϕ_e, and of measuring i_0, can be solved by combining electrochemical and radiotracer measurements (see Section III.5).

4. Conditions for the Accumulation of Ionic Intermediates in Solution

The treatment presented in this section leads to important conclusions concerning conditions that can lead to increase in the concentration of intermediates during electrolysis. This is of great significance, providing proof of a stepwise mechanism. Thus, if the transfer of the last electron is the limiting step ($m = n$), it follows from the equilibrium condition of the remaining steps [equation (18)] that at constant concentration of the initial and final products (M and M^{n+}), the concentration of all intermediates should increase when the potential shifts to more positive values for the case of anodic polarization, and, correspondingly, decrease in the case of cathodic polarization. On the other hand, when the step $M \rightarrow M^+ + e$ ($m = 1$) is rate-limiting, it can be expected from the equilibrium condition (19) that there will be an accumulation of intermediates as the potential shifts to more negative values, i.e., under cathodic polarization conditions.[47a]

Such effects can be used for providing experimental proof of the formation of low-valence intermediates under non-steady-state conditions, when, for example, there is a rapid transition from one potential to another. In the case of a stepwise process, such a transition will result in the consumption of a certain amount of electricity passing at the electrode interface for changing the concentration of the intermediates up to the level of a new quasiequilibrium concentration corresponding to the new value ϕ. Therefore, under nonstationary conditions, the current passing through each of the successive steps will not be the same. For example, for anodic polarization in the case when the last step is rate-limiting ($m = n$), a certain amount of additional charge will be required for creating new higher concentrations of all intermediates. Thus, the current $i_{a,m}$ for this step will be less than i_a/n, and only after reaching the steady state will the condition $i_a = ni_{a,m}$ [equation (16)] be obeyed.

That is, the overall current i_a will be equally distributed among all single-electron steps.*

The accumulation of low-valence intermediate ions (or their compounds) adsorbed on the electrode surface has been detected and also interpreted within the framework of stepwise mechanisms in the study of the discharge on mercury electrodes of cyanides of bivalent mercury[46] and nickel,[50] as well as in the investigation of charge-transfer processes involving copper[51,52] and iron.[53] If low-valence intermediates are formed during anodic dissolution† and cathodic formation of amalgams and solid metals, and can exist not only in the adsorbed state but also in the bulk of the solution, then indicator electrodes can be used for their detection and identification. Such electrodes can be made from inert metals[54–60] or from the same metal as the intermediates[61,62]; rotating ring-disk electrodes can also be used.[63–67]

The effect of the accumulation of low-valence intermediates in a stepwise mechanism can lead to distortion of the polarization data obtained by various nonstationary methods.[48] Indeed, if during the polarization measurements under steady-state conditions, the current is required only for process (1), then, during nonstationary measurements, a certain time-dependent fraction of the overall current should be required for the transient accumulation of intermediates. Therefore, the overall current will not precisely indicate the rate of reaction (1).‡ At very low concentrations of intermediates, this effect will, of course, be negligible and would not complicate the results of nonstationary measurements. However, if the concentration of the intermediates is such that the current required for the accumulation is comparable to the overall current, then nonstationary measurements will lead to erroneous results. Consequently, polarization curves for a given system obtained by

*For experimental proof of this effect, measurements of the effective valence of dissolving metal ions under nonstationary conditions can be employed.[48,49]

†*Editors' note*: Recently, solvated electrons have been suggested as intermediates in the dissolution of amalgams, but this suggestion is not consistent[60a] with the commonly accepted standard electrode potential of e_{aq}^- or with other electrode kinetic conditions (see Chapter 2).

‡This effect may be formally manifested in an increase in the measured capacity as compared with the double-layer capacity. It is of particular importance when the intermediates are adsorbed and has received detailed theoretical and experimental examination in various papers, e.g., Ref. 67a, b, c for such cases.

stationary and nonstationary methods may not coincide, except after appropriate corrections.

5. Combined Use of Electrochemical and Radiotracer Measurements

As remarked in Section III.3, it is necessary to supplement ordinary polarization measurements by direct measurements of the partial current of one of the two coupled reactions of process (1) (e.g., the anodic) near the equilibrium potential ϕ_e and, in particular, at ϕ_e (exchange current). In the present section, methods are discussed for solving this problem and a procedure is described, based on the combined use of electrochemical and radiotracer measurements. In order to overcome the distorting effect of the reverse reaction on the measurements of the forward reaction rate, in other words, to extend the Tafel section of the polarization curves into the equilibrium potential region, the modified equation (34) can be used,

$$i = i_a - i_k = i_0 \exp(\alpha_a F\, \Delta\phi/RT)[1 - \exp(-nF\, \Delta\phi/RT)] \tag{64}$$

Equation (64) is obtained from (34) by employing (28). Since $i_a = i_0 \exp(\alpha_a F\, \Delta\phi/RT)$ [equation (29)], equation (64) leads to

$$i_a = i/[1 - \exp(-nF\, \Delta\phi/RT)] \tag{65}$$

The denominator of the r.h.s. of this equation is the correction coefficient by which the value of the external current i should be divided in order to obtain the anodic partial current. By using the expression for i_a in equation (65), equation (40) becomes*

$$\phi = -(RT/\alpha_a F) \ln i_0 + (RT/\alpha_a F)\{\ln i - \ln[1 - \exp(-nF\, \Delta\phi/RT)]\} \tag{66}$$

The dependence of $\Delta\phi$ as a function of the term in curly brackets in equation (66) is linear up to ϕ_e. From the slope of this line, α_a can be determined and the intercept at $\Delta\phi = 0$ gives i_0. In an analogous way, equation (41) gives the corrected relation for the cathodic partial current. The disadvantage of this method is that it is necessary to utilize condition (28) in deriving equation (66); i.e., in

*This relation (with additional correction terms compensating for the effect of two diffusion steps) was first used by Esin.[41]

calculating the partial anodic current curve, it is necessary to make certain assumptions concerning the nature of the partial cathodic curve, and in some cases, such a procedure is invalid. Furthermore, it follows from equation (65) that as ϕ_e is approached, small errors in measuring $\Delta\phi$ will lead to increasingly higher errors in the i_a values. Finally, this method cannot be used for obtaining curves of $\Delta\phi$ versus $\ln i_a$ and $\Delta\phi$ versus $\ln i_k$ under conditions when the reverse reaction is predominant.

A method of measuring directly the partial rate of one of the coupled reactions which is free from the above disadvantages is based on combining electrochemical stationary polarization measurements with radiotracer measurements of the transfer rate of labeled metal species across the phase boundary.[36,37,68–73] Thus, for example, by carrying out polarization measurements on amalgams labeled by the radioisotope of the same metal, it is possible, using data for the solution activity, to directly determine the partial anodic current not only at ϕ_e (exchange current) and for anodic polarization, but also for cathodic polarization, that is, under conditions when the corresponding reverse reaction is predominant. Polarization measurements on a large amalgam electrode under conditions of intensive stirring are used in conjunction with radiotracer measurements (in the same cell) of the rate of metal ion transfer in one of the coupled reactions over a wide range of potentials. When the labeled amalgam is used, the partial anodic current i_a (at a given potential) and the exchange current i_0 are determined by periodically or continuously taking samples of the solution and determining their activities.[73a]

A significant advantage of the radiotracer method of studying the kinetics of process (1) (both on amalgams and at solid metals) is that the transfer of the labeled species takes place across the phase boundary during the electrode process. Because of this transfer, the initial substance and reaction product are in different phases. An analogous study of the kinetics of an electrochemical redox process in which both the initial substance and the final reaction product are in solution requires the preliminary separation of the labeled reaction product from the labeled initial substance before the concentration of the former in the solution can be determined. Such measurements can also be complicated owing to the effect of homogeneous isotopic exchange between the initial substance and the reaction product.

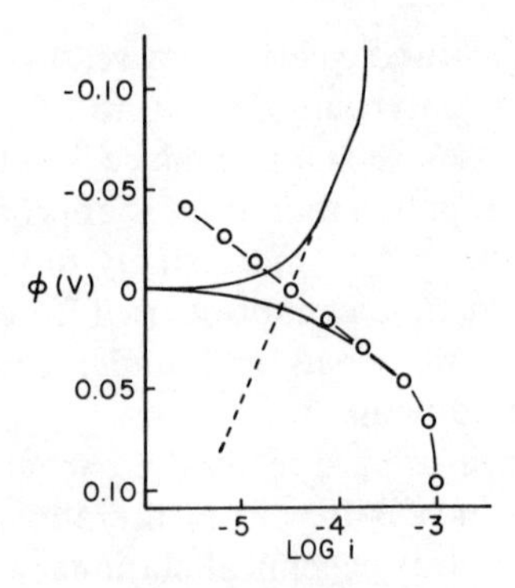

Figure 2. Typical results of polarization (lines) and radiotracer measurements (circles).

The distinguishing feature of the method is that radiotracer measurements are carried out during the electrolysis and not after the electrochemical measurements have been made.* Radiotracer measurements are carried out in a closed, all-glass system, and thus there is no exposed radioactive solution or amalgam and no need for making special samples for counting.

The advantages of the combined use of electrochemical and radiotracer measurements over the usual polarization method are evident from the diagram shown in Figure 2 for the case of the labeled amalgam. On this diagram, in accordance with equation (35) (that is, by assuming that both coupled reactions proceed independently) are shown the ordinary anodic and cathodic polarization curves; on the same diagram is shown the curve for the anodic partial current i_a (circles), which can be derived from the increase in the solution activity for various amalgam potentials. Thus, radiotracer measurements make it possible to obtain the plot of ϕ versus $\ln i_a$ directly in the region of ϕ_e and even for cases where the cathodic process predominates.†

A comparison of the plot of ϕ versus $\ln i_a$ with the ordinary anodic polarization curves shows that though the anodic limiting

*Joliot was the first to use this technique, in his work on the electrodeposition of metals from very dilute solutions.[74]

†Similar combined use of electrochemical and radiotracer measurements has also been applied to solid metals.[75–77]

current exceeds the exchange current by $1\frac{1}{2}$ orders of magnitude, the linear section on the ordinary anodic curve is practically absent. At the same time, the radiotracer curve (ϕ versus $\ln i_a$), unaffected by the reverse process, remains linear over a wide range of potentials. By using relation (35), $i = i_a - i_k$, it is possible to plot for the same range of potentials the curve of ϕ versus $\ln i_k$ (dashed line) and thus to obtain the complete polarization diagram for the amalgam. From the slopes of the ϕ versus $\ln i_a$ and ϕ versus $\ln i_k$ plots, the apparent transfer coefficients α_a and α_k can be determined and their sum at the constant potential obtained. A comparison of i_a and i_k at a given potential yields v for the electrode process.[78] The determination of the positions of the curves of ϕ versus $\ln i_a$ at different concentrations of metal in the amalgam and in the solution, and at different pH, etc., makes it possible to determine the corresponding anodic reaction orders and the dependence of i_a on various factors.

Valuable information can be obtained by comparing the radiotracer value of the exchange current with that obtained by polarization measurements. Measurement of the rates of direct and reverse reactions at equilibrium with the aid of labeled metal species is the most direct method of determining the exchange current. Other methods of determining the exchange current, such as those requiring stationary and nonstationary polarization, impedance, and Faradaic rectification measurements, involve extrapolations and assumptions concerning the applicability of various mathematical relations to the given electrode process. Therefore, the coincidence of the true exchange current obtained by the radiotracer technique with i_0 derived from simultaneous polarization measurements can serve as a convincing proof that such relations correctly describe the electrode process.* It should be noted that in the general case, such agreement may not be observed. Thus, far away from ϕ_e, the mechanism and, in particular, the nature of the limiting step may, in general, change. Evidence for this arises when radiotracer and extrapolated exchange current values do not agree. As will be shown in Section IV, the absence of agreement can be considered a criterion for the stepwise mechanism.

*One of the first attempts to compare the isotopic exchange rate with the exchange current of the corresponding electrode process determined by the extrapolation of the polarization curve was made by Levina.[79]

Finally, the combined use of electrochemical and radiotracer measurements is also convenient for determining the current efficiency of anodic and cathodic processes and for calculating the effective valence.[48,60,61,80,81] This is of significance for proving a stepwise mechanism and for plotting the partial curve of the given process when side processes are present or even predominant (for example, discharge or ionization of metals with simultaneous evolution of hydrogen or reduction of oxygen); also, low dissolution rates of amalgams and metals can be determined.

Calculation of the rate of one of the coupled processes (for example, the anodic i_a) from radiotracer data obtained during the passage of an external current can be carried out with the aid of an approximate form of the kinetic equation describing the transfer of the labeled species across the phase boundary in the electrochemical process not only at ϕ_e but at $\Delta\phi \gg 0$.[37,68,70] This equation is applicable under conditions when, owing to the low specific activity of metal ions in solution, the rate of the reverse transfer of labeled species into the amalgam is very small compared with the rate of transfer of these species from the amalgam into solution.

Under steady-state conditions, radiotracer measurements make it possible to determine only the rate of the *slowest* of the successive steps of the overall process. For processes having a high exchange current, even under conditions of vigorous stirring, the concentration of the labeled species in the amalgam at the electrode surface can become lower, as a result of isotopic exchange, than in the bulk of the labeled amalgam. In this case, the experimental data will be complicated by diffusion of the labeled species. Therefore, the system for investigation and the solution composition should be chosen in such a way that the electrode process on the amalgam is irreversible to the maximum degree. This will make it easier to overcome the effect of diffusion or to make a quantitative estimate of this effect, and will make it possible to carry out polarization and radiotracer measurements over a wide range of potentials.

At comparable rates of charge transfer and diffusion, an estimate of i_a based on the experimental apparent anodic partial rate (i_a') of the process can be made with the aid of the equation

$$\frac{1}{i_a'} = \frac{1}{i_a} + \frac{1}{i_d^a} + \frac{(i_a - i)}{i_a} \frac{1}{i_d^k} \tag{67}$$

where i_d^a and i_d^k are the anodic and cathodic diffusion-limiting currents, respectively. At equilibrium ($i = 0, i_a = i_0$), this equation becomes the usual relation

$$\frac{1}{i_0'} = \frac{1}{i_0} + \frac{1}{i_d^a} + \frac{1}{i_d^k} \tag{68}$$

by which the true i_0 can be calculated from the apparent exchange current i_0' derived from radiotracer measurements.[27,82]

The features characterizing the absence of diffusion distortion and agreement between radiotracer exchange rate and true exchange current are:

(a) The exchange rate is independent of the stirring velocity.
(b) There is a fractional dependence of the exchange rate on $[M]_{Hg}$ and $[M^{n+}]$ as defined by equation (38).*
(c) The measured values of the exchange rate are significantly lower than the values for the limiting current [see equation (68)].
(d) Abrupt change in the exchange rate following the introduction to the solution of inhibiting or accelerating additives. (Usually, such additives have no effect on the rate of the diffusion step).

6. Experimental Data for Zinc and Indium Amalgams

Use of electrochemical measurements in conjunction with radiotracer studies was first applied to zinc amalgams,[36,37,70,83] where the exchange rate is relatively high. Thus, to overcome the complicating effect of the diffusion steps on the radiotracer and polarization measurements, in the early part of the work with zinc amalgams,[36,37] i_0 was lowered by addition of tetrabutylammonium sulfate. Figure 3 shows the plot of the radiotracer data for i_a versus the amalgam potential, together with the usual polarization curves obtained in the same experiment.[36] The plot of ϕ versus $\ln i_a$ is linear in the potential range of 100 mV including the region of ϕ_e. The coupled cathodic process thus has no effect on i_a, as expected; similarly, the anodic partial process does not affect i_k.

*It can readily be seen that when diffusion is the limiting step, e.g., in the amalgam, the isotopic exchange rate should be proportional to the concentration of the amalgam and should be independent of the concentration of the solution. In this case, $i_d^a \ll i_0, i_d^k$, and equation (68) gives $i_0' = i_d^a \propto [M]$, since the anodic limiting current is proportional to the amalgam concentration.[28,82]

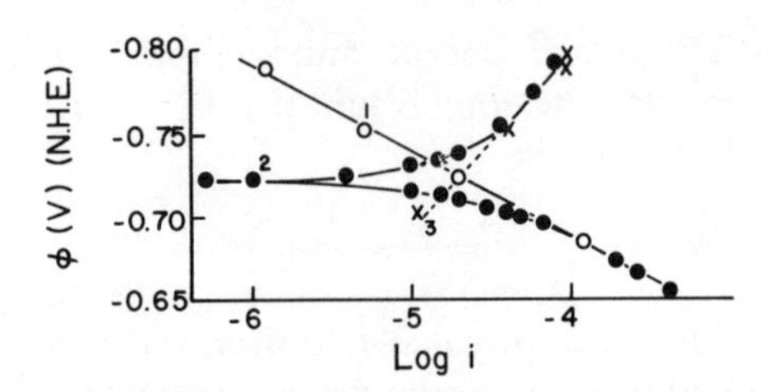

Figure 3. Polarization curves for a radio-labeled zinc amalgam (0.4 M) in 0.1 M $ZnSO_4$ + 0.01 N H_2SO_4: (1) anodic partial current curve; (2) ordinary polarization curves; (3) cathodic partial current curve.

The linear cathodic partial current curve (ϕ versus $\ln i_k$) shown in Figure 3 (broken line) was calculated with the aid of equation (34) ($i_k = i_a - i$) from the polarization curve and the ϕ versus $\ln i_a$ relation.

The slope b_a of the ϕ versus $\log i_a$ plot is independent of the zinc concentrations in the amalgam and in the solution; $b_a = 0.051$ V and hence corresponds to $\alpha_a = 1.16$ For the ϕ versus $\log i_k$ plot, the corresponding values are $b_k = 0.080$ V and $\alpha_k = 0.74$. Thus, $\alpha_a + \alpha_k$ near ϕ_e is close to 2 and their ratio is $\alpha_a/\alpha_k \sim 1.6$. On zinc amalgam, in the absence of surface-active compounds, the radiotracer method makes it possible to measure only the exchange current.[70,83] From the plot of i_0 versus amalgam concentration and the concentration of the solution, using equation (45), the transfer coefficients were found to be $\alpha_k = 0.5$ and $\alpha_a = 1.4$, i.e., $\alpha_a/\alpha_k = 2.8$. This is in good agreement with the conclusions reached from impedance and polarization data.[10,73a,83a–86b]

With increasing amalgam concentration, the anodic curves (ϕ versus $\ln i_a$) shift in the negative direction.[37] The plot of i_a values derived from these curves, as a function of the amalgam concentration c_a at constant potential, shows that the i_a is directly proportional to c_a. Change in concentration of $ZnSO_4$ from 0.02 to 0.3 N at constant c_a has no effect on the position of the anodic curve, indicating that the rate of the anodic process is independent of the concentration of Zn^{++} in the solution.

Analogous results[38,87] were obtained for indium amalgams in excess $NaClO_4$ (Figure 4). The plot of ϕ versus $\log i_a$ (circles) measured by the radiotracer technique is linear both near ϕ_e and at considerable cathodic $\Delta\phi$. Falling on the same line are the values of i_0 obtained at the same c_a in solutions having different concentrations c_s of In^{3+} ions. This result indicates that the partial rate of the anodic process measured away from ϕ_e is consistent with that measured at equilibrium. It also indicates that the coefficients α_a determined from the slopes of ϕ versus log i_a for constant c_a, c_s, and ϕ_e versus ln i_0 for constant c_a are the same. This agreement is one of the indications that the electrode process is limited by the charge-transfer step (see p. 329) and at the same time shows that i_a is independent of the In^{3+} ion concentration.

The slope of the ϕ versus log i_a plot is independent of c_a and c_s; its mean value is $b_a = 0.026(5)$ V and corresponds to $\alpha_a = 2.20$; α_k was determined from the plot of i_0 versus ϕ_e at different c_a and constant c_s (points in Figure 4). As can be seen in Figure 4, the plot of i_k derived from the exchange current and extrapolated into the region of cathodic potentials diverges significantly from the usual polarization curve at $\phi > \phi_e$. A detailed investigation of the kinetics of indium discharge under these conditions showed that when $\phi < \phi_e$, the cathodic process becomes limited not by the charge-transfer step, but by a preceding chemical reaction leading to the formation of hydrolyzed and partially dehydrated indium ions from simple $In(H_2O)_6^{3+}$ ions.[69,88,89] A detailed analysis[90]

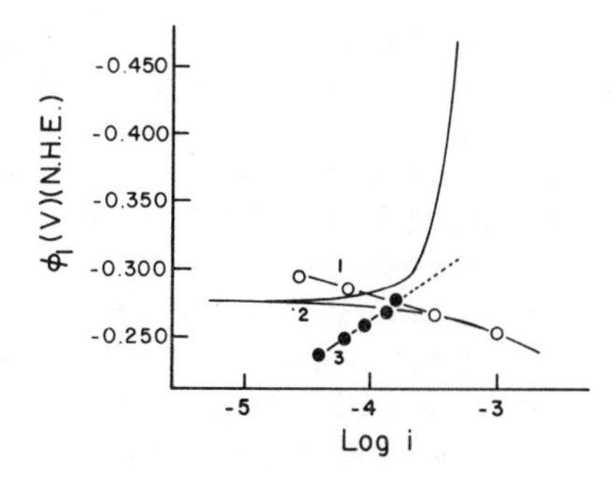

Figure 4. (1) Anodic partial current curve (open circles), (2) polarization curves for $c_a = 0.3\,M$ and $c_s = 0.01\,M$, and (3) the cathodic partial current curve based on the exchange currents at different c_a (darkened circles) for indium amalgam.

showed that the i_0 data and the ϕ versus log i_a plot were not invalidated by the presence of this chemical reaction. The plot of ϕ_e versus log i_0 is linear and has a slope $b_k = 0.065$ V and, according to equation (46), $\alpha_k = 0.91$, so that $\alpha_a/\alpha_k = 2.4$. As in the case of zinc, for the indium amalgam, the sum of the transfer coefficients is close to n ($\alpha_a + \alpha_k = 3.11$).

The anodic reaction order with respect to $c_{In^{3+}}$ ions is 0 but the order of the cathodic process with respect to In^{3+} ions is 1.[87] From the ln i_0 versus $1/T$ plot, the apparent activation energy of the electrode process was found to be 7.4 and 12.8 kcal mole^{-1} for zinc[70] and indium[87] amalgams, respectively.

As can be seen from Figures 3 and 4, at sufficiently anodic $\Delta\phi$, the points for i_a derived from radiotracer measurements coincide with the anodic polarization curve obtained from current measurements. Since it is assumed in the estimation of i_a values based on radiotracer measurements that the final products of the anodic dissolution of zinc and indium amalgams are stable cations of the highest valence (Zn^{2+} and In^{3+}, respectively), the agreement between the results of independent measurements shows that the anodic current efficiency is 100%. Thus, under the given conditions, electrochemical side reactions are insignificant and rates of parallel chemical reactions involving low-valence intermediates are negligible; hence, the subsequent oxidation of these intermediates proceeds mostly on the electrode.*

The shape of polarization curves (Figures 3 and 4) shows that the exchange current is usually significantly lower than the anodic and cathodic limiting currents. Furthermore, the dependence of radiotracer exchange current for zinc[70,83] and indium amalgams[38,87] on concentration is described by equation (38). Thus, diffusion complications are insignificant.

Radiotracer measurements showed that the partial rate of one of the coupled reactions for zinc and indium amalgams obeyed Tafel's law both at small as well as large $\Delta\phi$ when the reverse reaction is predominant, thus indicating directly the independence of the partial electrode processes in and the applicability of equation

*Generally, considerable deviation of the anodic current efficiency from 100% provides a basis for proof of a stepwise mechanism. This deviation may indicate the formation of low-valence intermediates; however, the interpretation of the polarization data under such conditions then becomes complicated.[48,60,64]

(34). The kinetic nature of the electrochemical equilibria is also directly demonstrated.

The coincidence of i_0 obtained from radiotracer measurements with i_0 obtained by extrapolation of the anodic polarization curve ($i^a_{0,\text{ext}}$) (Figures 3 and 4) indicates that as $\Delta\phi \to 0$ the mechanism of the electrode process remains the same. In accordance with equation (28), the sum of the transfer coefficients for zinc and indium amalgams is in agreement with the total number of electrons n transferred in the electrode process not only at constant potential, but also for values of α_a and α_k estimated from anodic and cathodic Tafel regions (except for i_k for In). Again, the mechanism of the overall process is therefore the same through the entire range of potentials employed.*

For zinc and indium amalgams, the ratio α_a/α_k considerably exceeds 1. As follows from Table 1, this may be an indication of a stepwise mechanism of the electrode process with the limiting transfer of the last electron. This conclusion accords with the results of radiotracer measurements. The ϕ–$\log i_a$ plot remains linear over a wide range of potentials including ϕ_e, and thus the remaining charge-transfer steps are practically at equilibrium at these potentials (see p. 332). Reversibility of the reaction $In \rightleftarrows In^+ + e$ is indicated by accumulation of In^+ ions upon electrolysis at the indium amalgam as well as at the indium metal (see Section III.4). This was established with the aid of indium and mercury indicator electrodes[56,57,60,61] and by means of the ring-disk technique.[64,65]

7. Experimental Data for Other Systems

As the literature shows, differences between anodic α_a and cathodic α_k transfer coefficients is often observed for multielectron processes of type (1). Usually, $\alpha_a > \alpha_k$.[91] For example, for cadmium amalgam, α_a lies in the range 1.36–1.78,[86b,92–96] while α_k is 0.24–0.28.[14,97]† For

*As will be shown in the following section, divergence between i_0 values obtained by two different methods and a deviation of $\alpha_a + \alpha_k$ from n, i.e., nonobservance of condition (28), are the criteria for comparable rates of *successive* charge-transfer steps, in other words, the absence of a single rate-limiting step.

†All values of transfer coefficients were recalculated in accordance with equation (28). The simultaneous determination of both transfer coefficients for a given system has, unfortunately, seldom been carried out. Thus, the above-mentioned conclusion concerning the value of the ratio α_a/α_k usually follows either from a comparison of α_k and α_a values determined for a given system by different authors, or by taking the difference of one of the coefficients from the sum $\alpha_a + \alpha_k = n$.

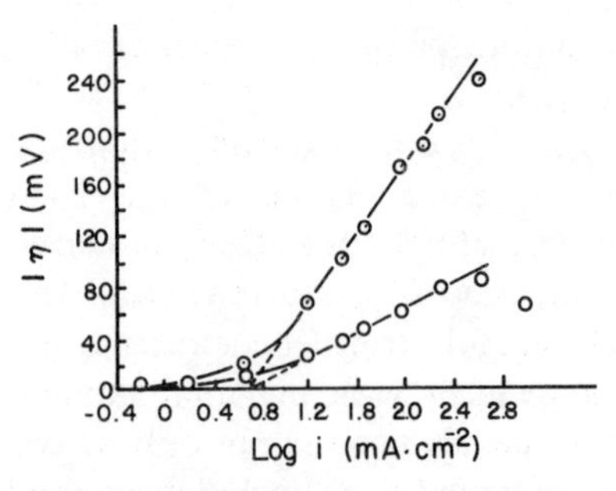

Figure 5. Anodic (circles) and cathodic (dotted circles) polarization curves for copper in 0.15 *N* $CuSO_4$ (Mattson and Bockris[44]).

copper amalgam, $\alpha_a = 1.60$ and $\alpha_k = 0.41$.[98] For amalgams of tin, lead, and gallium, the values of α_k are respectively 0.4–0.6,[99] 0.5,[14] and 0.23.[100]

As mentioned earlier, for solid electrodes, plots of anodic and cathodic polarization curves which are not complicated by preceding and subsequent chemical and diffusion steps, or by side reactions, are quite difficult to obtain. Nevertheless, for a number of solid metals, $\alpha_a > \alpha_k$; e.g., Mattson and Bockris[44] showed that in the case of copper in an acidic solution of $CuSO_4$, b_a obtained by the galvanostatic transient method was approximately three times smaller than b_k (Figure 5); extrapolation of each of these curves to ϕ_e gives the same value for i_0.[44] The transfer coefficients derived from slopes of these curves are $\alpha_k = 0.57$ and $\alpha_a = 1.42$ (i.e., $\alpha_a/\alpha_k \simeq 2.5$ and $\alpha_a + \alpha_k \simeq 2$).* Such a large difference between α_k and α_a values was explained on the basis of a stepwise mechanism with the rate-limiting step being (69),

$$Cu^{++} + e \rightarrow Cu^{+} \tag{69}$$

$$Cu^{+} + e \rightarrow Cu_{ads} \tag{70}$$

$$Cu_{ads} \rightleftarrows Cu \tag{71}$$

*The low b_a for dissolution of copper corresponding to $\alpha_a = 1.44$ was first established by Esin and Antropov.[101] The higher slope of the steady-state cathodic curve corresponding to $\alpha_k \simeq 0.5$ was also observed in earlier work.[102–104] However, none of these studies contains a comparison of the slopes of the anodic and cathodic curves.

Audubert[75] obtained a similar value for α_a (= 1.40) from the slope of the i_a curve at ϕ_e and for cathodic polarization using the radiotracer technique. The fact that this slope is identical with the slope of the usual anodic polarization curve indicates the absence of a break in the plot of i_a versus ϕ in the region of ϕ_e. This proves experimentally that the assumption of Mattson and Bockris concerning the reversibility of step (70) is correct, as is also supported by the potential–time behavior of the copper electrode after switching on anodic and cathodic galvanostatic transients.[51]* This conclusion is also confirmed by the potentiostatic current–time transients following application of an anodic pulse[52] and by measurements of $[Cu^+]$[47a,63,65,66,67] which indicate, in particular, an increase in the concentration of Cu^+ ions upon anodic polarization. A high anodic transfer coefficient is observed not only for cadmium amalgam (see p. 345) but also for solid cadmium[104a] ($\alpha_a = 1.54$).

From the slope of the anodic polarization curve for solid indium and the partial anodic current curve obtained from the exchange currents derived from radiotracer data at different concentrations of In^{3+} in perchlorate solution, the transfer coefficient was found[105] to be $\alpha_a = 1.85$, a value which satisfactorily accords with that given above ($\alpha_a = 2.20$) for indium amalgam. The calculated ratio $\alpha_a/\alpha_k = 1.6$, taking into account the condition $\alpha_a + \alpha_k = 3$, corresponds either to the rate-limiting step $In^+ \rightarrow In^{++} + e$ (Table 1) or the rate-limiting step $In^+ \rightarrow In^{3+} + 2e$.[1]

Low anodic slopes, corresponding to a ratio $\alpha_a/\alpha_k > 1$, have also been observed for all metals of the iron group (see, for example, Refs. 1 and 106).

8. Causes of Relative Slowness of the Last Charge-Transfer Step

Comparisons of α_a and α_k for amalgam electrode reactions may be interpreted in terms of a stepwise mechanism with rate-limiting transfer of the last electron. It is interesting to consider the causes of the small rate of this step. An analysis of kinetic data for reactions of type (1) clearly shows that the value of the exchange current tends

*The slope of these plots was much lower than that corresponding to the charging of the double layer. In this case, the dependence of the excess quantity of electricity on the potential can be quantitatively interpreted by assuming that this excess is required for bringing about a change in the concentration of Cu^+ ions at the electrode in accordance with the Nernst equation for step (70).

to decrease with increasing valence of the cations, this decrease being accompanied by a considerable increase in the activation energy of the electrode process and in the free energy of hydration of the corresponding cations.[91,107] This correlation was first noted by Pleskov and Miller.[108]

The hydration energy ΔH_s of ions of a given element in different oxidation states ($Cu^+ \rightarrow Cu^{2+}$, $Fe^{2+} \rightarrow Fe^{3+}$, $Tl^+ \rightarrow Tl^{3+}$, $Ce^{3+} \rightarrow Ce^{4+}$) increases as the charge of the cation increases. Since ΔH_s is approximately proportional to the square of the ionic charge, the change in ΔH_s for a given ion upon an increase of charge by one is the greater, the greater the charge of the initial cation.[109] Though the activation energy is only a small fraction of ΔH_s for the corresponding cation, it is known that the activation energy is largely due to the change in the structure of the hydrate shell following the change in the charge of the reacting particle.[26b,110,111]

Thus, the fact that in the stepwise process $M \rightleftarrows M^{n+} + ne$, the limiting stage is the final step $M^{(n-1)+} \rightleftarrows M^{n+} + e$, is due apparently to the very large difference in ΔH_s for cations taking part in the last reaction. For earlier steps, this difference should be considerably smaller. Since univalent ions are more weakly hydrated, the step $M \rightleftarrows M^+ + e$ usually requires only a relatively low activation energy.[26b] Thus, in comparison with subsequent charge-transfer steps, this step is usually practically at equilibrium.

The assumption that the rate of process (7) is limited by the last electron transfer accords with measurements of i_0 on redox reactions involving highly charged metal cations at the mercury electrode; for example, in the reactions $V^{2+} \rightleftarrows V^{3+} + e$,[43] $Eu^{2+} \rightleftarrows Eu^{3+} + e$,[42] and $Cr^{2+} \rightleftarrows Cr^{3+} + e$,[43a] the rate constants are much lower than for reactions of the type $M \rightleftarrows M^+ + e$. This observation makes it possible to explain the far-reaching parallelism between the kinetics of a given process on an amalgam and that on the corresponding solid metal, e.g., as noted above for Cu, Cd, and In, where $\alpha_a > \alpha_k$. Since the rate-limiting step is in fact an ionic redox reaction for which the electrode serves only as a source or sink of electrons, the kinetics of this step and the corresponding kinetic parameters for the solid and the amalgam electrode can be very close to one another.

With respect, however, to the step $M \rightleftarrows M^+ + e$, its kinetic parameters can be quite different for the solid and amalgam surfaces owing to the dissimilar interaction conditions under which the discharge takes place. However, owing to its high relative rate, the step is often practically reversible and hence should have no effect on the kinetics of the overall process. This assumption is supported by numerous experimental data for very high exchange currents between M^+ cations and the corresponding amalgams[17,112–14] and between M^+ cations and solid metals.[77,115]

That the redox step plays the determining role in the kinetics of processes on amalgams and corresponding solid metals is also confirmed by the fact that the effect of acidity on the cathodic reduction of In^{3+} and that on the ϕ_e are almost identical for both electrodes.[73,116]

The kinetic analysis presented here for the case of a stepwise mechanism with a single rate-limiting step makes it possible to differentiate this case from that involving several successive steps having comparable rate constants (see Section IV) and to establish the limiting step. However, proof of a stepwise mechanism when it is based on a comparison of α_a and α_k on the plots of partial anodic curves and evaluation of reaction orders is not always unequivocal. The same experimental results can also be consistent with simultaneous transfer of all electrons in the limiting step, but this assumption usually requires the assignment of unreasonable values for the symmetry factors.

As will be shown in the following section, unambiguous criteria can be established in the case where successive charge-transfer steps have comparable rates, as well as in the case of more complex mechanisms.

IV. PROCESSES WITH COMPARABLE RATE CONSTANTS FOR SUCCESSIVE STEPS

1. Rate Equations for the Process with One-Electron Steps

When the rate constants of successive charge-transfer steps are comparable, the passage of current disturbs the equilibrium not only of the slowest step, but also of the remaining steps. In this case, rate equations for the multistep process (7) have a complex form because they include the kinetic parameters of all steps. Thus, it is more convenient to analyze a simpler process such as

$$M \rightleftarrows M^{3+} + 3e \tag{72}$$

consisting of three successive one-electron steps

$$M \rightleftarrows M^{+} + e \tag{73}$$

$$M^{+} \rightleftarrows M^{++} + e \tag{74}$$

$$M^{++} \rightleftarrows M^{3+} + e \tag{75}$$

The derivation of the rate equation follows the method of Vetter,[2,3] so that detailed descriptions of the steps involved[117] will not be given. In deriving the equations, the assumptions stated at the beginning of the previous section will be adopted. Double-layer conditions will be assumed invariant. For anodic polarization,

$$i_1 = i_{a,1} - i_{k,1} = k_{a,1}[M] \exp(\beta_{a,1} F\phi/RT) - k_{k,1}[M^{+}] \times \exp(-\beta_{k,1} F\phi/RT) \tag{76}$$

$$i_2 = i_{a,2} - i_{k,2} = k_{a,2}[M^{+}] \exp(\beta_{a,2} F\phi/RT) - k_{k,2}[M^{++}] \times \exp(-\beta_{k,2} F\phi/RT) \tag{77}$$

$$i_3 = i_{a,3} - i_{k,3} = k_{a,3}[M^{++}] \exp(\beta_{a,3} F\phi/RT) - k_{k,3}[M^{3+}] \times \exp(-\beta_{k,3} F\phi/RT) \tag{78}$$

In this case, the concentrations of the intermediate particles in solution, M^{+} and M^{2+} (unlike M and M^{3+}), vary with the passage of the current.

At equilibrium, $i_1 = i_2 = i_3 = 0$, and from equations (76)–(78), the expressions for the exchange currents of corresponding steps ($i_{0,1}, i_{0,2}, i_{0,3}$) are obtained and contain the equilibrium concentrations of the intermediates M^{+} and M^{++}, which obey the Nernst equation for the individual steps.

The overall current i equals the sum $i_1 + i_2 + i_3$. Since each of the three steps is a one-electron step, and in the steady state $i_1 = i_2 = i_3$, the overall current is equal to three times the current for a given step; for example, $i = 3i_1$. By using these conditions, the concentrations of intermediates can be eliminated from equations (76)–(78), and the constants $k_{a,1}$, $k_{a,2}$, etc. can be expressed[117] through $i_{0,1}$, $i_{0,2}$, and $i_{0,3}$ with the aid of the equilibrium conditions for steps (73)–(75). The final expression for the steady-state polarization curve is

$$i = 3\left\{i_{0,1}i_{0,2}i_{0,3}\left[\exp\frac{(\beta_{a,1} + \beta_{a,2} + \beta_{a,3})F\,\Delta\phi}{RT} - \exp - \frac{(\beta_{k,1} + \beta_{k,2} + \beta_{k,3})F\,\Delta\phi}{RT}\right]\right\}$$
$$\times\left\{i_{0,2}i_{0,3}\exp\left[\frac{(\beta_{a,2} + \beta_{a,2})F\,\Delta\phi}{RT}\right] + i_{0,1}i_{0,3}\exp\left[\frac{(\beta_{a,3} - \beta_{k,1})F\,\Delta\phi}{RT}\right] + i_{0,1}i_{0,2}\exp\left[\frac{(\beta_{k,1} + \beta_{k,2})F\,\Delta\phi}{RT}\right]\right\}^{-1} \tag{79}$$

where $\Delta\phi > 0$ for anodic polarization. This expression can also be derived directly from the general equation of the Temkin theory[118] of steady-state reactions. From equation (79),

$$i = 3\left[\exp\frac{(\beta_{a,1} + \beta_{a,2} + \beta_{a,3})F\,\Delta\phi}{RT} - \exp - \frac{(\beta_{k,1} + \beta_{k,2} + \beta_{k,3})F\,\Delta\phi}{RT}\right]$$
$$\times\left\{\frac{1}{i_{0,1}}\exp\left[\frac{(\beta_{a,2} + \beta_{a,3})F\,\Delta\phi}{RT}\right] + \frac{1}{i_{0,2}}\exp\left[\frac{(\beta_{a,3} - \beta_{k,1})F\,\Delta\phi}{RT}\right] + \frac{1}{i_{0,3}}\exp\left[\frac{(\beta_{k,1} + \beta_{k,2})F\,\Delta\phi}{RT}\right]\right\}^{-1} \tag{80}$$

or, more conveniently, for anodic polarization

$$i = 3i_{0,1}\left(\exp\frac{\beta_{a,1}F\,\Delta\phi}{RT}\right)\left[1 - \exp\left(-\frac{3F\,\Delta\phi}{RT}\right)\right]$$

$$\times\left\{1 + \frac{i_{0,1}}{i_{0,2}}\exp\left[-\frac{(\beta_{k,1} + \beta_{a,2})F\,\Delta\phi}{RT}\right]\right.$$

$$\left. + \frac{i_{0,1}}{i_{0,3}}\exp\left[-\frac{(1 + \beta_{k,1} + \beta_{a,3})F\,\Delta\phi}{RT}\right]\right\}^{-1} \tag{81}$$

and for cathodic polarization

$$i = -3i_{0,3}\left(\exp -\frac{\beta_{k,3}F\,\Delta\phi}{RT}\right)\left[1 - \exp\left(\frac{3F\,\Delta\phi}{RT}\right)\right]$$

$$\times\left\{1 + \frac{i_{0,3}}{i_{0,1}}\exp\left[\frac{(1 + \beta_{k,2} + \beta_{a,3})F\,\Delta\phi}{RT}\right]\right.$$

$$\left. + \frac{i_{0,3}}{i_{0,2}}\exp\left[\frac{(\beta_{k,2} + \beta_{a,3})F\,\Delta\phi}{RT}\right]\right\}^{-1} \tag{82}$$

For sufficiently high $\Delta\phi$, equations (81) and (82) become

$$i = i_a = 3i_{0,1}\exp(\beta_{a,1}F\,\Delta\phi/RT) \tag{83}$$

$$i = i_k = -3i_{0,3}\exp(-\beta_{k,3}F\,\Delta\phi/RT) \tag{84}$$

which are Tafel relations, the extrapolation of which to ϕ_e give *different* values for the exchange current ($3i_{0,1}$ and $3i_{0,3}$) corresponding to the different rate-controlling steps of the overall process operative at various potentials,[2,3] i.e., at sufficiently high anodic polarization, the first electron transfer [equation (73)], and at high cathodic polarization, transfer of the last electron [equation (75)].[119] The above criterion of different extrapolated anodic and cathodic i_0 values (*the Vetter criterion*) is indefinite in the case where the values of the exchange currents for steps (73) and (75) are equal or close to one another. In this case, extrapolations from anodic and cathodic regions give practically identical values for the exchange current. Similar agreement is obtained for a stepwise process with a single slow step and for a simple, single-step process.

An unequivocal criterion of the consecutive process, the steps of which have comparable rate constants, can be obtained from a comparison of the Tafel slopes corresponding to equations (83) and (84). In the case where a single step is rate-controlling at all potentials, as in the case of "simultaneous" transfer of all electrons, the sum of the apparent transfer coefficients should equal n [equation (28)]; that is, in the present case,

$$\alpha_a + \alpha_k = 3 \tag{85}$$

As follows from equations (83) and (84) in the case of steps having comparable rate constants, the sum of the apparent transfer coefficients determined from b_a and b_k at potentials displaced from the equilibrium value is given by

$$\alpha_a + \alpha_k = \beta_{k,3} + \beta_{a,1} < 2 \tag{86}$$

since $\beta_{k,3} < 1$ and $\beta_{a,1} < 1$. Thus, for example, if the values of the symmetry coefficients are close to 0.5, equation (86) gives the condition $\alpha_a + \alpha_k \sim 1$, which is quite different from that in equation (85).

Observance of condition (86) is an unequivocal criterion of a consecutive process, the steps of which have comparable rate constants. This condition, which may be called *the criterion of the transfer coefficient sum*, is applicable, in particular, to the case where the i_0 values obtained by extrapolation from the anodic and cathodic regions are identical. The criterion (86), with definitions of b_a and b_k in terms of transfer coefficients, can be written in the following form:

$$\alpha_a + \alpha_k = (2.3RT/b_k F) + (2.3RT/b_a F) < 2 \tag{87}$$

from which is obtained*

$$(1/b_a) + (1/b_k) < 2F/2.3RT \tag{88}$$

Also, in the case of a single slow step, equation (43) gives

$$(1/b_a) + (1/b_k) = 3F/2.3RT \tag{89}$$

Equations (88) and (89) make it possible to establish the existence of a

*It can be readily shown that the conditions (86) and (88) should be observed for all values of n and, in particular, for $n = 2$.

process, the steps in which have comparable rate constants, directly from b_a and b_k determined at $|\Delta\phi| \gg 0$.

An important case where the exchange current, e.g., $i_{0,3}$, of one of the steps is much lower than those of the remaining steps will now be considered. At a certain $\Delta\phi$, equations (81) and (82) lead, respectively, to

$$i = 3i_{0,3} \exp[(2 + \beta_{a,3})F \, \Delta\phi/RT] \tag{90}$$

$$i = -3i_{0,3} \exp[-\beta_{k,3}F \, \Delta\phi/RT] \tag{91}$$

In this case, extrapolation of both Tafel regions to ϕ_e gives the same value for i_0 corresponding to the step (75), and the transfer coefficients $\alpha_a = 2 + \beta_{a,3}$ and $\alpha_k = \beta_{k,3}$ can be determined from the slopes of these regions. Thus, in this case, as in the case of a single slow step, the sum $\alpha_a + \alpha_k = 3$. That is, the criterion of the transfer coefficient sum does not indicate that the process has a stepwise nature.*

However, unlike the case of the single slow step, it follows from equation (81) that for sufficiently large $\Delta\phi$, there should appear on the anodic polarization curve a transition from a linear region with $b_a = 2.3RT/(2 + \beta_{a,3})F$ to a region with a higher slope $2.3RT/\beta_{a,1}F$, which corresponds to the step (73). The appearance of this transition region is due to the different dependences of the rates of different successive steps on ϕ.[98] With a shift of potential in the positive direction, the transfer rate for the first electron increases only owing to a decrease in the activation energy [equation (76)], whereas the transfer rate for the third electron increases not only owing to the same decrease in the activation energy, but also to an increase in the concentration of M^{2+} ions, a factor which is included in the rate equation for this step [equation (78)]. Thus, for a sufficiently large, positive $\Delta\phi$, a transition takes place from kinetically limiting transfer of the last electron to a limiting transfer of the first electron.

It can be readily shown that in some cases, for example, in the case where $i_{0,3} < i_{0,2} < i_{0,1}$ and at $\Delta\phi$ values corresponding to the

*The example cited (for $n = 2$ and $i_{0,2} < i_{0,1}$) is schematically illustrated in Fig. 1, where the polarization curve together with the i_a and i_k curves for both steps are shown.

inequalities

$$\frac{i_{0,1}}{i_{0,2}} \exp\left[-\frac{(\beta_{k,1} + \beta_{a,2})F\,\Delta\phi}{RT} \right] \gg 1$$

$$\frac{i_{0,1}}{i_{0,2}} \exp\left[-\frac{(\beta_{k,1} + \beta_{a,2})F\,\Delta\phi}{RT} \right] \gg \frac{i_{0,1}}{i_{0,3}} \times \exp\left[-\frac{(1 + \beta_{k,1} + \beta_{a,3})F\,\Delta\phi}{RT} \right] \quad (92)$$

there can appear on the anodic curve, between the above-mentioned two linear sections, one more linear section corresponding to step (74) with i given by

$$i = 3i_{0,2} \exp[(1 + \beta_{a,2})F\,\Delta\phi/RT] \quad (93)$$

Table 2 gives the slopes of different Tafel regions corresponding to equation (80). Also given are the transfer coefficients [obtained from equations (21) and (23)] of the corresponding limiting steps. The two lower rows in Table 2 refer to the mechanisms involving the "simultaneous" transfer of two electrons in the limiting step. The data contained in Table 2 make it possible to determine the limiting step of a three-electron process from the anodic and cathodic transfer coefficients for a given range of potentials.

For significantly different rates of the successive steps, the change in the limiting step as a function of ϕ becomes evident in the appearance of a break in the polarization curve when $\Delta\phi$ is appreciable. At comparable rates of these steps, this effect is already seen in an analogous break in curves of ϕ versus $\log i_a$ and ϕ versus $\log i_k$ near ϕ_e (see Section IV.2). This change in the kinetically limiting step with change of ϕ is an unequivocal criterion of the stepwise nature of the electrode process (*the "break" criterion*[121a]) and cannot be explained in terms of simultaneous transfer of electrons in the electrode process.

Examples are shown in Figure 6 for computer-calculated curves using equation (80) with $i_{0,3} < i_{0,2} < i_{0,1}$ and arbitrarily chosen values of the kinetic parameters. At small $\Delta\phi$, the rate of the electrode process is determined by the transfer of the last electron. In accordance with the data in Table 2, $b_a = b_k/5$. At more positive potentials, the anodic curve shows breaks and two additional linear

Table 2
Slopes and Transfer Coefficients for Various Stepwise Mechanisms

	Anodic Process			Cathodic Process		
Limiting step	Slope	Transfer coefficient	Symmetry coefficient	Slope	Transfer coefficient	Symmetry coefficient
$M \rightarrow M^{+} + e$	$2.3RT/F < b_a < \infty$	$0 < \alpha_a < 1$	$\beta_{a,1} = \alpha_a$	$2.3RT/3F < b_k < 2.3RT/2F$	$2 < \alpha_k < 3$	$\beta_{k,1} = \alpha_k - 2$
$M^{+} \rightarrow M^{++} + e$	$2.3RT/2F < b_a < 2.3RT/F$	$1 < \alpha_a < 2$	$\beta_{a,2} = \alpha_a - 1$	$2.3RT/2F < b_k < 2.3RT/F$	$1 < \alpha_k < 2$	$\beta_{k,2} = \alpha_k - 1$
$M^{++} \rightarrow M^{3+} + e$	$2.3RT/3F < b_a < 2.3RT/2F$	$2 < \alpha_a < 3$	$\beta_{a,3} = \alpha_a - 2$	$2.3RT/F < b_k < \infty$	$0 < \alpha_k < 1$	$\beta_{k,1} = \alpha_k$
$M^{+} \rightarrow M^{3+} + 2e$	$2.3RT/3F < b_a < 2.3RT/F$	$1 < \alpha_a < 3$	$\beta_{a,2} = (\alpha_a - 1)/2$	$2.3RT/2F < b_k < \infty$	$0 < \alpha_k < 2$	$\beta_{k,2} = \alpha_k/2$
$M \rightarrow M^{++} + 2e$	$2.3RT/2F < b_a < \infty$	$0 < \alpha_a < 2$	$\beta_{a,1} = \alpha_{a/2}$	$2.3RT/3F < b_k < 2.3RT/F$	$1 < \alpha_k < 3$	$\beta_{k,1} = (\alpha_k - 1)/2$

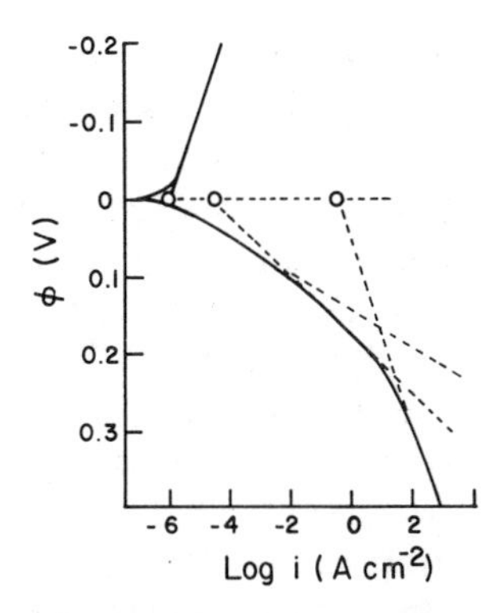

Figure 6. Calculated polarization curves for a three-step mechanism; $\beta_{k,1} = \beta_{k,2} = \beta_{k,3} = 0.5$; $\beta_{a,1} = \beta_{a,2} = \beta_{a,3} = 0.5$; $i_{0,1} = 0.1$; $i_{0,2} = 1.10^{-5}$; $i_{0,3} = 3.3 \times 10^{-7}$ A cm^{-2}.

sections corresponding to rate-limiting steps of transfer of the second and first electrons.

The appearance of several Tafel sections on the anodic polarization curve should be accompanied by the appearance of corresponding linear sections on the cathodic polarization curve in the same potential regions. In this case, the sum of the apparent transfer coefficients determined from b_a and b_k at a given ϕ should be equal to the number of electrons transferred in the overall reaction, i.e., condition (85) should be obeyed for all ϕ.

The observation of several Tafel regions and breaks on steady-state polarization curves follows from the Vetter theory of stepwise processes[2,3] and has been noted in various papers.[40,119–121]

As may be seen from Figure 6, in order to obtain reliable experimental breaks and linear sections corresponding to different limiting steps, it is necessary to carry out measurements sufficiently far from ϕ_e and over a wide range of potentials up to current densities exceeding the exchange current of the overall process by several orders of magnitude. For electrodes with high i_0, this may be difficult. If the exchange currents of successive steps are of comparable magnitudes, the linear sections become short. If the transition region between these regions is near ϕ_e, the break and the two linear sections on the polarization curve are obscured owing to the effect of the reverse reaction. This disadvantage can be partially overcome by using the i_a curve on which it is possible to detect the break in the immediate vicinity of ϕ_e (see p. 335 and Section IV, 2).

Another possibility for determining Tafel regions corresponding to different limiting steps arises from the difference in the dependence of the i_0 of each step on [M] and $[M^{n+}]$. In the equations for the i_0's of various steps [obtained from equations (76)–(78) when there is no external polarization], ϕ_e and the concentration of intermediate particles M^+ and M^{2+} can be expressed through the concentrations of the initial and final substances with the aid of Nernst equations for the overall process and for individual steps. Then,[117]

$$i_{0,1} = k_1[\mathrm{M}]^{(2+\beta_{k,1})/3}[\mathrm{M}^{3+}]^{\beta_{a,1}/3} \tag{94}$$

$$i_{0,2} = k_2[\mathrm{M}]^{(1+\beta_{k,2})/3}[\mathrm{M}^{3+}]^{(1+\beta_{a,2})/3} \tag{95}$$

$$i_{0,3} = k_3[\mathrm{M}]^{\beta_{k,3}/3}[\mathrm{M}^{3+}]^{(2+\beta_{a,3})/3} \tag{96}$$

from which the i_0 values as a function of the concentrations for each step are (cf. Refs. 2 and 51)

$$\left\{\frac{\partial(\ln i_{0,1})}{\partial(\ln[\mathrm{M}])}\right\}_{[\mathrm{M}^{3+}]} = \frac{2+\beta_{k,1}}{3}; \quad \left\{\frac{\partial(\ln i_{0,1})}{\partial(\ln[\mathrm{M}^{3+}])}\right\}_{[\mathrm{M}]} = \frac{\beta_{a,1}}{3} \tag{97}$$

$$\left\{\frac{\partial(\ln i_{0,2})}{\partial(\ln[\mathrm{M}])}\right\}_{[\mathrm{M}^{3+}]} = \frac{1+\beta_{k,2}}{3}; \quad \left\{\frac{\partial(\ln i_{0,2})}{\partial(\ln[\mathrm{M}^{3+}])}\right\}_{[\mathrm{M}]} = \frac{1+\beta_{a,2}}{3} \tag{98}$$

$$\left\{\frac{\partial(\ln i_{0,3})}{\partial(\ln[\mathrm{M}])}\right\}_{[\mathrm{M}^{3+}]} = \frac{\beta_{k,3}}{3}; \quad \left\{\frac{\partial(\ln i_{0,3})}{\partial(\ln[\mathrm{M}^{3+}])}\right\}_{[\mathrm{M}]} = \frac{2+\beta_{a,3}}{3} \tag{99}$$

These relations show that with increasing concentration of the amalgam (other conditions being kept constant), i_0 for the first step increases to the greatest extent, whereas that of the third step increases least. As $[M^{3+}]$ is varied, there is a maximum change in i_0 for the third step and a negligible change in that of the first step. Thus, by varying [M] and $[M^{3+}]$ in opposite directions, it is possible to obtain a significant change in the ratio of i_0's of separate steps. This creates favorable conditions for the detection of linear Tafel regions corresponding to different rate-limiting steps,[117] so that the respective i_0 values and transfer coefficients can be determined.

Additional mechanistic information can be obtained by comparing the i_0's of differential steps determined by the above method with the i_0 of the overall process determined by radiotracer or impedance measurements, or from the slope of the *initial* section of the same polarization curve. The true exchange current of the three-electron process is related in the usual way[2] to R_d determined near ϕ_e,

$$R_d = [\partial(\Delta\phi)/\partial i]_{i=0} = (RT/3F)(1/i_0) \tag{100}$$

Differentiation of equation (80) with respect to ϕ gives the transfer resistance as a function of the exchange currents of the steps (73)–(75), i.e.,[121]

$$R_d = \left(\frac{\partial(\Delta\phi)}{\partial i}\right)_{i=0} = \frac{1}{3}\left(\frac{RT}{3F}\right)\left(\frac{1}{i_{0,1}} + \frac{1}{i_{0,2}} + \frac{1}{i_{0,3}}\right) \tag{101}$$

From equations (100) and (101), we obtain the relation between the true exchange current and the exchange currents of separate steps [cf. Refs. 3 and 123 in relation to Ref. 122 regarding the correct form of equation (101)][117] as

$$\frac{1}{i_0} = \frac{1}{3}\left(\frac{1}{i_{0,1}} + \frac{1}{i_{0,2}} + \frac{1}{i_{0,3}}\right) \tag{102}$$

By substituting into equation (102) the i_0 values for steps from equations (94)–(96), the dependence of the true exchange current on the concentrations of metal in the amalgam and in the solution is derived. The agreement between the values of i_0 determined from independent measurements (e.g., by the radiotracer technique) and those calculated from experimental i_0 values from equation (102) shows that equation (80) describes the process (72) both for small and large $\Delta\phi$.

In the case of a single slow step, and when the i_0 for one of the steps is significantly below that of the remaining steps (e.g., $i_{0,1} \ll i_{0,2} < i_{0,3}$), equation (102) yields $i_0 \simeq i_{0,\text{ext}} = 3i_{0,1}$. That is, the true exchange current coincides with the extrapolated exchange current $i_{0,\text{ext}}$ of the rate-limiting step. It is readily seen that when the steps have comparable i_0's, the true value i_0 obtained from equation (102) should be smaller than the exchange current $i_{0,\text{ext}}$ obtained by extrapolation according to equations (83) and (84), e.g., when $i_{0,1} = i_{0,2} = i_{0,3}$, equation (102) gives $i_0 = i_{0,1} = i_{0,\text{ext}}/3$. The difference between the true and extrapolated exchange currents is hence another criterion for a process having comparable rate constants for the successive steps, and may be referred to as *the criterion of the true exchange current.* The application of this criterion makes it possible to detect a stepwise mechanism when the rate constants for the individual steps are close to one another and the Vetter criterion is not unequivocal. Herein lies the advantage of the criterion of the true exchange current.

Thus, for the stepwise process having comparable rates of successive steps, the following criteria apply:

(a) *The Vetter criterion* [equations (83) and (84)]: divergence of the values for $i_{0,\text{ext}}$ from data obtained at appreciable anodic or cathodic polarization, $i^{a}_{0,\text{ext}} \neq i^{k}_{0,\text{ext}}$.

(b) *Criterion of the transfer coefficient sum*: deviation of the sum of the transfer coefficients measured at appreciable $\Delta\phi$ from the overall number of electrons transferred in the electrode process ($\alpha_a + \alpha_k < 2$).

(c) *The break criterion*[121a]: appearance of a break on the ordinary polarization curve at appreciable $\Delta\phi$ or on the partial current curve for a given process near ϕ_e (see Section IV.2).

(d) *The criterion of the true exchange current*: divergence between the true and the extrapolated exchange currents ($i_0 < i_{0,\text{ext}}$).

2. Equations for the Partial Anodic Current

As shown in the previous subsection, one of the main criteria[121a] for a stepwise mechanism in electrode processes having comparable rate constants for successive steps is the appearance of a break in the Tafel relations when $\Delta\phi$ is appreciable (p. 355). On the partial current curve of the corresponding process (for example, the anodic i_a) obtained by the radiotracer method under conditions where no distortion arises due to the reverse process, the break can also be detected near ϕ_e. This considerably broadens the range of applicability of the break criterion. The conclusion was reached in the analysis of a two-step process[36,37] and the effect was first found in experiments with bismuth amalgam.[72,124] In addition, such measurements permit the determination of the slopes of the i_a and i_k curves for processes taking place at the same ϕ. Thus, it is possible to calculate the sum of the transfer coefficients corresponding to this potential.

It is now of interest to analyze the kinetics of the anodic partial process where the rate constants in each of the successive steps in the simple case of a two-electron process are comparable [equations (73) and (74)].[125] With the assumptions made in Section IV.1, equations (76) and (77) apply. The concentration of the intermediate M^+ is determined by using the condition for the steady state

$$i = 2i_1 = 2i_2 = 2(i_{a,1} - i_{k,1}) = 2(i_{a,2} - i_{k,2}) \tag{103}$$

and substituting the corresponding values from equations (76) and (77); this gives

$$[M^+] = \frac{k_{a,1}[M]\exp(\beta_{a,1}F\phi/RT) + k_{k,2}[M^{++}]\exp(-\beta_{k,2}F\phi/RT)}{k_{a,2}\exp(\beta_{a,2}F\phi/RT) + k_{k,1}\exp(-\beta_{k,1}F\phi/RT)} \tag{104}$$

From this equation, we can derive the relation for an i_a curve when different i_0 ratios for the two steps apply. For example, let $i_{0,2} < i_{0,1}$; then, in the steady state, the anodic component of the overall current is described in the first approximation by the partial current of the slower step through the equation

$$i_a = 2i_{a,2} \tag{105}$$

By using equations (77) and (104) and substituting the rate constants by the respective exchange currents for each step, equation (105) gives,[125] for both positive or negative $\Delta\phi$, an expression for the anodic partial current, viz.

$$i_a = 2k_{a,2}[M^+]\exp\left(\frac{\beta_{a,2}F\phi}{RT}\right)$$

$$= 2\frac{\exp[(\beta_{a,1} + \beta_{a,2})F\,\Delta\phi/RT] + (i_{0,2}/i_{0,1})\exp[(\beta_{a,2} - \beta_{k,2})F\,\Delta\phi/RT]}{(1/i_{0,2})\exp(-\beta_{k,1}F\,\Delta\phi/RT) + (1/i_{0,1})\exp(\beta_{a,2}F\,\Delta\phi/RT)} \tag{106}$$

and at small $\Delta\phi$,

$$i_a = 2i_{0,2}\exp[(1 + \beta_{a,2})F\,\Delta\phi/RT] \tag{107}$$

At such potentials (region II in Figure 7) there is a linear section on the ϕ versus $\log i_a$ curve with a low slope of $2.3RT/(1 + \beta_{a,2})F$, and the rate of the overall process is limited by the second step.

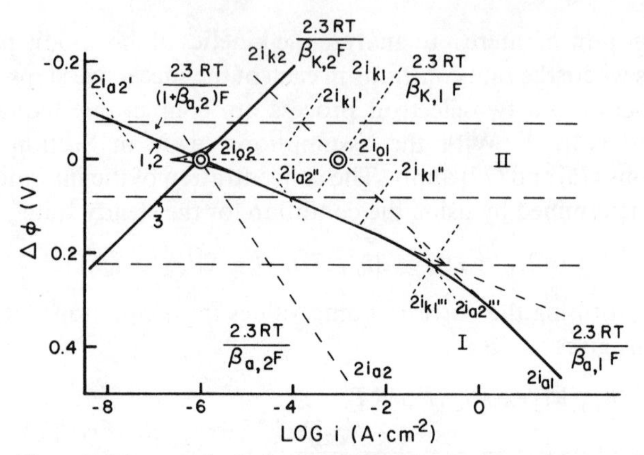

Figure 7. Diagram of calculated anodic and cathodic partial current curves: (1) curve plotted from exchange current values; (2, 3) curves corresponding to polarization with the parameters $\beta_{k,1} = 0.4$, $\beta_{a,1} = 0.6$, $\beta_{k,2} = 0.6$, $\beta_{a,2} = 0.4$, $i_{0,1} = 1000 i_{0,2}$.

For sufficiently high anodic polarization, equation (106) gives

$$i_a = 2i_{0,1} \exp(\beta_{a,1} F \, \Delta\phi/RT) \tag{108}$$

so that a break and transition to a second linear region with a slope of $2.3RT/\beta_{a,1}F$ arises, corresponding to the first step becoming limiting (region I in Figure 7).

As can be seen in Figure 7, at $i_{0,1} = 1000 i_{0,2}$, on the anodic branch of the ordinary polarization curve, there is also a break and a linear section having a low slope. When the successive steps have comparable rate constants, this linear section on the usual anodic curve cannot be detected (Figure 8). The reason is that the break lies near ϕ_e, where the slope of the anodic curve is distorted owing to the effect of the reverse process.

Important conclusions can also be drawn from the analysis of the rate equation in the region of cathodic polarization. For a sufficiently large, negative $\Delta\phi$, the first term in the numerator and the second term in the denominator of equation (106) can be disregarded, so that

$$i_a = 2i_{0,2}(i_{0,2}/i_{0,1}) \exp[(\beta_{a,2} + \beta_{k,1} - \beta_{k,2})F \, \Delta\phi/RT] \tag{109}$$

Hence, on cathodic polarization, the curve of ϕ versus $\log i_a$ should exhibit an additional break and a third linear section with

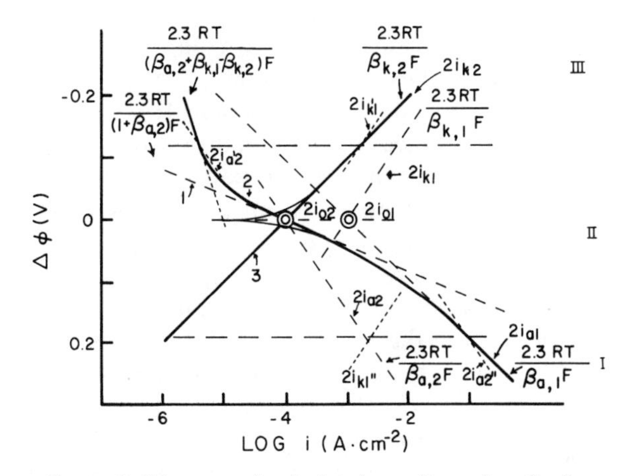

Figure 8. Diagram of calculated anodic and cathodic partial current curves: (1) curve plotted from exchange current values; (2, 3) curves corresponding to polarization: $\beta_{k,1} = 0.4$, $\beta_{a,1} = 0.6$, $\beta_{k,2} = 0.6$, $\beta_{a,2} = 0.4$, $i_{0,1} = 10i_{0,2}$.

$b_c \doteqdot 2.3RT/(\beta_{a,2} + \beta_{k,1} - \beta_{k,2})F$ (region III, Figure 8). This effect is due to the diminished reversibility of the first step in region III [the boundary of this region is determined by the intersection of the straight lines $i_{k,2}$ and $i_{a,1}$ (Figure 8)] so that the concentration of M^+ ions becomes practically independent of $\Delta\phi$. Thus, for significant negative shifts of potential, equation (104) gives

$$[M^+] = (k_{k,2}/k_{k,1})[M^{++}]\exp[-(\beta_{k,2} - \beta_{k,1})F\phi/RT] \quad (110)$$

whence $[M^+]$ is practically independent of ϕ since, usually, $\beta_{k,1} \sim \beta_{k,2}$. In this case, i_a varies with ϕ only as a result of the usual change in activation energy with ϕ.

Hence, in region III as well as in region I, both steps proceed irreversibly, whereas in region II, only the second step is irreversible and the first step proceeds near equilibrium. The intersection points of the horizontal dashed lines in Figures 7 and 8 with the lines i_a and i_k indicate, in accordance with equation (103), the degree of irreversibility of these steps. From Figure 7, it can be seen that in region II at a given potential, the equilibrium in the first step is almost undisturbed since the rates of the direct and reverse processes of this step are practically the same ($i_{a,1} \sim i_{k,1}$), whereas the second step is irreversible ($i_{a,2} \gg i_{k,a}$ at $\Delta\phi > 0$ and $i_{a,2} \ll i_{k,2}$ when $\Delta\phi < 0$).

Figure 8 shows that in regions I and III, both steps are irreversible, since at a given potential, $i_{a,1} \gg i_{k,1}$ and $i_{a,2} \gg i_{k,2}$ (region I) and, accordingly, $i_{a,1} \ll i_{k,1}$ and $i_{a,2} \ll i_{k,2}$ (region III).

For nonstationary measurements in the region of irreversibility of both steps (e.g., region I in Figure 8), some characteristic features also arise. When the potential is changed in anodic polarization and the single limiting step is retained (e.g., the second), accumulation of M^+ ions to a level corresponding to a new steady-state concentration (p. 334) should take place gradually. If the new value of ϕ lies in region I, where the first step is also irreversible, the concentration of M^+ ions should pass through a maximum because, during the initial period, practically all the current is consumed in producing a new and higher concentration of the intermediate ($i_{a,1} > i_a/2$). Thus the first step proceeds initially at higher overvoltage, to which corresponds a higher $[M^+]$ as compared with that at steady state; after this state is reached. The current is equally distributed between both steps ($i_{a,1} = i_{a,2} = i_a/2$). As the M^+ concentration passes through a maximum, there should appear a maximum in the $\Delta\phi$–time curve[126] and in the i–time curve (at constant $\Delta\phi$).[126a]

The break in the ϕ–$\log i_a$ curve, where regions II and III are adjacent, that is, for cathodic polarization, can be detected only by the radiotracer technique.[125] Its appearance is an unambiguous criterion of a stepwise mechanism.

The simultaneous plotting of the radiotracer curve for ϕ versus $\log i_a$ and the usual cathodic polarization curve in regions II and III makes it possible to compare the apparent transfer coefficients derived from the slopes of these curves at the same potential. In region II (Figure 8), $\alpha_a + \alpha_k$ (at constant ϕ) equals 2 (as in the case of an electrode process with a single limiting step), whereas in region III, as a result of irreversibility of both steps, this sum is close to unity; i.e., $\alpha_a + \alpha_k = \beta_{k,2} + (\beta_{a,2} + \beta_{k,1} - \beta_{k,2}) = \beta_{k,1} + \beta_{a,2} \simeq 1$. In region I, both steps are also irreversible and the sum $\alpha_a + \alpha_k$ obtained from the slopes of curves of ϕ versus $\ln i_k$ and ϕ versus $\ln i_a$ is also close to unity ($\alpha_a + \alpha_k = \beta_{k,2} + \beta_{a,1} \simeq 1$). This deviation from $n = 2$ of the sum $\alpha_a + \alpha_k$ measured at a constant potential in regions III and I is an additional criterion of the stepwise mechanism in a process having comparable rate constants for its successive steps.

This shows that it is possible to utilize the break criterion as well as the criterion of the sum of the transfer coefficients at constant potential, even when the rates of the successive steps (measured near $\Delta\phi = 0$) are close to one another.

3. Conditions for the Applicability of the "Break" Criterion

The appearance of the break on both the polarization curve and the partial current curve for a given process cannot be interpreted in terms of "simultaneous" transfer of electrons in the electrode process. Apparently this effect is the most direct proof of a stepwise mechanism, for the appearance of two linear sections with different slopes on the curve of one of the coupled processes makes it possible to follow the transition of the limiting step in a relatively narrow range of potentials without the need for extrapolation. The break criterion can also be used when, for some reason, it is impossible to study the kinetics of the reverse process and carry out measurements in the region of ϕ_e. In order to use this criterion, it is necessary to analyze thoroughly the conditions under which it is applicable and to consider other possible causes for the appearance of the break and the two Tafel regions on the polarization curves. Possible sources of error involved in the use of this criterion are as follows:

(a) Distortions of the shape of the polarization curve could be caused by electrochemical side reactions taking place simultaneously (for example, hydrogen evolution). These complications can be detected by current efficiency measurements and overcome by obtaining the partial curves of the process under investigation, e.g., by means of radiotracers.

(b) The presence of a slowly preceding diffusion or chemical step, an Ohmic potential drop in the solution, or electrode passivation could also lead to the appearance of a break on the polarization curve. However, in these cases, the section of the polarization curve following the break should not obey the Tafel relation.

(c) If the appearance of one of two linear regions is due to the presence of succeeding diffusion or chemical steps[90,127,128] (see also Section IV.4, cases c and d), the first (i.e., that nearest to ϕ_e) should have $b = 2.3RT/nF$. This section should shift in the direction of lower $\Delta\phi$ when the stirring rate is increased for the case when a slow diffusion step is involved.

(d) If the break is due to transition of the limiting step to a "barrierless" condition ($\beta_{k,m} = 1$), or to "activationless" discharge

($\beta_{k,m} = 0$), the slopes of the corresponding sections of the polarization curve should be $2.3RT/(m - 1)F$, where m is a whole number ($0 \leq m \leq n$) the magnitude of which depends on the number of preceding equilibrium steps.[129]

(e) A break accompanied by a transition to a Tafel region having a higher slope may be due to the attainment of the region of limiting surface coverage by low-valence intermediates formed in the preceding equilibrium step.[130] Although such a break is a result of the stepwise mechanism of the process, it is not related to the change in the nature of the limiting step, which remains the same for both sections. Unlike the cases of breaks analyzed earlier, for this section, the rate of the overall process in the region of high slope should not vary with an increase in the concentration of initial substance, since the concentration of particles participating in the limiting step remains constant.

So far it has been assumed that the overall process takes place through one path. If this process can take place simultaneously along several parallel paths, the rate equations will become considerably more complex. In this case, there is also the possibility of the appearance on the polarization curves of Tafel sections with different slopes. Thus, for example, the appearance of two linear sections can be due to the parallel occurrence of the electrode process through two paths: the stepwise and the simple path (Figure 9).[131] To the simple single-step reaction correspond anodic and cathodic curves with the same slopes. For a symmetrical energy barrier there should be little difference between these slopes. For example, for the reaction $M \rightleftarrows M^{++} + 2e$, the slopes should be ($b_a \sim b_k \sim 2, 3.2\, RT/F$) (see p. 354).

This means that, all other conditions being the same, for the process of transfer in the limiting step of only one charge, the decrease in the activation energy upon the shift of the potential from ϕ_e will be considerably less than for the process involving simultaneous transfer of all charges in one step. Thus, at very large $\Delta\phi$ the overall reaction can proceed predominantly through direct transfer of the cation (e.g., M^{2+}) across the phase boundary. As can be seen from Figure 9, this effect should lead to the appearance of breaks in the polarization curves in the far away anodic and cathodic regions. In the case of a stepwise reaction with a single limiting step, the polarization curve will have only one break, located in the cathodic or anodic region, depending on the magnitude of the ratio α_a/α_k. When $\alpha_a > \alpha_k$, the break can occur only in the cathodic region. In the case of comparable rates of simple and stepwise reactions proceeding simultaneously, these breaks apparently can be determined experimentally.

The characteristic feature of such breaks is that with a shift from ϕ_e the slope of the polarization curve in the region of the break should decrease[131] whereas in the case of the single-path process with successive charge transfer steps, the slope of the curve in the region of break should increase.

4. Processes Involving Higher-Order Reactions

In the stepwise electrode process, low-valence intermediates are formed which can undergo not only further electrochemical oxidation or reduction but also a chemical reation between the intermediates themselves. There is little likelihood that reactions will take place between cations of intermediate valence (for example, disproportionation and dimerization) owing to their low concentration. At such concentrations, reactions of the second or third order are less likely and the high activation energy of such reactions, especially disproportionation, does not favor them. Nevertheless, it is of interest to establish the criteria for the mechanism of reactions which may involve such steps.[132]

The simplest process of this type is

$$M \rightleftarrows M^{++} + 2e \tag{111}$$

If the intermediate particles can undergo disproportionation, the overall process is then described by a combination of three steps,

$$M \underset{k_{k,1}}{\overset{k_{a,1}}{\rightleftarrows}} M^{+} + e \tag{112}$$

$$M^{+} \underset{k_{k,2}}{\overset{k_{a,2}}{\rightleftarrows}} M^{++} + e \tag{113}$$

$$2M^{+} \underset{k''_d}{\overset{k'_d}{\rightleftarrows}} M + M^{++} \tag{114}$$

with $i_{0,1}$, $i_{0,2}$, and $i_{0,d}$. Various cases can be considered.

(*a*) $i_{0,2} \ll i_{0,1} < i_{0,d}$. The step (113) is rate-determining, and the process passes through a limiting step (112) followed by fast disproportionation. In this case, $\nu = 2$. Since n also equals 2, the rate equations for the anodic and cathodic process[8a] become simplified to

$$i_a = k_{a,1}[\mathrm{M}]\exp(\beta_{a,1}nF\phi/\nu RT) = k_{a,1}[\mathrm{M}]\exp(\beta_{a,1}F\phi/RT) \quad (115)$$

$$\begin{aligned} i_k &= k_{k,1}[\mathrm{M}^+]\exp(-\beta_{k,1}nF\phi/\nu RT) \\ &= k_{k,1}[\mathrm{M}^+]\exp(-\beta_{k,1}F\phi/RT) \end{aligned} \quad (116)$$

For quasiequilibrium in (114), the relation

$$i_k = k_{k,1}K^{-1/2}[\mathrm{M}]^{1/2}[\mathrm{M}^{++}]^{1/2}\exp(-\beta_{k,1}F\phi/RT) \quad (117)$$

is obtained from (116). Thus, i_k depends on the concentration of metal species in both solution and amalgam. At ϕ_e, $i_a = i_k = i_0$, and on the basis of (115) and (117), it follows that

$$\phi_e = \text{const} + [RT/2(\beta_{k,1} + \beta_{a,1})F]\ln([\mathrm{M}^{++}]/[\mathrm{M}]) \quad (118)$$

By comparing this equation with the Nernst relation for the overall process (111), it is found that $\beta_{a,1} + \beta_{k,1} = 1$. Equations (115) and (117) show that the symmetry coefficients of the limiting step, $\beta_{k,1}$ and $\beta_{a,1}$, coincide with the transfer coefficients (α_a and α_k) derived with the aid of equations (39a) and (39b) from the b values for ϕ versus ln i_a and ϕ versus ln i_k curves. Thus, contrary to the case of processes which do not include step (114),

$$\alpha_a + \alpha_k = 1 \quad (119)$$

By using Tafel slopes, it follows from (119) that

$$(1/b_a) + (1/b_k) = F/2.3RT \quad (120)$$

From (115) and (117), at ϕ_e, the equations for the exchange current and the polarization curve result:

$$i_0 = i_0^\circ[\mathrm{M}]^{(1+\alpha_k)/2}[\mathrm{M}^{++}]^{\alpha_a/2} \quad (121)$$

$$i = i_a - i_k = i_0[\exp(\alpha_a F\Delta\phi/RT) - \exp(-\alpha_k F\Delta\phi/RT)] \quad (122)$$

respectively. From (122), R_d is given by

$$R_d = \left[\frac{\partial(\Delta\phi)}{\partial i}\right]_{i=0} = \frac{RT}{F}\frac{1}{i_0} \quad (123)$$

It follows from (122) that the $i_{0,\text{ext}}$ obtained by anodic and cathodic extrapolations agree when such a mechanism applies.

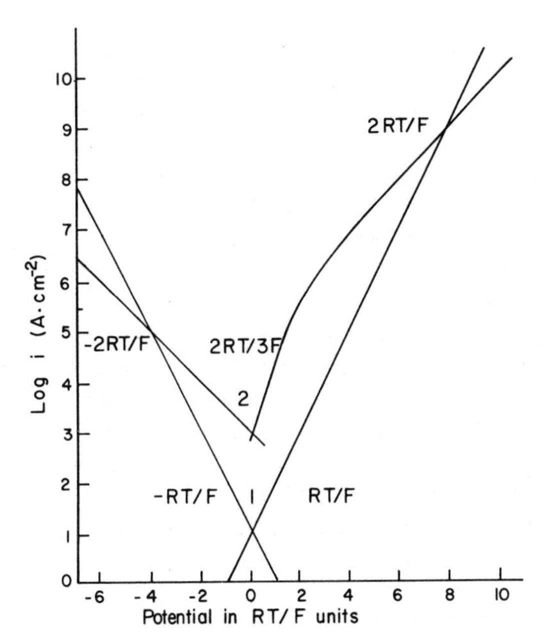

Figure 9. Theoretical curves for the reaction $M \rightleftarrows M^{++} + 2e$ in the case of (1) a simple, single-step and (2) a stepwise mechanism (after Heusler[131]).

For processes which include the fast disproportionation step (114), an additional criterion can be formulated which is based on the comparison of exchange currents determined from electrochemical (i_0) and radiotracer measurements ($i_{0,r}$).[132] Thus, for the process (111), together with the usual "electrochemical" path of isotopic exchange, there is possible a transfer of radioactive particles from one phase to the other under equilibrium conditions also through the disproportionation step, bypassing the electrochemical step, that is, by "nonelectrochemical" exchange. For a radioactive amalgam, isotopic exchange can proceed either through the charge-transfer steps [equations (112) and (113)] at the rate i_0, or through the "chemical" mechanism (two successive disproportionation steps)

$$\overset{*}{M} + M^{++} \rightarrow M^{+} + \overset{*}{M}^{+} \qquad \text{and} \qquad M^{+} + \overset{*}{M}^{+} \rightarrow M + \overset{*}{M}^{++}$$

at a rate equal to $\frac{1}{2}i_{0,d}$ [since the step (114) occurs twice], in which case

$$i_{0,r} = \tfrac{1}{2}i_{0,d} + i_0 \tag{124}$$

This equation makes it possible to determine experimentally the rate of exchange arising from the disproportionation step (114). If the rate of step (114) is significantly higher than that of the steps (112) and (113), it should be found that $i_{0,r} > i_0$.* Thus, the divergence between $i_{0,r}$ and i_0 may serve as a criterion of a stepwise mechanism for the process (111) and as an indication of the possible presence of a fast disproportionation step; (cf. Ref. 133 for analogous cases in heterogeneous gas reactions). This criterion, which may be referred to as the *criterion* of *nonelectrochemical exchange*,[132] was examined by using reactions involving hydrogen evolution proceeding through the discharge and electrochemical desorption steps[134–136] and the process $2Hg \rightleftarrows Hg_2^{++} + 2e$.[28]

Thus, the features of mechanism (*a*) which distinguish it from the previously analyzed cases involving successive charge transfer steps with $\nu = 1$ are: (i) a fractional dependence of the rate of the cathodic process on the concentration of M and M^{2+}; (ii) a stoichiometric number differing from unity ($\nu = 2$) [owing to which, the sum $\alpha_a + \alpha_k$ is low; equation (119)†] and R_d has an unusually high value [equation (123)]; (iii) a considerably stronger dependence of i_0 on [M] than on $[M^{n+}]$ [equation (121)]; and (iv) the presence of nonelectrochemical exchange.

(*b*) $i_{0,1} \ll i_{0,2} < i_{0,d}$. This case is analogous to case (a). However, the first step in the anodic polarization is the rapid disproportionation reaction proceeding now from right to left, and the limiting step is (113). In this case, the rate of the anodic process is proportional to $[M]^{1/2}[M^{2+}]^{1/2}$, and i_0 depends to a greater extent on $[M^{2+}]$ than on [M]. This case and case (d) analyzed below are of special interest, since when M^{2+} ions are absent from the solution, the initial stage of anodic dissolution of metal should proceed through steps (112) and (113); only after the accumulation of M^{2+} ions does the process take place through steps (114) and (113). This peculiarity can be employed

*The same effect can also be observed for the process which includes only successive one-electron steps, if the particles M^+ formed in the equilibrium step (112) are capable of entering into a fast homogeneous exchange with M^{2+} ions, bypassing the slow step (113). Thus, if condition $i_{0,r} = i_0$ holds, such homogeneous exchange is practically absent.[105]

†This feature of the mechansim involving a fast disproportionation step is ambiguous since the low value for the sum $\alpha_a + \alpha_k$ is also obtained when successive steps have comparable rate constants (Section IV.1), i.e., for the case where a single limiting step is absent; hence, ν becomes physically meaningless[137] for this situation.

for determining the mechanism by means of nonstationary measurements.[132]

(*c*) $i_{0,2} \ll i_{0,d} < i_{0,1}$. In this case, the limiting process is the disproportionation step with $\nu = 1$, which yields a mechanism previously analyzed (p. 316). The kinetic equations describing this case are

$$i_a = k'_d[\mathrm{M}^+]^2 = k'_d K_{(1)}[\mathrm{M}]^2 \exp(2F\phi/RT) \tag{125}$$

$$i_k = k''_d[\mathrm{M}][\mathrm{M}^{++}] \tag{126}$$

which distinguish case (c) from other mechanisms.

From equations (125) and (126), the relation describing the polarization curve is obtained:

$$i = i_a - i_k = i_0[\exp(2F\,\Delta\phi/RT) - 1] \tag{127}$$

and

$$R_d = \left[\frac{\partial(\Delta\phi)}{\partial i}\right]_{i=0} = \frac{RT}{2F}\frac{1}{i_0} \tag{128}$$

The anodic polarization curve should exhibit a Tafel region with $b = 2.3RT/2F$ and on the cathodic curve there will be a kinetic limiting current $i = i_0$.

(*d*) $i_{0,1} \ll i_{0,d} < i_{0,2}$. This case is analogous to case (c), so that

$$i_a = k''_d[\mathrm{M}][\mathrm{M}^{++}] \tag{129}$$

$$i_k = k'_d[\mathrm{M}^{++}]^2 \exp(-2F\phi/RT) \tag{130}$$

That is, $\alpha_k = 2$ and $\alpha_d = 0$.

(*e*) $i_{0,1} \simeq i_{0,2} \ll i_{0,d}$. In this case, the two mechanisms (a) and (b) can occur simultaneously, i.e., the process takes place along two *parallel* paths and the special criteria for this situation then apply.

Since case (e) involves a combination of mechanisms (a) and (b) and since for these mechanisms, the dependences of i_a and i_k on the metal concentration in the amalgam and in solution are different, by changing the metal concentration in one of the phases, it is possible to go over from case (e) to a region in which mechanisms (a) and (b) are predominant and thus differentiate case (e) from other cases.

5. Summary of Various Criteria for Stepwise Mechanisms

It is of interest to combine the criteria for stepwise mechanisms analyzed in Sections III and IV. Table 3A summarizes the kinetic parameters and Table 3B illustrates the applicability of the criteria for stepwise mechanisms in different cases of the simple stepwise process with two one-electron steps (I–III). For purposes of comparison, a single-step case is included, i.e., a case where the transfer of both electrons takes place in "one step" (IV). Rows a–e of Table 3A present the conclusions for various cases involving one-electron steps and the step involving disproportionation of intermediates.

It can be readily seen that more definite criteria of a stepwise mechanism can be obtained in the case when successive steps have comparable rate constants (III) than in the case where a single step is rate-limiting (I–II). In fact, the occurrence of case III can be established with the aid of the Vetter criterion as well as with the aid of the criteria of the break, the sum of transfer coefficients, and the true exchange current (p. 360). For cases I and II, however, application of these criteria, together with the determination of R and ν, and the comparison of the true exchange current with the isotopic exchange rate (the criterion of nonelectrochemical exchange) do not permit differentiation of the stepwise mechanism from the simple, single-step mechanism. In this case, the questions of whether the process is stepwise and which is the rate-limiting step can be answered only by considering the transfer coefficients (Section III.3).

The presence in the overall process of a fast or a slow disproportionation step involving low-valence intermediates (cases a–e) can be reliably established by analyzing the reaction orders, the stoichiometric number, the transfer coefficients, and the dependence of the exchange current on concentration, and determining if nonelectrochemical exchange occurs. Thus, by applying the criteria in Table 3B to the experimental data presented in Section III for electrochemical processes on zinc and indium amalgams (the reaction orders, values of α_a and α_k, coincidence of radiotracer and extrapolated exchange currents, and other effects), it can readily be shown that these processes do not involve disproportionation steps.

For a reliable proof of the stepwise mechanism for a given electrode process, the analysis should not be limited to the application of any one criterion of the stepwise mechanism, but all the criteria applicable under given experimental conditions should be examined.[138]

Table 3A
Kinetic Parameters for Different Mechanisms in the Process $M \rightleftarrows M^{++} + 2e$

			Reaction order				Transfer coefficients		
			Anodic process		Cathodic process				
Case	Ratio of step rate constants	ν	[M]	$[M^{++}]$	[M]	$[M^{++}]$	α_k	α_a	$\alpha_a + \alpha_k$
I	$i_{0,2} \ll i_{0,1}$	1	1	0	0	1	$1 > \alpha_k > 0$	$2 > \alpha_a > 1$	2
II	$i_{0,1} \ll i_{0,2}$	1	1	0	0	1	$2 > \alpha_k > 1$	$1 > \alpha_a > 0$	2
III	Comparable $i_{0,1}$ and $i_{0,2}$	1	1	0	0	1	$1 > \alpha_k > 0$	$1 > \alpha_a > 0$	$\beta_{k,2} + \beta_{a,1}$
IV	—	1	1	0	0	1	$2 > \alpha_k > 0$	$2 > \alpha_a > 0$	2
a	$i_{0,2} \ll i_{0,1} < i_{0,d}$	2	1	0	$\frac{1}{2}$	$\frac{1}{2}$	$1 > \alpha_k > 0$	$1 > \alpha_a > 0$	1
b	$i_{0,1} \ll i_{0,2} < i_{0,d}$	2	$\frac{1}{2}$	$\frac{1}{2}$	0	1	$1 > \alpha_k > 0$	$1 > \alpha_a > 0$	1
c	$i_{0,2} \ll i_{0,d} < i_{0,1}$	1	2	0	1	1	0	2	2
d	$i_{0,1} \ll i_{0,d} < i_{0,2}$	1	1	1	0	2	2	0	2
e	$i_{0,1} \simeq i_{0,2} \ll i_{0,d}$	2	1	0	0	1	$1 > \alpha_k > 0$	$1 > \alpha_a > 0$	$2 > \alpha_k + \alpha_a > 1$

Table 3B
Criteria for Different Mechanisms in the Process $M \rightleftarrows M^{++} + 2e$

Case	Ratio of step rate constants	Exchange current*		Rate of isotopic exchange $i_{0,r}$	The Vetter criterion†	Relation of i_0 and $i_{0,ext}$‡	Presence of break in polarization line
		x	y				
I	$i_{0,2} \ll i_{0,1}$	$\alpha_a/2$	$\alpha_k/2$	i_0	$i^a_{0,ext} = i^k_{0,ext}$	$i_0 = i_{0,ext}$	No
II	$i_{0,1} \ll i_{0,2}$	$\alpha_a/2$	$\alpha_k/2$	i_0	$i^a_{0,ext} = i^k_{0,ext}$	$i_0 = i_{0,ext}$	No
III	Comparable $i_{0,1}$ and $i_{0,2}$	—	—	i_0	$i^a_{0,ext} \neq i^k_{0,ext}$	$i_0 < i_{0,ext}$	Yes
IV	—	$\alpha_a/2$	$\alpha_k/2$	i_0	$i^a_{0,ext} = i^k_{0,ext}$	$i_0 = i_{0,ext}$	No
a	$i_{0,2} \ll i_{0,1} < i_{0,d}$	$\alpha_a/2$	$(1 + \alpha_k)/2$	$i_0 + (1/2)i_{0,d}$	$i^a_{0,ext} = i^k_{0,ext}$	$i_0 = i_{0,ext}$	No
b	$i_{0,1} \ll i_{0,2} < i_{0,d}$	$(1 + \alpha_a/2)$	$\alpha_k/2$	$i_0 + (1/2)i_{0,d}$	$i^a_{0,ext} = i^k_{0,ext}$	$i_0 = i_{0,ext}$	No
c	$i_{0,2} \ll i_{0,d} < i_{0,1}$	1	1	i_0	$i^a_{0,ext} = i^k_{0,ext}$	$i_0 = i_{0,ext}$	No
d	$i_{0,1} \ll i_{0,d} < i_{0,2}$	1	1	i_0	$i^a_{0,ext} = i^k_{0,ext}$	$i_0 = i_{0,ext}$	No
e	$i_{0,1} \simeq i_{0,2} \ll i_{0,d}$	—	—	$i_0 + (1/2)i_{0,d}$	$i^a_{0,ext} \neq i^k_{0,ext}$	$i_0 > i_{0,ext}$	Yes

*x and y are exponents in the equation $i_0 = i_0^\circ[M^{++}]^x[M]^y$.
†$i^a_{0,ext}$ and $i^k_{0,ext}$ are the exchange currents obtained by extrapolating the linear sections of anodic and cathodic branches of polarization curves.
‡i_0 is the true exchange current, $i_{0,ext}$ is that derived by extrapolation.

6. Experimental Results for Bismuth and Copper Amalgams

An electrode process of the type $M \rightarrow M^{n+} + ne$ having comparable rate constants for successive charge-transfer steps was first observed in the case of bismuth amalgam in a perchloric acid solution.[72,124,139] Figure 10 shows the dependence of i_a, obtained from radiotracer measurement points, on the potential (curve 1), together with the usual polarization curves (curves 2 and 3). Here, the anodic curve has two linear sections. The section with $b_a = 0.033$ V ($\alpha_a = 1.76$), which lies in the region of ϕ_e, can be reliably determined only by means of radiotracer measurements. The reason is that on the usual polarization curve, this section is observed only over a narrow potential range.

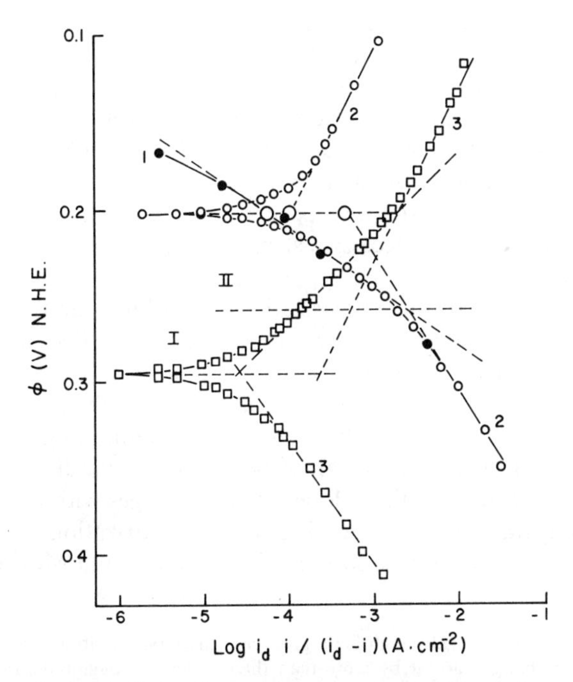

Figure 10. (1) Anodic partial current curve and (2) polarization curves for bismuth amalgam, $c_a = 0.12\,M$ and $c_s = 7.8 \times 10^{-4}\,M$ ($i_d^a = 0.83$, $i_d^k = 2.3 \times 10^{-3}$ A cm^{-2}); (3) polarization curves at $c_a = 10^{-4}\,M$ and $c_s = 10^{-2}\,M$ ($i_d^a = 6.6 \times 10^{-4}$, $i_d^k = 2.4 \times 10^{-2}$ A cm^{-2}).

For positive $\Delta\phi$ on curve 1 (Figure 10), a break and a second linear section appear with a higher slope $b_a = 0.074$ V corresponding to $\alpha_a = 0.78$. The extrapolation of this section to the equilibrium potential gives the value for the exchange current $i^a_{0,\text{ext}}$, which exceeds the true radiotracer exchange current $i_{0,\text{ex}}$ by nearly five times.*

The cathodic polarization curve 2 has only one short linear section with a high slope. For the purpose of investigating the kinetics of the cathodic process over a wider range of potentials, plots were made at low amalgam concentration and for a higher concentration of Bi^{3+} ions in the solution. The result is that on the cathodic curve 3 (Figure 10), two linear sections appear with slopes $b_k = 0.049$ V and $b_k = 0.101$ V. The mean values of α_k derived from these slopes are 1.18 and 0.57, respectively.

The presence of breaks and two linear sections on the curve of ϕ versus $\ln i_a$ and on the cathodic polarization curve, the divergence of the extrapolated exchange currents ($i^a_{0,\text{ext}} \neq i^k_{0,\text{ext}}$), the difference between $i^a_{0,\text{ext}}$ and the radiotracer value $i_{0,r}$, as well as the deviation of the sum of coefficients $\alpha_a + \alpha_k$ (derived at $\Delta\phi \gg 0$ on linear sections of the polarization curve) from $n = 3$ ($\alpha_a + \alpha_k = 0.78 + 0.57 = 1.35$), all indicate a stepwise process with comparable values of the rate constants of successive steps (see Tables 3A, B). The divergence between $i^a_{0,\text{ext}}$ and $i^k_{0,\text{ext}}$ was also noted by Hamelin and Valette,[140] who obtained polarization curves for bismuth amalgam by the potentiostatic impulse method. They applied the Vetter criterion and reached the conclusion that the electrode process was a stepwise one. However, their conclusion that in the anodic region the two-electron step $Bi \rightleftarrows Bi^{++} + 2e$ was rate-limiting is not fully substantiated, since the linear sections on the anodic curves are arbitrarily drawn and their slope actually changes with variation of the amalgam concentration. Breaks on polarization curves for bismuth amalgam were also noted by Lovrecek and Mekjavic.[140a]

*It should be noted that although, in the given case, the anodic limiting current exceeds the true exchange current by more than three orders of magnitude, the reliable determination of two Tafel sections on the same curve and of the values of two exchange currents corresponding to the Tafel sections are possible only by combining electrochemical and radiotracer measurements. The result illustrates the greater potentialities of the radiotracer method as compared with the stationary polarization curve method for examining processes near ϕ_e.

Thus, measurements over a wide potential range give three different regions of potentials on the polarization curves (Figure 10): (I) $\alpha_a = 0.78$; (II) $\alpha_a = 1.76$, $\alpha_k = 1.18$, and $\alpha_a + \alpha_k = 2.94$; and (III) $\alpha_k = 0.57$. The two linear sections on the curve of ϕ versus log i_a match the corresponding linear sections on the cathodic curves in the same region of potentials. The kinetics of the cathodic process in region I were investigated by using polarization measurements at different concentrations of metal in the amalgam and in the solution. For this purpose, polarization curves were obtained at very low amalgam concentrations and at a high concentrations of bismuth ions in solution (Figure 11). The results show that the entire anodic curve is shifted into region I and on the cathodic curve, there appears an additional break and a third linear section (in region I) with slope $b_k = 0.027$ V ($\alpha_k = 2.18$). That is, for this region, $\alpha_a + \alpha_k = 2.18 + 0.78 = 2.96$. Thus the sum of the transfer coefficients is very close to 3, both in regions II and I. The existence of two Tafel regions with different slopes on the anodic and cathodic curves shows that the rates of the cathodic and anodic processes are limited in regions I and II by different single-electron steps. On the basis of the values for α_a and α_k for these regions, the conclusion can be

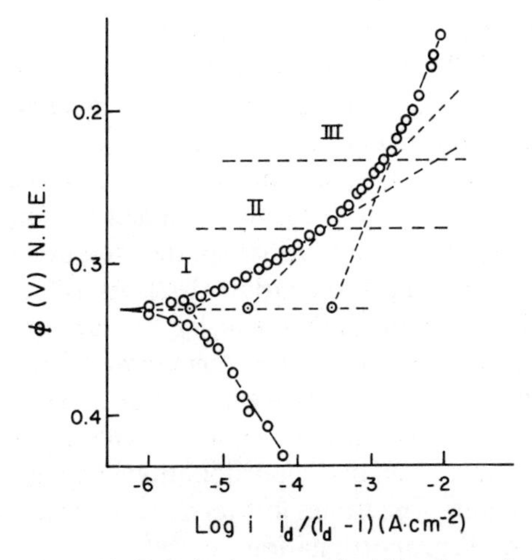

Figure 11. Polarization curves for bismuth amalgam at $c_a = 5 \times 10^{-6}$ M and $c_s = 0.1$ M ($i_d^a = 4.2 \times 10^{-5}$, $i_d^k = 0.3$ A cm^{-2}).

drawn (cf. Table 2) that in region I, the reaction $Bi \rightarrow Bi^{+} + e$ is the rate-limiting step, while that in region II is the step $Bi^{+} \rightarrow Bi^{++} + e$. The value of $\alpha_k = 0.57$ indicates that in region III, the process $B^{++} \rightarrow Bi^{3+} + e$ is the rate-limiting step. Extrapolation of the Tafel regions of the cathodic curve (Figure 11) from regions III, II, and I to ϕ_e gives the values for the i_0's of each of the three one-electron steps.

As can be seen from Figures 10 and 11, different Tafel sections of the polarization curves are divided by relatively large transition regions. Also, in the vicinity of ϕ_e, there arises the distorting effect of the reverse process. At the highest current densities, there may also be present an influence of diffusion. In addition, when values of i_0 for different steps are similar, the linear Tafel regions become shortened and are sometimes absent. As a result, the derivation of the kinetic parameters (exchange current for each step and values of transfer coefficients) on the basis of the Tafel region alone becomes unreliable. Thus, in deriving kinetic parameters, use should be made not only of the linear sections, but of the entire polarization curve. Computer calculations were made[117,141] on the basis of equation (80) for the three-step successive process. Into this equation were introduced the usual correction terms to allow for diffusion effects both in the amalgam and solution. (The anodic and cathodic limiting currents were determined experimentally.) The polarization curves calculated on the basis of equation (80), using optimal values for the kinetic parameters, are in good agreement with the experimental data for bismuth amalgam.

Analogous problems have been solved by using a computer for single-step[142] and two-step electrode processes.[143] On the basis of optimal values for the i_0's for each step, the true exchange current was calculated according to equation (102). Its value agreed with that obtained in the same experiment by the radiotracer method. This agreement and the value of the stoichiometric number ($\nu = 1$) derived from the plot of $\Delta\phi$ versus $\ln(i_a/i_k)$ (see Tables 3A, B),[31,136] as well as the reaction orders (see below), indicate the absence in the overall process of a fast disproportionation step involving low-valence bismuth ions. The values for the reaction orders indicate the absence of a slow disproportionation step.[132]

In agreement with the simple scheme of the process (see p. 319) used for deriving rate equations, the anodic and cathodic processes on bismuth, as well as those on zinc and indium amalgams (see Section III.6), are found to exhibit first-order kinetics with respect to the bulk concentration of the initial reactants and the rates are independent of the concentration of the corresponding reaction products. Thus, for example, the rate of anodic dissolution of bismuth amalgam is proportional to its concentration and is independent of the concentration of bismuth ions in the solution.[139] Since particles at the surface of the electrode participate in the electrode process, it follows that there is proportionality between the bulk and the surface concentrations of these species.

The activation energy of the overall electrode process on bismuth amalgam as determined from the temperature dependence of the radiotracer exchange currents was found to be 9 kcal $mole^{-1}$.[144] Since the processes taking place have comparable rate constants for the successive steps, the usual linear dependence between the logarithm of the exchange current for the overall process, i_0, and the reciprocal of temperature may not hold. Thus, owing to different temperature dependences of the exchange currents for different steps, the limiting step in the overall reaction can change with temperature. Nevertheless, the experimental dependence of $\ln i_0$ on $(1/T)$ was found to be linear. Apparently this can be explained by the fact that in this case, all three one-electron steps have practically the same activation energies.

In the study of bismuth discharge and ionization on a rotating disk electrode in $2M$ $HClO_4$, the steady-state polarization curves (see Figure 12) obtained[145] were similar to the curves described above for bismuth amalgams. As in the case of bismuth amalgam, the anodic curve for solid bismuth exhibits a break and two Tafel regions; the exchange currents derived by extrapolation of the linear sections of the curves from the anodic and cathodic potential regions far away from equilibrium differ from one another; the sum of apparent transfer coefficients derived from these sections (for curves 1 and 2, the sums $\alpha_a + \alpha_k$ are, respectively, 1.27 and 2.15) differ from the value of n $(= 3)$. The cathodic curves for solutions having a high concentration of bismuth ions also exhibit two linear regions with slopes differing from one another by nearly a factor of three. All these observations indicate the stepwise nature of the process with comparable rate constants for the successive steps.

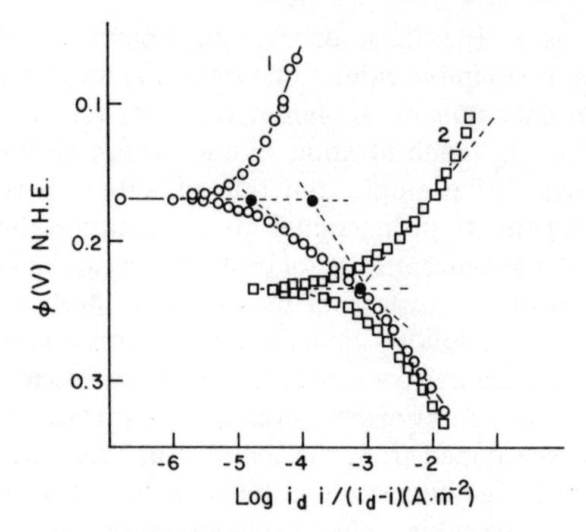

Figure 12. Polarization curves for solid bismuth at different concentrations of $Bi(ClO_4)_3$: (1) 4.6×10^{-5} *M* ($i_d^k = 8 \times 10^{-5}$ A cm^{-2}); (2) 0.1 *M* ($i_d^k = 0.37$ A cm^{-2}).

On the whole, the results of the investigations on bismuth amalgam and the comparison of these results with the experimental data for zinc and indium amalgams confirm the conclusion that for the case of comparable rate constants for the successive steps, more definite criteria of the stepwise mechanism can be used (Section IV.5). Thus, one of possible ways of proving the stepwise mechanism is to change the relative rates of the successive steps in the direction of their convergence in order to use these criteria.

As shown in Section III.7, under the usual conditions, the transfer of the last electron is apparently the rate-limiting step for the amalgams. The kinetic analysis shows that the rates of successive steps become closer when the potential is shifted in the positive direction (p. 354). In order to achieve this, polarization measurements should be carried out either far away from the equilibrium potential in the positive direction or in accordance with equations (97)–(99) at very low amalgam concentrations and at high concentrations of the metal ions in solution. It is useful now to examine the application of this method, together with the other criteria (see Table 3B), to the copper electrode as an example. Judging by the magnitude of the ratio $\alpha_a/\alpha_k \simeq 2.5$ (Figure 5), in this case, the rate of the process through the entire range of potentials under investigation is limited by the step $Cu^{2+} + e \rightarrow Cu^{+}$ (see p. 346). The ratio α_a/α_k close to 3 and $\alpha_a + \alpha_k = 2$, is also observed for copper amalgam in contact with solutions of low Cu^{++} ion concentration[98]; under these conditions, the steady-state polarization curves have the same forms as those for the solid electrode. Owing to the complicating effect of the preceding diffusion step in the amalgam, it is impossible to obtain the anodic polarization curve at higher anodic overvoltages. However, when the concentration of copper in the solution is abruptly raised and the amalgam concentration is lowered, the cathodic curve has, besides the usual linear section with a high slope ($\alpha_k = 0.41$), a second linear section with a low slope ($\alpha_k = 1.40$) at $\phi > 0.26$ V (Figure 13). Apparently, in the potential region more positive than 0.26 V, the cathodic process is limited by the discharge of Cu^{+} ions, whereas under these conditions, the step $Cu^{2+} + e \rightarrow Cu^{+}$ becomes practically reversible. As a result, a break appears on the cathodic curve near $\phi \simeq 0.26$ V between two linear sections. Extrapolation of these sections to ϕ_e gives, of course, the values for the exchange currents of these steps; they differ from one another by two orders of magnitude. In the case of solid copper electrodes, the break on the cathodic curve has not yet been sub-

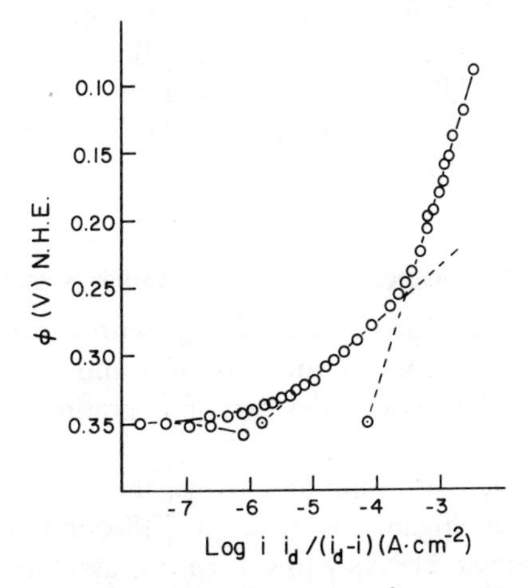

Figure 13. Polarization curves for copper amalgam ($\sim 3 \times 10^{-6}$ *M*) in 2×10^{-3} *M* $Cu(ClO_4)_2$ + 5 *M* $HClO_4$ ($i_d^k = 2.6 \times 10^{-3}$ A cm^{-2}). (Circled points show $i_{0,\text{ext}}$ values.)

stantiated (although some indications of the appearance of a break on the solid-copper cathodic curve and on the curve $\ln i_0$ vs. $\ln[Cu^{++}]$[51] have been reported) since this requires a large shift of the equilibrium potential of copper in the positive direction. In the case of an amalgam, such a shift can be effected by lowering its concentration. This peculiarity of the amalgam electrode, together with its other advantages, makes it more convenient for experimentally proving stepwise mechanisms than in the case of the corresponding solid electrode.

Analysis of the reaction order for the copper amalgam process and the values of the sum $\alpha_a + \alpha_k$, using the data in Tables 3A, B, reveals the absence of a disproportionation step involving Cu^+ ions ($2Cu^+ \rightleftarrows Cu^{++} + Cu$) in the overall process.[98] This is confirmed by the direct experimental proof of the absence of homogeneous disproportionation of these ions even when their concentration is high.[146]

A characteristic feature of the copper electrode is that it is possible to have relatively high concentrations of monovalent ions in the solution. However, as shown in the previous section (see p. 335), this can distort results of nonstationary polarization measurements. Apparently it is because of this effect that the experimental polarization data for copper amalgams obtained by potentiostatic transient methods[147] differ from the results obtained by stationary polarization measurements as described above.

For bismuth and zinc amalgams, the results of stationary and transient measurements are practically the same. This means that, for these systems, the concentrations of the intermediates are very low.

7. Experimental Results for Other Systems

The case of an electrode process having comparable rate constants for the successive steps (as with bismuth and copper amalgams) can apparently also arise under certain conditions in a number of other systems.

For example, with zinc amalgams, the α_a/α_k ratio, as in the case of copper amalgams, is close to 3 (Section III.6). Since the i_0 for this system is relatively low, it can be assumed that if a sufficiently positive $\Delta\phi$ is applied, it should be possible to detect experimentally the change in the limiting step, e.g., by observing an increase in α_a. The data of Hush and Blackledge[148] apparently prove the significance of this effect. These authors carried out impedance measurements on zinc amalgam. At equilibrium, they obtained a high value for the transfer coefficient $\alpha_a = 1.38$ (this value has been recalculated by the present author for the condition $\alpha_a + \alpha_k = 2$). However, in their calculation from the slope of the anodic curve, Hush and Blackledge obtained a considerably lower value ($\alpha_a = 0.68$). In accordance with these results, the impedance and faradaic rectification measurements showed an increase in the cathodic transfer coefficient α_k when the potential was shifted in the positive direction.[149,149a] A similar effect was observed on cadmium amalgam.[149a]

The possibility of a stepwise mechanism in the electrode process on solid zinc in an acid perchlorate solution is indicated by: the high b_a and b_k values on the steady-state polarization curves (~ 125 mV), the difference between the values of extrapolated exchange currents ($i^{a}_{0,\mathrm{ext}} > i^{k}_{0,\mathrm{ext}}$), the appearance on the anodic curve of a second linear section with a lower slope (~ 40 mV) in the vicinity of ϕ_e in solutions with a low concentration of zinc ions, and by the results of nonstationary measurements.[149b] Analogous results were obtained on solid cadmium.[149c]

Table 3A shows that determination of ν does not permit differentiation of the case of the stepwise process having comparable rate constants of successive charge-transfer steps (case III) from the simple, single-step process (case IV). Therefore, the conclusion drawn by Horiuti and Matsuda[78] about the single-step mechanism of the electrode process on zinc amalgam, based only on the value of $\nu\,(=1)$, is unconvincing.

Judging from the magnitude of the ratio α_a/α_k for indium amalgam in the region of ϕ_e and at anodic polarizations, the transfer of the last electron is the rate-limiting step (see Section III.6). In polarization measurements on 0.3 wt.% indium amalgam at high current densities (up to 500 mA cm^{-2}), using the galvanostatic transient method, Markovac and Lovrecek[150] observed on the anodic curve, in addition to the usual Tafel section with a low slope (Figure 4), a break and a second linear section with a high slope. They related this to the transition from one of the limiting charge-transfer steps to the other. Breaks on the anodic curves were also observed by Kozin and co-workers[151] for concentrated indium amalgams (20 at.%). Markovac and Lovrecek attributed the increase in the slope of the anodic curve (from 0.021 to 0.060 V in a sulfate solution) to a sharp increase in the concentrations of In^+ and In^{2+} as the c.d. increased. Under these conditions, the removal of these species by disproportionation becomes possible, leading to the formation of In^{3+} ions. As a result, the rate of the step $In \rightarrow In^+ + e$ begins to determine the velocity of the overall process.

It can readily be shown that this explanation is erroneous, for it is incompatible with the authors' own assumption concerning the accumulation of In^+ ions during anodic polarization. Indeed, if the step $In \rightarrow In^+ + e$ tended to be slower than the following disproportionation step, the concentration of In^+ ions during anodic polarization (in solution with a constant concentration of In^{3+} ions) could not significantly exceed the equilibrium value which is determined by the equilibrium constant for the disproportionation process. However, direct experiments prove that during anodic polarization of In amalgam, an accumulation of In^+ ions actually takes place. This accumulation indicates a higher rate constant for the $In \rightarrow In^+ + e$ step as compared with those of the subsequent steps.[56,57,60] Furthermore, the mechanism proposed by Lovrecek and Markovac seems to correspond to an improbably high value for the true transfer coefficient of the first step $\beta_{a,1} = \alpha_a \doteqdot 1$ (Table 2). The experiments carried out with a ring-disk electrode[64] also indicate the absence of disproportionation of In^+ ions in the anodic polarization of indium. Nevertheless, the rise in the slope of the anodic curve at high current densities observed by Lovrecek and Markovac may indicate a change in the limiting step of the process.

These authors also observed two linear sections and a break on the cathodic curve upon increasing the current density.[47,150] They related this to the transition from a limiting $In^{3+} + e \rightarrow In^{2+}$ step to the limiting step $In^{3+} + 2e \rightarrow In^+$. As detailed investigations of the cathodic process showed,[69,88–90] the first linear section with a high slope corresponds, in fact, to the limiting preceding chemical step (see p. 343), and the second section corresponds to the limiting step $In^{3+} + e \rightarrow In^{2+}$. Thus, the break on the cathodic curve is not related to a change in the limiting charge-transfer step.

Similarly, in studies on Cd amalgam by the galvanostatic transient method, Lovrecek and Marincic[152] observed Tafel regions on the anodic and cathodic branches of the polarization curve with high and similar slopes (~ 180 mV), which correspond to $\alpha_a = \alpha_k = 0.32$. That is, the sum $\alpha_a + \alpha_k$ is significantly less than 1.[152] [These data are not, however, in accord with results of other workers, according to whom $\alpha_a/\alpha_k > 3$ and $\alpha_a + \alpha_k \simeq 2$ (p. 345)]. Extrapolations of these sections to the equilibrium potential give identical values for the exchange current. Since the exchange current is very high, these Tafel regions were obtained only for a narrow range of current densities (250–500 mA cm^{-2}), and the measurements required large Ohmic corrections. The authors concluded that the limiting steps in the overall anodic and cathodic processes are, respectively, $Cd \rightarrow Cd^+ + e$ and $Cd^{2+} + e \rightarrow Cd^+$, and in both cases, the slow step is followed by the fast disproportionation of Cd^+ ions. As follows from the data in Tables 3A, B, for such a mechanism (case e), the condition $\alpha_a + \alpha_k > 1$ should hold and one of the branches of the polarization curve should have a break. This, however, is not in agreement with the experimental observations of Lovrecek and Marincic.[151] Furthermore, when there is a fast disproportionation step, nonelectrochemical exchange should be observed (see p. 369); this, however, contradicts the results of Baticle,[153] who obtained good agreement between values for the exchange current on cadmium amalgam obtained by means of radiotracer and electrochemical measurements. Thus, the mechanism proposed by Lovrecek and Marincic is still in doubt.

It can be readily seen that these data cannot alternatively be explained on the basis of another mechanism involving fast disproportionation. Thus, for example, for the cases (a) and (b) (Table 3A), the condition $\alpha_a + \alpha_k = 1$ must be obeyed. Also, nonelectrochemical exchange ($i_{0,r} > i_0$) should be observed and the participation of these mechanisms can be readily established by determination of reaction orders (see Tables 3A, B).

Besides the amalgams, a liquid gallium electrode is a convenient means of studying the kinetics of metal discharge-ionization processes. On the anodic and cathodic branches of the polarization curve for liquid gallium in alkaline solutions of $GaCl_3$, Bockris and Enyo[154] obtained distinct Tafel regions. The ratio of the transfer coefficients was $\alpha_a/\alpha_k \simeq 3$; however, their sum was considerably less than $n = 3$ ($\alpha_a + \alpha_k = 0.45 + 1.3 = 1.75$). Extrapolation of each of the Tafel regions to ϕ_e gave identical values for i_0. The authors concluded that the anomalously low value for the sum $\alpha_a + \alpha_k$ could be explained in terms of the following mechanism:

$$H_2GaO_3^- + e \rightarrow HGaO_2^- + OH^- \quad (131)$$

$$HGaO_2^- + e \rightarrow GaO^- + OH^- \quad (132)$$

$$HGaO_2^- + GaO^- + H_2O \rightarrow Ga + H_2GaO_3^- + OH^- \quad (133)$$

The rate-limiting step in this process is (131). The assumption of monovalent gallium as an intermediate in the electrode process is supported by data on the anodic dissolution of liquid gallium in alkaline solutions.[155] It would be desirable to confirm such a mechanism (which is similar to case b in Table 3A) by determining reaction orders[40a] since the rate of the anodic process should be proportional to the concentration of the final product, i.e., $H_2GaO_3^-$ ions. Furthermore, in this case, as in the case of mechanisms involving a fast disproportionation step (Tables 3A, B), nonelectrochemical exchange should again be observed.

The low value of $(\alpha_a + \alpha_k)$ can be caused by the fact that the overall process takes place through three successive one-electron steps with comparable rate constants for steps (131) and (132). Indeed, if step (132) is rate-limiting in the anodic polarization and (131) is the limiting step in the cathodic polarization, the sum $\alpha_a + \alpha_k$ determined from corresponding slopes should not be 3, but 2. In this case, when the concentration of gallium ions is significantly decreased, the values of $i^a_{0,\mathrm{ext}}$ and $i^k_{0,\mathrm{ext}}$ should become different (p. 358) and on the anodic curve there may appear near ϕ_e a second Tafel region with a lower slope corresponding to the limiting step (131). In accordance with this assumption, anodic dissolution of liquid gallium in alkaline chloride solutions free of gallium ions apparently gives Tafel regions having low slopes.[156]

In conclusion, it should be noted that stepwise mechanisms involving the formation of low-valence intermediates and the criteria which indicate them are more and more frequently invoked for interpreting polarization measurements in electrochemical dissolution and deposition of metals on solid electrodes such as iron,[23,24,45,157–159] cobalt,[160–162] nickel,[128,163,164,164a] copper,[44] tellurium,[165] and bismuth.[145,166] An analysis of all these contributions would be outside the scope of the present chapter. Thus, here, only those new problems and difficulties will be listed which are encountered as the study of reaction (1) on *amalgams* is extended to that on *solid* electrodes.

Electrolytic deposition of the majority of metals, especially metals of the iron group, is accompanied by hydrogen evolution. On this account, it is necessary to obtain the cathodic partial curves of the process under investigation and to take into account the possible effects of hydrogen penetration into the surface of the cathode on the kinetics of this process.

For a number of metals, the rate of the electrode process sharply decreases with decreasing *p*H of the solution.[1,106] Thus, in the workup of experimental data, allowances must be made for the possible difference of the *p*H value at the surface of the electrode and in the bulk of the solution.[45] In the case of indium amalgam (in contrast to solid metal electrodes), the effect of *p*H could be explained only on the basis of a stepwise mechanism with participation of hydrolyzed particles in a rate-limiting step such as $InOH^+ \rightarrow InOH^{2+} + e$ followed by the chemical step $InOH^{2+} + H^+ \rightarrow In^{3+} + H_2O$.[69,88] The significance of such a step is indicated by analysis of the anodic[90,127] and cathodic[89] polarization curves. The formation of $InOH^{2+}$ intermediates (together with low-valence In^+ intermediates) was proved directly by means of an auxiliary mercury electrode.[57,60]

In the case of some metals, interpretation of the results of polarization measurements requires consideration of the adsorption of initial reactants and absorption of reaction products. Account should also be taken of the adsorption of low-valence intermediates and derived compounds on the electrode surface and the change in the surface coverage by such species as the electrode potential is changed.[51,53,157,158,161]

In those cases where the metal under investigation can form an amalgam, valuable information concerning the mechanism of the electrode process can be obtained by comparing the electrochemical behavior of the solid metal with that of its amalgam. Thus, the agreement of α_a/α_k values for a metal and its amalgam (Section III, 7) apparently shows that in both cases the mechanism is analogous. The far-reaching parallelism between the electrochemical behavior of copper, zinc, indium and bismuth, and that of the corresponding amalgams, confirms the conclusion about the critical role of redox steps in the kinetics of both types of electrode reaction.

V. APPLICATION OF THE CRITERIA FOR STEPWISE MECHANISMS IN SOME ELECTRODE REDOX REACTIONS

Any stepwise process of metal discharge or ionization includes electrochemical redox steps involving cations of different valencies [equation (6)]. Thus, it is desirable to conclude by examining briefly the experimental proof of stepwise mechanisms in such redox reactions occurring at inert electrodes.

Esin and Loshkarev[9] found that when the reaction $Sn^{2+} \rightarrow Sn^{4+} + 2e$ occurs in acid chloride solutions at a streaming mercury electrode, the slope of the anodic polarization curve is considerably lower than that of the cathodic curve; extrapolation of Tafel regions to ϕ_e gave values of i_0 that are close to one another. In this case, $\alpha_a/\alpha_k \simeq 4$ and the sum $\alpha_a + \alpha_k \simeq 2$. From Table 3A, these results may indicate the presence of a two-step process under conditions when the exchange current of the step $Sn^{3+} \rightarrow Sn^{4+} + e$ is considerably lower than that of the first step (case I). This conclusion is confirmed by the results obtained for the same system at electrodes of pyrolytic graphite.[167] The anodic polarization curve has a break and two distinct Tafel sections with slopes of ~40 and 135 mV (Figure 14). Extrapolation to ϕ_e gave i_0 values which differed by three orders of magnitude. The experimental anodic curve over a wide range of potentials coincides with the theoretical curve calculated from an equation similar to (80) (for the two-step process with corrections for the effect of diffusion) using kinetic parameters ($i_{0,1}$, $i_{0,2}$, $\beta_{a,1}$, and $\beta_{a,2}$), derived from these Tafel regions. The authors concluded that the electrode reaction proceeds stepwise.

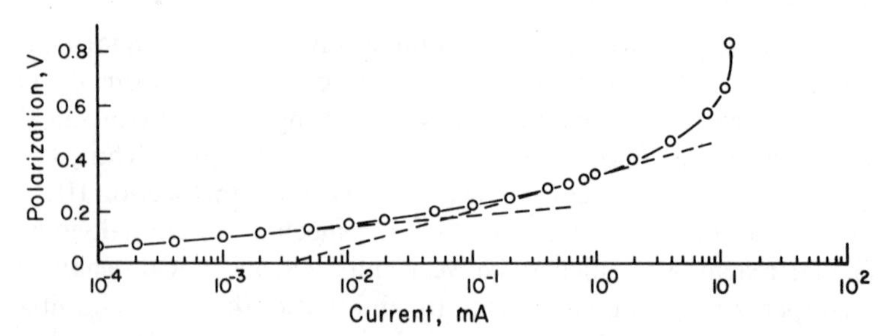

Figure 14. Anodic polarization curve for the solution 0.104 M $SnCl_2$ + 0.081 M $SnCl_4$ + 4 N HCl on pyrolytic graphite (after Lerner and Austin[168]).

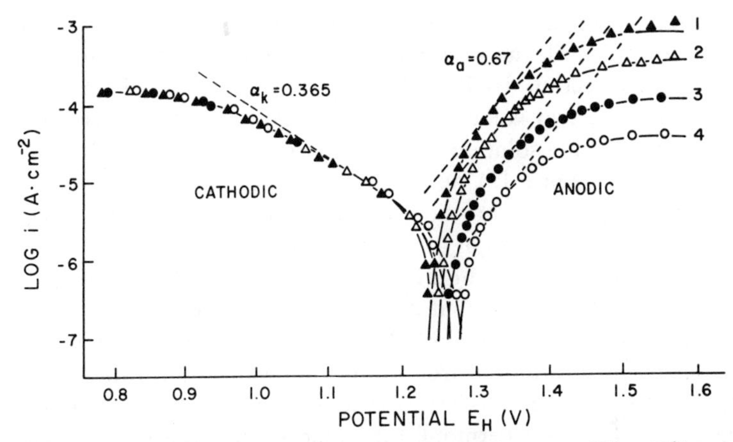

Figure 15. Anodic and cathodic polarization curves for the process $Tl^+ \rightleftarrows Tl^{3+} + 2e$ on a smooth platinum electrode in 15*N* H_2SO_4 at constant concentration of Tl^{3+} (10^{-3} *M*) and different concentrations of Tl^+: (1) 3×10^{-3} *M*; (2) 10^{-3} *M*; (3) 3×10^{-4} *M*; (4) 10^{-4} *M* (after Vetter and Thiemke[170]).

A similar conclusion was reached for the anodic oxidation of As^{3+} to As^{5+} at Pt in 1 *M* $HClO_4$.[169] In this work, a polarization curve with two Tafel slopes (~45 and 95 mV) was also obtained. A polarization curve with a break and two Tafel sections (with α_k 0.67 and 0.25–0.33) was obtained for the cathodic reduction $CO_2 \xrightarrow{+2e,+2H^+} HCOOH$ on Hg[169a] and interpreted in terms of a stepwise mechanism with two one-electron steps.

A typical example of a simple redox process is

$$Tl^+ \rightleftarrows Tl^{3+} + 2e \tag{134}$$

which was shown by Vetter and Thiemke[170] to proceed by a stepwise mechanism. These authors studied the kinetics of reaction (134) in 15 *N* H_2SO_4 on a platinum electrode by carrying out polarization measurements under conditions of vigorous stirring over a wide range of potentials and at different concentrations of Tl^+ and Tl^{3+}. After introducing corrections for diffusion, the polarization curves exhibit Tafel regions with high slopes (dashed lines in Figure 15). Extrapolations of these regions to ϕ_e give different $i^a_{0,\text{ext}}$ and $i^k_{0,\text{ext}}$, indicating a stepwise mechanism for the process with comparable rate constants of successive one-electron steps (the Vetter criterion). At equal concentrations of Tl^+ and Tl^{3+}, $i^a_{0,\text{ext}} > i^k_{0,\text{ext}}$; that is, the transfer of the last electron $Tl^{2+} \rightarrow Tl^{3+} + e$ is the slower step. This accords with the results of the analysis of data obtained for amalgams (Section III,7). When the concentration of Tl^{+} ions decreases, the two values for the exchange current approach one another and, for a solution having a concentration of 1.10^{-4} M Tl^{+}, $i^{a}_{0,\text{ext}} = i^{k}_{0,\text{ext}}$ (Figure 15), an effect which is caused by the different dependences on concentration of the exchange currents of the steps according to equations of the type (97)–(99).

The conclusion about the stepwise mechanism of reaction (134) is confirmed by the fact that it is consistent with the two criteria of the stepwise mechanism (Table 3A, case III): the sum $\alpha_a + \alpha_k$ is close to unity; $i_0 < i^{a}_{0,\text{ext}}$ [the true exchange current i_0 can be readily calculated from the transfer resistance cited in Ref. 170 according to equation (100)]; furthermore, the experimental values for i_0 agree with values calculated from $i^{a}_{0,\text{ext}}$ and $i^{k}_{0,\text{ext}}$ according to equation (102). The reaction orders also correspond to case III and are incompatible with the mechanisms (a)–(d) (Table 3A), including the disproportionation step. In order to exclude case (e), it would be necessary to carry out measurements at concentrations of Tl^{+} and Tl^{3+} that differ considerably from each other (see p. 371) and to establish the absence of nonelectrochemical exchange.

Vetter and Thiemke[170] used experimental values of the kinetic parameters of both steps and for the diffusion-limiting currents in their calculation of the entire polarization curve according to an equation of the type (80) (solid lines in Figure 15). The calculated values are in good agreement with the experimental polarization data. The homogeneous isotopic exchange between Tl^{+} and Tl^{3+} ions apparently also proceeds stepwise involving Tl^{2+} intermediates, as is indicated by a large increase in the exchange rate which takes place upon the illumination of the system, due to a chain mechanism involving these ions.[171]

In their study of the process $Tl^{+} \rightleftarrows Tl^{3+} + 2e$ in 1 M $HClO_4$, Catherino and Jordan[172] concluded that in the anodic and cathodic regions, the reaction rate is limited by the respective steps $Tl^{+} \rightarrow Tl^{2} + e$ and $Tl^{3+} + e \rightarrow Tl^{2+}$, and $k_a/k_k \doteqdot 10^{3}$. This conclusion accords with the data of Vetter and Thiemke[170] cited above. However, the suggestion by these authors, that in both cases disproportionation involving Tl^{2+} ions is the fast following step, is unconvincing. Indeed, when a disproportionation step occurs, which is reversible and independent of the potential, the cathodic process would have taken place not through the slow-transfer step $Tl^{3+} + e \rightarrow Tl^{2+}$, but through a parallel, more facile disproportionation $Tl^{3+} + Tl^{+} \rightarrow 2Tl^{2+}$ with a subsequent step $Tl^{2+} + e \rightarrow Tl^{+}$. That is, the rate-limiting step both in the anodic and cathodic regions would be the same (Table 3A, case a). Further investigation of the process (134) showed that the b values are strongly affected by oxidation of the platinum surface as well as by the adsorption of Tl^{+} and traces of Cl^{-} ions,[173,174] and this may be the cause of the observed divergence of the anodic and cathodic rate constants. In more concentrated H_2SO_4 solutions, these anomalies are less evident[173] and it can be assumed that the experimental data of Vetter and Thiemke[170] for 15 N H_2SO_4 solution were not complicated by such effects. These results demonstrate the difficulties which are encountered in the study of the kinetics of complex stepwise electrode reactions on solid electrodes and, in particular, on platinum. The classical benzoquinone–hydroquinone redox couple[123,175,176] is another example which illustrates the large effect of the state of the platinum surface, the adsorption effects, and the effect of the technique of polarization measurement on the kinetic parameters evaluated. A stepwise mechanism with comparable rate constants of successive one-electron steps was established for the duroquinone–durohydroquinone couple on Au.[176a] In this case, it was quantitatively proved that disproportionation of intermediates proceeds slowly and can be neglected.

The results of polarization measurements for the reaction

$$Pb^{++} \rightleftarrows PbO_2 + 2e \tag{135}$$

in concentrated $HClO_4$ solutions indicate a stepwise mechanism for this reaction.[177,178] In this case, the Tafel slope of the anodic branch of the polarization curve is several times higher than that of the cathodic branch. The sum of the values of transfer coefficients obtained from these slopes was significantly smaller than 2 ($\alpha_a + \alpha_k = 0.86 + 0.24 = 1.1$) and the $i_{0,\mathrm{ext}}$ values did not coincide with one another ($i^k_{0,\mathrm{ext}} < i^a_{0,\mathrm{ext}}$). Using the Vetter criterion and that of the transfer coefficient sum, Hampson *et al.*[178] concluded that reaction (135) occurs by successive one-electron steps having comparable rate constants (case III, Table 3A):

$$Pb(2+) \overset{i_{0,1}}{\rightleftarrows} Pb(3+) + e \tag{136}$$

$$Pb(3+) \overset{i_{0,2}}{\rightleftarrows} Pb(4+) + e \tag{137}$$

where Pb(2+) and Pb(4+), respectively, represent the hydrolyzed forms of bivalent and the tetravalent lead formed, as a result of preceding chemical reactions from Pb^{2+} and PbO_2. With increasing concentration of Pb^{2+} ions, $i_{0,1}$ increases more slowly than $i_{0,2}$. This is inconsistent with the equations for the concentration dependence of i_0 values of different steps (see p. 358). It is possible that this unexpected result is due to a considerable deviation from 1 of the order of the anodic reactions with respect to Pb^{2+} ions observed by the authors.[178] The assumption of a stepwise mechanism for reaction (135) which involves the formation of trivalent lead intermediates was first made by Varypaev and Fedote'v,[179] who obtained anodic polarization curves with high slopes ($\alpha_a = 0.41$–0.46) in nitric acid solutions. The experimental data[178] can be described by equation (47) for the simple, single-step process by means of a computer, only if it is assumed that a substantial change in the transfer coefficients occurs with variation of the potential in the vicinity of ϕ_e. However, this is unlikely, thus indicating the inapplicability of the single-step mechanism.

The i_0 for the overall process calculated from experimental values of $i^k_{0,\mathrm{ext}}$ and $i^a_{0,\mathrm{ext}}$ according to equation (102) was found to be 2–3 times smaller that that determined directly from the initial slope of the polarization curve.[177] In order to explain this difference, it was assumed that near ϕ_e, the process takes place in a single-step mechanism. Such a change in the mechanism of the electrode process within a narrow potential range near equilibrium is questionable. In fact, analysis of reaction (135), proceeding by two parallel mechanisms (simple and stepwise) with comparable rates in both paths (Section IV,3), shows that if, near ϕ_e, the reaction predominantly occurs by the simple mechanism, this would also be the pathway at appreciable $\Delta\phi$ owing to the larger change in the activation energy with ϕ for the simple mechanism than for the stepwise one. Then, at least for one of the branches of the polarization curve, this predominance should be evident at all $\Delta\phi$.

REFERENCES

[1] M. V. Simonova and A. L. Rotinyan, *Usp. Khim.* **34** (1965) 734.
[2] K. Vetter, *Elektrochemische Kinetik*, Springer-Verlag, Berlin, 1961.
[3] K. Vetter, *Z. Naturforsch.* **7a** (1952) 328.
[4] D. C. Grahame, *Ann. Rev. Phys. Chem.* **6** (1955) 337.
[5] N. N. Semyonov, *O nekotoryck problemack Khim. kinetiki*, Chapter V, 1958.
[6] A. N. Frumkin, *Tr. soveshch. po elektrokhimii*, p. 34, 1953.
[7] K. Vetter, in *Trans. Symp. on Electrode Processes* (*Electrochem. Soc.*), 1958; p. 47, John Wiley and Sons, New York, 1961.
[8a] P. Delahay, *Double-Layer and Electrode Kinetics*, Interscience, New York, 1965.
[8b] B. E. Conway, *Theory and Principles of Electrode Processes*, Ronald Press, New York, 1964.
[9] O. Esin and M. Loshkarev, *Zh. Fiz. Khim.* **13** (1939) 794; *Acta Physicochem. USSR* **10** (1939) 513.
[10] O. Esin, *Zh. Fiz. Khim.* **17** (1943) 159; *Acta Physicochem. URSS* **16** (1942) 103.
[11] H. Gerischer, *Z. Physik. Chem.* **202** (1953) 292.
[12] A. G. Stromberg, *Zh. Fiz. Khim.* **29** (1955) 409.
[13] G. C. Barker, in *Trans. Symp. on Electrode Processes* (*Electrochem. Soc.*), 1958, p. 325, John Wiley and Sons, New York, 1961.
[14] J. E. B. Randles, in *Trans. Symp. on Electrode Processes* (*Electrochem. Soc.*), p. 209, John Wiley and Sons, New York, 1961.
[15] T. P. Hoar, in *Modern Aspects of Electrochemistry*, Eds., J. O'M. Bockris and B. E. Conway, Vol. 2, Chapter IV, Butterworths, London, 1959.
[16] T. Hurlen, *Electrochim. Acta* **7** (1962) 653.
[17] J. Heyrovsky, *Disc. Faraday Soc.* **1** (1947) 212.
[18] M. A. Loshkarev and A. A. Krjukova, *Zh. Fiz. Khim.* **26** (1952) 731; **30** (1956) 2336.
[19] A. N. Frumkin, *Voprosy Khim. Kinetiki* **1955**, 402.
[20] A. N. Frumkin, *Dokl. Akad. Nauk SSSR* **85** (1952) 373.
[21] J. Heyrovsky, *Sb. tschekhoslov. khim. rabot*, Suppl. 2, **1954**, 58; *Chem. Zvesty* **8** (1954) 617.
[22] J. Heyrovsky and J. Kuta, *Principles of Polarography*, Chapter XVI, Prague, 1965.
[23] B. N. Kabanov and D. I. Leikis, *Dokl. Akad. Nauk SSSR* **58** (1947) 1685.
[24] B. Kabanov and R. Burstein, and A. Frumkin, *Disc. Faraday Soc.* **1** (1947) 259.
[25] A. N. Frumkin, *Zh. Fiz. Khim.* **10** (1937) 568; *Acta Physicochim.* **7** (1937) 475.
[26a] M. A. Vorotynzev and A. M. Kuznetsov, *Elektrokhimia* **6** (1970) 208.
[26b] B. E. Conway and J. O'M. Bockris, *Proc. Roy. Soc.* **A248**, (1958) 394; *Electrochim. Acta* **3** (1961) 340.
[27] V. V. Losev and G. M. Budov, *Zh. Fiz. Khim.* **37** (1963) 578.
[28] V. V. Losev and A. I. Molodov, *Zh.Fiz. Khim.* **35** (1961) 2289.
[28a] D. M. Mohilner, *J. Phys. Chem.* **73** (1969) 2652.
[29] R. Parsons, *Ann. Rep. Progr. Chem.* **61** (1965) 80.
[29a] A. I. Molodov and V. V. Losev, *Elektrokhimia* **7** (1971) 818.
[29b] I. Kiss and J. Farkas, *Acta Chim. Acad. Sci. Hung.* **64** (1970) 241; **66** (1970) 33.
[30a] A. N. Frumkin, V. S. Bagotzky, S. A. Iofa, and B. N. Kabanov, *Kinetika elektrodnykh prozessov*, Izd. MGU, Moscow, 1952.
[30b] A. N. Frumkin, *Z. phys. Chem.* **AI64** (1933) 121.
[31] J. Horiuti and M. Ikusima, *Proc. Imp. Akad. Tokio* **15** (1939) 39; J. Horiuti, *J. Res. Inst. Catalysis* **1** (1948) 8.

[32]R. Parsons, in *Adv. in Electrochem.*, Vol. 1, p. 54, Eds., C. Tobias and P. Delahay, Interscience, New York, 1961.
[33]V. V. Scorcheletti, *Teoret. Elektrokhimiya*, Leningrad, 1959.
[34a]A. N. Frumkin, *Z. phys. Chem.* **AI60** (1932) 116.
[34b]C. Wagner and W. Traud, *Z. Elektrochem.* **44** (1938) 391.
[34c]Ya. M. Kolotyrkin and A. N. Frumkin, *Dokl. Akad. Nauk SSSR* **33** (1941) 445.
[35]A. C. Wahl and N. A. Bonner, *Radioactivity Applied to Chemistry*, 1951.
[36]V. V. Losev, *Dokl. Akad. Nauk SSSR* **100** (1955) 111.
[37]V. V. Losev and A. M. Khopin, *Tr. 4 Soveshch. po elektrokhimii, 1956*, p. 116, Moscow, 1959.
[38]G. M. Budov and V. V. Losev, *Dokl. Akad. Nauk SSSR* **129** (1959) 1321.
[39]B. Lovrecek, *J. Phys. Chem.* **63** (1959) 1795.
[40]D. M. Mohilner, *J. Phys. Chem.* **68** (1964) 623.
[40a]B. E. Conway and E. J. Rudd, *Disc. Faraday Soc.* **45** (1968) 87; B. E. Conway and M. Salomon, *Electrochim. Acta* **9** (1964) 1599.
[41]O. Esin, *Zh. Fiz. Khim.* **14** (1940) 731; *Acta Physicochim. URSS* **13** (1940) 429.
[42]L. Gierst and P. Cornelissen, *J. Electroanal. Chem.* **12** (1966) 524.
[43]J. E. B. Randles, *Can. J. Chem.* **37** (1959) 238; J. E. B. Randles and D. R. Whitehouse, *Trans. Faraday Soc.* **64** (1968) 1376.
[43a]R. Parsons and E. Passeron, *J. Electroanal. Chem.* **12** (1966) 524.
[43b]H. Gerischer, *Z. Elektrochem.* **54** (1950) 366.
[44]E. Mattson and J. O'M. Bockris, *Trans. Faraday Soc.* **55** (1959) 1586.
[45]J. O'M. Bockris, D. Drazic, and A. R. Despic, *Electrochim. Acta* **4** (1961) 325.
[46]M. Perez-Fernandez and H. Gerischer, *Z. Elektrochem.* **64** (1960) 477.
[47]B. Lovrecek and V. Markovac, *J. Electrochem. Soc.* **109** (1962) 727.
[47a]A. I. Molodov, G. N. Markosyan, and V. V. Losev, *Elektrokhimia* **7** (1971) 263.
[48]A. I. Molodov, V. I. Barmashenko, and V. V. Losev, *Elektrokhimia* **7** (1971) 18.
[49]A. I. Molodov, *Elektrokhimia* **6** (1970) 365.
[50]A. Vleck, *Coll. Czech. Chem. Commun.* **22** (1957) 948, 1736.
[51]J. O'M. Bockris and M. Enyo, *Trans. Faraday Soc.* **58** (1962) 1187.
[52]O. R. Brown and H. R. Thirsk, *Electrochim. Acta* **10** (1965) 383.
[53]J. O'M. Bockris and H. Kita, *J. Electrochem. Soc.* **108** (1961) 676.
[54]W. Kangro and E. Weingartner, *Z. Elektrochem.* **58** (1954) 505; see also G. Biedermann and T. Wallin, *Acta Chem. Scand.* **14** (1960) 594.
[55]V. M. Novakovsky and A. S. Schubin, *Tr. Ural. Nauchnoissled. khim. inst.* **9** (1961) 25.
[56]R. E. Visco, J. Electrochem. Soc. **112** (1965) 936.
[57]A. I. Molodov and V. V. Losev, *Elektrokhimia* **4** (1968) 835.
[58]A. I. Molodov, G. N. Markosyan, A. P. Pchelnikov, and V. V. Losev, *Elektrokhimia* **4** (1968) 1370.
[59]V. V. Gorodetzki, V. V. Losev, and L. I. Fedorzev, *Elektrokhimia* **5** (1969) 1271.
[60]V. V. Losev, *Electrochim. Acta* **15** (1970) 1095.
[60a]B. E. Conway and D. J. MacInnon, *J. Phys. Chem.*, **74** (1970) 3663.
[61]A. P. Pchelnikov and V. V. Losev, *Zashchita metallov* **1** (1965) 482.
[62]V. V. Losev and A. P. Pchelnikov, in *Proc. 3rd Congr. on Metallic Corrosion, 1966*, p. 101, 1968.
[63]L. N. Nekrasov and N. P. Berezina, *Dokl. Akad. Nauk SSSR* **142** (1962) 855.
[64]B. Miller and R. E. Visco, *J. Electrochem. Soc.* **115** (1968) 251.
[65]L. Kiss, *Mady. Kem. Foly.* **72** (1966) 191; L. Kiss, A. Körösi, and J. Farkas, *Ann. Univ. Budapest, sect. Chim.* **9** (1967) 3; L. Kiss and J. Farkas, *Acta Chim. Acad. Sci. Hung.* **64** (1970) 241; **65** (1970) 7, 141; **66** (1970) 33; *Mag. Kem. Foly.* **76** (1970) 389; *Kemiai Kozlemények* (1970) 261.

[66]J. Jacq, B. Cavalier, and O. Bloch, *Electrochim. Acta* **13** (1968) 1119.
[67]G. W. Tindall and S. Bruckenstein, *Anal. Chem.* **40** (1968) 1051, 1637.
[67a]B. E. Conway and E. Gileadi, *Trans. Faraday Soc.* **58** (1962) 2493.
[67b]E. Gileadi and B. E. Conway, in *Modern Aspects of Electrochemistry*, Vol. 3, Chapter 3, Eds., J. O'M. Bockris and B. E. Conway, Butterworths, London, 1964.
[67c]R. Parsons, in *Advances in Electrochemistry and Electrochemical Engineering*, Vol. 7, Chapter 3, Ed., P. Delahay, Interscience, New York, 1970.
[68]V. V. Losev, *Tr. Inst. Fiz. Khim. Akad. Nauk SSSR* **6** (1957) 20.
[69]V. V. Losev and A. I. Molodov, *Electrochim. Acta.* **6** (1962) 81.
[70]G. M. Budov and V. V. Losev, *Zh. Fiz. Khim.* **37** (1963) 1023.
[71]V. V. Losev, M. A. Dembrovski, and A. I. Molodov, *Problemy Fiz. Khimii* **3** (1963) 43; *Zh. Fiz. Khim.* **37** (1963) 1904.
[72]V. V. Losev, M. A. Dembrovski, A. I. Molodov, and V. V. Gorodetzki, *Electrochim. Acta* **8** (1963) 387.
[73]A. P. Pchelnikov and V. V. Losev, *Elektrokhimia* **1** (1965) 1058.
[73a]N. B. Miller and V. A. Pleskov, *Dokl. Akad. Nauk SSSR* **74** (1950) 323; *Tr. 3 Soveshch. Elektrokhimii, 1950*, Moscow (1953) 165.
[74]F. Joliot, *J. Chim. Phys.* **27** (1930) 119.
[75]R. Audubert, *Compt. Rend.* **238** (1954) 1997.
[76]S. Raschkov, K. Hamparzumyan, and N. Pangarov, *Isv. Inst. Fiz. Khimii BAN* **4** (1964) 97; K. Hamparzumyan and G. Raichevski, *Isv. Inst. Fiz. Khimii BAN* **6** (1967) 131.
[77]H. Gerischer and R. Tischer, *Z. Elektrochem.* **58** (1954) 819; **61** (1957) 1159.
[78]J. Horiuti and A. Matsuda, *J. Res. Inst. Catalysis* **7** (1959) 19.
[79]S. Levina, *Acta Physicochim. URSS* **14** (1941) 294.
[80]V. V. Gorodetzki, L. I. Fedortzev, and V. V. Losev, *Elektrokhimia* **4** (1968) 967.
[81]A. I. Molodov, G. N. Markosyan, and V. V. Losev, *Elektrokhimia* **5** (1969) 918.
[82]A. M. Baticle and Y. Thouvenin, *Compt. Rend.* **248** (1959) 794, 1330.
[83]G. M. Budov and V. V. Losev, *Dokl. Akad. Nauk SSSR* **122** (1958) 90.
[83a]B. V. Ershler and K. I. Rosental, *Tr. 3 Soveshch. Elektrokhimii, 1950*, Moscow (1953) 446.
[84]H. Gerischer, *Z. Phys. Chem.* **202** (1953) 302.
[85]J. H. Sluyters and I. I. C. Omen, *Rec. trav. chim.* **79** (1960) 1101.
[86]A. M. Baticle, *Compt. Rend.* **254** (1962) 668.
[86a]N. Tanaka, Y. Aoki, and A. Yamada, *Electrochim. Acta* **14** (1969) 1155.
[86b]G. Salie, *Z. phys. Chem.* **244** (1970) 1.
[87]G. M. Budov and V. V. Losev, *Zh. Fiz. Khim.* **37** (1963) 1230.
[88]V. V. Losev and A. I. Molodov, *Dokl. Akad. Nauk SSSR* **135** (1960) 1432.
[89]A. I. Molodov and V. V. Losev, *Elektrokhimia* **1** (1965) 651.
[90]A. I. Molodov and V. V. Losev, *Elektrokhimia* **1** (1965) 1253.
[91]V. V. Losev and G. M. Budov, *Zh. Fiz. Khim.* **37** (1963) 1461.
[92]H. Gerischer, *Z. Elektrochem.* **57** (1953) 604.
[93]H. Gerischer and M. Krause, *Z. phys. Chem., N.F.*, **10** (1957) 264.
[94]W. Vielstich and P. Delahay, *J. Am. Chem. Soc.* **79** (1957) 1874.
[95]N. A. Hampson and D. Larkin, *J. Electroanal. Chem.* **18** (1968) 401.
[96]D. J. Kooijman and J. H. Sluyters, *Electrochim. Acta* **12** (1967) 693.
[97]T. Biegler and H. Laitinen, *J. Electrochem. Soc.* **113** (1966) 855.
[98]V. V. Losev, L. E. Srybni, and A. I. Molodov, *Elektrokhimia* **2** (1966) 1431.

[99]S. Meibuhr, E. Yeager, A. Kozawa, and F. Hovorka, *J. Electrochem. Soc.* **110** (1963) 190.
[100]K. Asada, P. Delahay, and A. K. Sundaram, *J. Am. Chem. Soc.* **83** (1961) 3396.
[101]O. Esin and L. Antropov, *Zh. Obshch. Khim.* **7** (1937) 2719.
[102]L. I. Antropov and S. Ya. Popov, *Zh. Prikl. Khim.* **27** (1954) 55.
[103]E. Mattsson and R. Lindstrom, *Proc. 6th Mtg. CITCE*, p. 263, 1955.
[104]H. Fischer and R. Sroka, *Z. Elektrochem.* **60** (1956) 109.
[104a]N. A. Hampson, R. J. Latham, and D. Larkin, *J. Electroanal. Chem.* **23** (1969) 211.
[105]V. V. Losev and A. P. Phelnikov, *Elektrokhimia* **6** (1970) 41.
[106]G. M. Florianovich, L. A. Sokolova, and Ya. M. Kolotyrkin, *Elektrokhimia* **3** (1967) 1027; *Electrochim. Acta* **12** (1967) 879.
[107]J. O'M. Bockris, in *Modern Aspects of Electrochemistry*, Vol. 1, Chapter IV, Eds., J. O'M. Bockris and B. E. Conway, Butterworths, London, 1954.
[108]V. A. Pleskov and N. B. Miller, *Tr. 3 Soveshch. po elektrokhimii, 1950*, p. 165, 1953.
[109]N. A. Izmailov, *Elektrokhimia rastvorov*, 1959.
[110]V. G. Levich, in *Advances in Electrochemistry and Electrochemical Engineering*, Vol. 4, p. 249, 1966.
[111]R. R. Dogonadze, A. M. Kusnetzov, and A. A. Chernenko, *Usp. Khim.* **34** (1965) 1779.
[112]O. Esin, M. Loshkarev, and K. Sofyski, *Zh. Fiz. Khim.* **10** (1937) 132; *Acta Physicochim. URSS* **7** (1937) 433.
[113]J. E. B. Randles and K. W. Somerton, *Trans. Faraday Soc.* **48** (1952) 951.
[114]H. Imai and P. Delahay, *J. Phys. Chem.* **66** (1962) 1183.
[115]A. R. Despic and J. O'M. Bockris, *J. Chem. Phys.* **32** (1960) 389.
[116]V. V. Losev and A. P. Pchelnikov, *Elektrokhimia* **4** (1968) 264.
[117]V. V. Losev and V. V. Gorodetzki, *Electrokhimia* **3** (1967) 1061; V. V. Losev, V. V. Gorodetzki and A. I. Molodov, *Coll. Czech. Chem. Comm.* **32** (1967) 2917.
[118]M. I. Temkin, *Dokl. Akad. Nauk SSSR* **152** (1963) 156; see also *Cons. naz. ricer, Fond corsie seminarie di chim.*, **5** (1966) 159; and *J. Res. Inst. Catalysis* **16** (1968) 355.
[119]H. Mauser, *Z. Elektrochem.* **62** (1958) 419.
[120]J. O'M. Bockris, *J. Chem. Phys.* **24** (1956) 817.
[121]R. M. Hurd, *J. Electrochem. Soc.* **109** (1962) 327.
[121a]R. Parsons, *J. Chim. Phys.* **49** (1952) 82; B. E. Conway and P. L. Bourgault, *Can. J. Chem.* **38** (1960) 1557.
[122]K. Vetter, *Z. Naturforsch.* **8a** (1953) 823.
[123]K. J. Vetter, *Z. Electrochem.* **56** (1952) 797.
[124]V. V. Gorodetzki and V. V. Losev, *Dokl. Akad. Nauk SSSR* **151** (1963) 361.
[125]V. V. Losev and V. V. Gorodetski, *Elektrokhimia* **4** (1968) 1103.
[126]H. Plonski, *J. Electrochem. Soc.*, **116** (1969) 944, 1688; **117** (1970) 1048.
[126a]K. J. Bachmann and U. Bertocci, *Electrochim. Acta* **15** (1970) 1877; K. J. Bachmann, *J. Electrochem. Soc.* **118** (1971) 227.
[127]V. V. Losev, A. I. Molodov, and V. V. Gorodetzki, *Elektrokhimia* **1** (1965) 572; *Electrochim. Acta* **12** (1967) 475.
[128]N. Sato and G. Okamoto, *J. Electrochem. Soc.* **111** (1964) 897.
[129]L. I. Krishtalik, *Usp. Khim.* **34** (1965) 1831; L. I. Krishtalik, in *Advances in Electrochemistry and Electrochemical Engineering*, Vol. 7, Chapter 4, P. Delahay, ed., Interscience, New York, 1970.
[130]J. Horiuti, A. Matsuda, M. Enyo, and H. Kita, in *Proc. 1st Austral. Cong. Electrochem., 1963*, Pergamon, London, 1965.

[131]K. Heusler, *Habilitationsschrift*, 1966.
[132]V. V. Losev and A. I. Molodov, *Elektrokhimia* **4** (1968) 1366.
[133]V. A. Evropin, N. V. Kulkova, and M. I. Temkin, *Zh. Fiz. Khim.* **30** (1956) 348; see also M. I. Temkin, *Sb. Nauchn. osnovy podbora i proisvodstva katalisatorov*, p. 46, Novosibirsk, 1964.
[134]P. D. Lukovtzev, *Zh. Fiz. Khim.* **21** (1947) 589; P. D. Lukovtzev and S. D. Levina, *Zh. Fiz. Khim.* **21** (1947) 599.
[135]A. N. Frumkin, *Dokl. Akad. Nauk SSSR* **119** (1958) 318.
[136]A. Matsuda and J. Horiuti, *J. Res. Inst. Catalys.* **10** (1962) 14.
[137]L. I. Krishtalik, *Electrokhimia* **1** (1965) 346.
[138]B. E. Conway, *Progress in Reactor Kinetics*, Vol. 4, p. 399, Ed., G. Porter, Pergamon Press, London, 1967.
[139]V. V. Gorodetzki and V. V. Losev, *Elektrokhimia* **3** (1967) 1192.
[140]A. Hamelin and G. Valette, *J. Chim. Phys.* **N10** (1966) 1285.
[140a]B. Lovrecek and I. Mekjavic, *Electrochim. Acta* **14** (1969) 301.
[141]V. V. Gorodetzki, K. A. Mishenina, V. V. Losev, A. N. Grimberg, and G. M. Ostrovski, *Elektrokhimia* **4** (1968) 46.
[142]I. P. G. Farr, N. A. Hampson, and M. E. Williamson, *J. Electroanal. Chem.* **13** (1967) 433, 462.
[143]J. Jacq, I. Vignes, and O. Bloch, *Compt. Rend.* **260** (1965) 3061; *Electrochim. Acta* **11** (1966) 93.
[144]V. V. Gorodetzki and V. V. Losev, *Elektrokhimia* **7** (1971) 631.
[145]V. V. Gorodetzki, A. G. Alenina, and V. V. Losev, *Elektrokhimia* **5** (1969) 227.
[146]I. A. Altermatt and S. E. Manahan, *Anal. Chem.* **40** (1968) 655.
[147]F. Chao and M. Costa, *Bull. Soc. chim. France* **N10** (1968) 4015.
[148]N. S. Hush and J. Blackledge, *J. Electroanal. Chem.* **5** (1963) 420.
[149]M. Sluyters-Rehbach, S. M. C. Brenkel, and J. H. Sluyters, *J. Electroanal. Chem.* **19** (1968) 85.
[149a]G. Salie, *Z. phys. Chem.* **239** (1968) 411; **244** (1970) 1.
[149b]L. Gaiser and K. E. Heusler, *Electrochim. Acta* **15** (1970) 161.
[149c]K. E. Heusler and L. Gaiser, *J. Electrochem. Soc.* **117** (1970) 762.
[150]V. Markovac and B. Lovrecek, *J. Electrochem. Soc.* **112** (1965) 520; **113** (1966) 838.
[151]L. F. Kozin, E. E. Kobrand, and I. A. Sheka, *Ukrain. Khim. Zh.* **32** (1966) 154.
[152]B. Lovrecek and N. Marincic, *Electrochim. Acta* **11** (1966) 237.
[153]A. Baticle, *Electrochim. Acta* **8** (1963) 595.
[154]J. O'M. Bockris and M. Enyo, *J. Electrochem. Soc.* **109** (1962) 48.
[155]J. D. Corbett, *Inorg. Chem.* **2** (1963) 634; *J. Electrochem. Soc.* **109** (1962) 1214.
[156]T. Hurlen, *Electrochim. Acta* **9** (1964) 1449.
[157]K. E. Heusler, *Z. Elektrochem.* **62** (1958) 582.
[158]Ya. D. Zytner and A. L. Rotinyan, *Elektrokhimia* **2** (1966) 1371.
[159]G. M. Florianovich, L. A. Sokolova, and Ya. M. Kolotyrkin, *Elektrokhimia* **3** (1967) 1359.
[160]V. I. Kravtzov and O. G. Lokshtanova, *Zh. Fiz. Khim.* **36** (1962) 2363.
[161]K. E. Heusler, *Z. Elektrochem.* **66** (1962) 177; *Ber. Bunsenges.* **71** (1967) 620.
[162]M. V. Simonova and A. L. Rotinyan, *Zh. Prikl. Khim.* **37** (1964) 1951; A. L. Rotinyan and M. V. Simonova, *Elektrokhimia* **1** (1965) 1407; Ya. D. Zytner and A. L. Rotinyan, *Zh. Prikl. Khim.* **40** (1967) 89.
[163]N. P. Fedotyev and S. I. Dmitresheva, *Zh. Prikl. Khim.* **30** (1957) 221.
[164]M. Hollnagel and R. Landsberg, *Z. Phys. Chem.* **212** (1959) 94.
[164a]K. E. Heusler and L. Gaiser, *Electrochim. Acta* **13** (1968) 59.

[165]V. M. Komandenko and A. L. Rotinyan, *Zh. Prikl. Khim.* **39** (1966) 123; *Elektrokhimia* **3** (1967) 723.
[166]M. S. Grilikhes, B. S. Krasikov, and N. S. Solotzkaya, *Elektrokhimia* **5** (1969).
[167]H. Lerner and L. G. Austin, *J. Electrochem. Soc.* **112** (1965) 636.
[168]H. A. Catherino, *J. Phys. Chem.* **70** (1966) 1338; **71** (1967) 268.
[169]W. Paick, T. N. Andersen, and H. Eyring, *Electrochim. Acta* **14** (1969) 1217.
[170]K. J. Vetter and G. Thiemke, *Z. Elektrochem.* **64** (1960) 805.
[171]D. R. Stranks and J. R. Yandell, *Proc. Symp. on Exchange Reactions, Vienna*, p. 83, 1965.
[172]H. A. Catherino and J. Jordan, *Talanta* **11** (1964) 159.
[173]S. D. James, *Electrochim. Acta* **12** (1967) 939.
[174]R. Greef, *Disc. Faraday Soc.* **45** (1968) 61.
[175]M. A. Loshkarev and B. I. Tomilov, *Zh. Fiz. Khim.* **34** (1960) 1753; **36** (1962) 132, 1902.
[176]Jao-Lu-An, Yu. B. Vasilyev, and V. S. Bagotzky, *Zh. Fiz. Khim.* **38** (1964) 205; *Elektrokhimia* **1** (1965) 170.
[176a]J. K. Dohrmann and K. J. Vetter, *Ber. Bunsenges. phys. Chem.* **73** (1969) 1068.
[177]N. A. Hampson, P. C. Jones, and R. F. Philips, *Can. J. Chem.* **45** (1967) 2045.
[178]N. A. Hampson, P. C. Jones, and R. F. Philips, *Can. J. Chem.* **46** (1968) 1325.
[179]V. N. Varapaev and N. P. Fedote'v, *Tr Leningr. Tekhnolog. Inst. im Lensoveta* **46** (1958) 103.

Index

W

Z